CIRCUIT DESIGN IDEA HANDBOOK

CIRCUIT DESIGN IDEA HANDBOOK

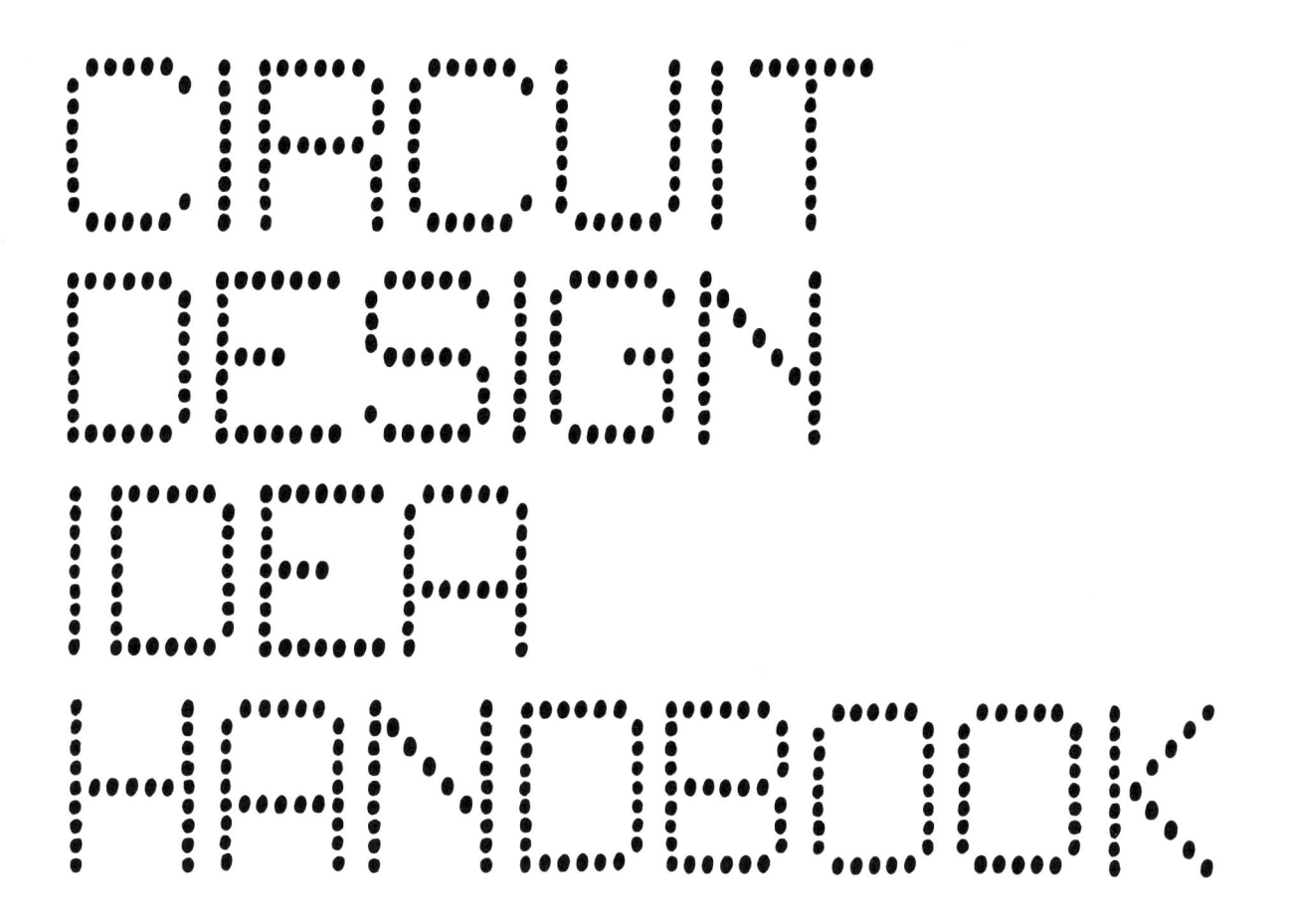

Edited by Bill Furlow Associate Editor, EDN magazine

Cahners Books, 89 Franklin Street, Boston, Massachusetts

International Standard Book Number: 0-8436-0205-8
Library of Congress Catalog Card Number: 73-76440
New Material Copyright © 1974 by Cahners Publishing Company, Inc.
© 1972, 1971, 1970, 1969, 1968, 1967, 1966, 1965 by Cahners Publishing Co., Inc.
All rights reserved.
Printed in the United States of America.
Halliday Lithograph Corporation, West Hanover, Massachusetts, U.S.A.

Contents

Preface	xiii
Section I Analog, Linear and Communications Designs	**1**
Vertical synch separator has no integrator	3
Precision clipper operates from mV to volts	3
Digital ICs serve as audio filters	4
An error-stabilized analog divider	5
Integrate and dump circuit	5
Phase-locked frequency discriminator	6
Exclusive-OR gates simplify modem designs	6
Circuit selects larger of two signals	7
Current to voltage converter	8
Set-reset latch uses optical couplers	8
Product detector uses FET tetrode	9
Low-noise preamplifier uses FET	9
Comparator with noninteracting adjustments	10
High-voltage source follower	11
Sample-hold circuit	11
Diodes provide switching and isolation for absquare circuit	12
Improved absolute-value circuit	13
Fixed bias extends zener range	13
Gating circuit detects positive or negative peaks	14
True RMS measurements	15
Fast-rise current switch	16
Analog switch controls 80V rms	16
Analog arithmetic unit offers good accuracy	17
Analog monitor has threshold and hysteresis controls	18
IC op amps make inexpensive instrumentation amplifiers	18
Voltage-controlled current source	19
Linear signal compressor has wide dynamic range	20
IC functions as sampling amplifier or tone-burst gate	21
Use opto-isolation in video amps	21
Op-amp in active filter provides gain	22
Transistor bridge detector	22
Op-amp comparator with latching function	23

Phase locked looped stereo decoder is aligned easily	23
Reduce integrator transients with synchronized gate signals	24
Constant-voltage current sink	25
Modified emitter follower has no offset	25
Differentiator circuit	26
Circuit yields linear pulse-width modulation	26
Current boosters for IC op amps	27
One transistor, 50-db dynamic range compression amplifier	27
Optical sensor has built-in hysteresis	28
Lamp test becomes circuit test	28
Multi-aperture solar cell amplifier	29
Coaxial cable driver circuit	29
High-speed circuit detects phase-difference	30
Linear modulator has excellent temperature stability	31
LED modulator	31
JFET switched resistor controls gain of op amps	32
6-bit D/A converter uses inexpensive components	32
Linear circuit integrates pulse trains	33
Low-cost UJT raster generators	33
Isolated current source	34
Squaring circuit makes efficient frequency doubler	34
Buffer/filter extracts WWV time codes	35
Transistor circuit speeds synch time of phase-locked oscillator	35
Higher-efficiency chokeless vertical sweep	36
Phase modulator has broad bandwidth	37
Active bandpass filter with adjustable center frequency and constant bandwidth	37
A polar clamp	38
Compensated adjustable zener	38
Adjustable level-detector	38
High-speed synchronous detector	39
Improved circuit for constant-current source	39
PIN-diode turnoff for microwave VCO	40
Amplitude switching for large analog signals	40
Multiplexing without FETs	41
Double-balanced diode mixer	41
Linear period-to-voltage converter with low ripple	42
Achieve 0.5 $\mu V/°C$ drift in IC video amps	42
Variable dc offset using a current source	43
Track-and-hold amplifier	44
Reversible linear counter	45
Low-power, multiple-input comparator for ac/dc inputs	46

Section II Digital and Pulse Circuits — 47

End edge ambiguities with two ICs	49
One shot multivibrator has delayed output	49
Control signal determines modulo of IC counter	50
Single transistor improves CMOS one shot	51

Simple bi-quinary to BCD converter	51
One shot triggers on both edges	52
Phase shifter for phase-locked loops	53
Flip flop operated by input signal NOR	53
Improved tunnel-diode threshold circuit has adjustable hysteresis	54
Timer produces time delays from secs to hours	55
2 TTL packages convert BCD up-counter for down counting	56
Simple FET timer	56
Single-digit BCD adder uses 3 ICs	57
Digital phase-locked loop with los-of-lock monitor	58
Two TTL gates drive very long coax lines	59
Double-edge pulser	60
Latching circuit provides noise immunity	60
High speed circuit converts binary to BCD	61
Flip-flop has improved rise time and stability	62
DC to DC one-shot starting circuit	62
J-FET generates clear or preset commands for low-power TTL	63
Improved one-shot output circuit	63
A synchronized phase locked loop	64
Fast-recovery one-shot multi gives 10:1 width control	65
Optical tape-marker detector	65
Low-cost manual pulser	66
Combination Schmitt trigger-monostable multivibrator	66
Wide range monostable multivibrator	67
Variable Schmitt, amplitude comparator	67
Improved one-shot multivibrator using ICs only	68
Low-hysteresis trigger circuit	68
A programmable combinational logic circuit	69
Extraneous pulse detector	70
Variable delay blanking-pulse generator	70
Anti-coincidence circuit prevents loss of data	71
Stable threshold circuit with low hysteresis	71
Pulse generator-to-CCSL interface	72
Eliminating false triggering in monostable multis	73
Pulse-peak indicator	73
Flip-flop with pulse and level outputs	74
Quad NAND full adder uses two ICs	74
Unbalanced to balanced level-shifter	75
Long-duration one-shot	75
Synchronized one shot	75
Divide-by-N circuit has 50/50 duty cycle	76
Complementary series multivibrator	77
Current source improves immunity of one-shot	77
Simplified Schmitt yields fast rise time	78
Clock driver for MOS shift registers	79
Level detecting flip flop with adjustable hysteresis	80
Silicon unilateral switches detect initial event	81
Simple pulse phase-splitter	82
Retriggerable monostable	82

High-speed, one-IC one-shot	83
Circuit shifts asynchronous high frequency clock sources	83
Pulse-train detector and counter	84
Resettable one-shot with high noise immunity	84
Three-state indicator	85
Double ended limit detector	85
Transfer parallel information without a clock	86
Constant-current generator speeds up wired-OR circuits	87
Non-inverting pulse amplifier uses one power supply	87
LED driver is TTL compatible	88
Low cost digital filter	88
Circuit remembers random data within periodic field	89
Circuit triggers one shot on both edges of square wave	90
Three IC's accurately sense pulse rate	90
Frequency comparator	91
Photo deteetors drive digital circuits	92
Single-voltage circuit generates "power-on" reset pulse	92
3 ICs monitor pulse width	93
Variable-threshold hybrid one-shot	93
Divider circuit maintains pulse symmetry	94
Single IC compares frequencies and phase	94
Pulse-width discriminator	95
One transistor provides ECL to LED interface	96
Time-delayed Schmitt sensor	96
Pulse-width discriminator	97
Digital comparator is self-adjusting	97
CMOS and LPTTL gates make low power Schmitt trigger	98
Clock pulse generator has addressable output	98

Section III Signal Sources   99

Triangular and square wave generator has wide range	101
Inexpensive UJT-SCR intervalometer	101
Linear bidirectional ramp generator	102
Op amp and one transistor produce ramp function	103
Wide range ramp generator has programmable outputs	103
FET stabilizes sine-wave oscillator	104
Zener-diode controls Wienbridge oscillator	105
Staircase generator uses current-regulating diode	106
Simple stair step generator uses 1 IC and 3 transistors	106
Op amp makes variable-frequency triangular wave generator	107
Triangle wave circuit has wide range controls	108
IC op amp makes gated oscillator	108
High efficiency relaxation oscillator	109
CMOS linear ramp generator has amplitude control	110
Stepped-sawtooth tone generator	111
Positive or negative slope generator	111
Cascade UJT oscillator generates linear frequency sweeps	112

Recycling timing with variable duty-cycle	113
PUT delivers ultra-low-power, high-energy pulses	113
Sequential bipolar multivibrator	114
Series-approximation sine-wave generation	115
2 gates make quartz oscillator	116
Monolithic timer generates 2-phase clock pulses	117
Current-controlled triangular/square-wave generator	117
Gated clock generates pulse train or single pulse	118
Crystal-controlled relaxation oscillator	119
TTL inverter makes stable Colpitts oscillator	119
Simple op amp relaxation oscillator generates linear ramp output	120
Astable operation of IC timers can be improved	121
Ultra low distortion oscillator	122
CMOS circuits generate arbitrary periodic wave forms	122
Improved multi with continuously variable rep rate	123
Precision oscillator is versatile	124
A wideband, linear VCO	125
PUT oscillator has 4-decade frequency range	125
Simple sinewave oscillator	126
Modified UJT oscillator has no timing error	126
Zero V=Zero frequency VCO	127
Ramp generator has adjustable nonlinearity	128
Pulse generator offers wide range of duty cycles	129
Self-completing gated astable	130
Clock-pulse generator	130
Low-power two-phase clock	132
Voltage-controlled oscillators	132
Triggered sweep features low dc offset	133
Keyed multivibrator produces symmetrical ac output	133
Stable low-distortion bridge oscillator	134
Gated 60 Hz clock avoids glitches	135
Wide-range pulse generator uses single IC	135
Emitter-coupled astable with saturated output	136
Crystal controls rep rate of simple IC pulse generator	137
Operational amplifier makes a simple delayed pulse generator	138
Crystal controlled multivibrator	138
Adjustable rectangular-wave oscillator interfaces with IC logic	139

Section IV Power Supply Circuits   141

Current transformer gives fast overload protection	143
One transistor improves IC voltage regulator	143
Simple ± 15V regulated-supply provides tracking	144
AC power interlock	144
Under-voltage sensing circuit	145
Spare IC gate serves as regulator	146
High voltage dc-to-dc converter	146
Resistor reduces switching regulator losses	147

Diodes avoid accidental damage from power supply sensing	147
IC timer makes transformerless power converter	148
Opto-isolators provide foldover indication	148
Versatile circuit behaves like SCR	149
Logic-supply crowbar	150
Adjustable-overvoltage circuit breaker	150
Window detector uses one IC regulator	151

Section V Switching Techniques — 153

High-voltage triacs reverse capacitor motor	155
Simple zero-crossing solid-state switch	156
Digital interlocking switch	156
Monolithic timer makes convenient touch switch	157
Triac gating circuit	157
Set-reset latch uses optical couplers	158
Noise-rejecting SCR trigger circuit	158
Putting the "thumb" on thumbwheel switch multiplexing	159
Battery-saving remote-command detector	160
600V switching circuit	160
Remote-controlled exclusive OR gate	161
An improved rotary-switch interlocking circuit	162
Make-and-break bounceless switching	163
Low-cost bistable relay circuits	163
Efficient and simple zero-crossing switch	164
Battery saver has automatic turn-off	165
Low-power timer drives stepping relay	165
Bistable switch with zero standby drain	166
Solid-state relay	167

Section VI Test Circuits — 169

One CMOS package makes universal logic probe	171
Technique locates shorted gates	171
Simple DC voltmeter uses single op amp	172
Simple jig determines transistor gain-bandwidth product	172
High-gain ac/dc oscilloscope amplifier	174
Logic probe tests three-state logic circuits	174
555 timer makes simple capacitance meter	175
In-circuit capacitor tester	176
IC power supply provides test spikes or level shifts	176
Bias supply circuit provides constant current	177
2 CMOS gates convert counter into capacitance meter	177
Temperature-coefficient measuring circuit	178
FET probe drives 50Ω load	179
Test box indicates logic levels for entire IC	180
Low-speed logic probe	181
Short-circuit alarm	181

Section VII Miscellaneous — 183

Novel clock circuit provides multiplexed display	185
60-Hz frequency discriminator	186
Convert-wrap-around code to sign-plus-magnitude	186
Fail-safe temperature sensor	187
Phase-interlocked gyro motor driver	187
Thirty-second timer uses IC one-shot	188
Super-stable reference-voltage source	189
Adjustable, low-impedance zener	190
A digital dead band comparator	190
Automatic scaling circuit for optical measurements	192
SCR's form electronic combination lock	193
Pulse catcher uses two IC's	193
Pulse-level discriminator and fault indicator	194
Low pass digital filter	194
Thermistor circuit senses air temperature and velocity	195
DPM makes self-contained digital thermometer	196
E-cell controls battery charging rate	197
Proportional oven-temperature controller	197
TTL/DTL interface to FET analog switch	198
FET provides automatic meter protection	198
Two TTL gates drive very long coax lines	199
Voltage- or pot-variable 400-Hz delay	200
Double duty photo alarm	200
Current-pulse generator for LED's	201
Photocells allow a free pendulum to operate silently	202
Low-cost, long-delay timer	203
Leakage testing diodes & JFET's	203
Universal transformers	204
Low cost IR system detects intruders	205
Optimum zener decoupling	206

Index — 207

Preface

"Electronic cookbooks" are always popular with electronic designers because they give you an instant way to compare problems and solutions with your compatriots.

There are 287 such solutions in this book, gathered from the "Design Awards" sections of *EDN* and *EEE* magazine. They may fit your problem exactly, or you may have to modify them; at any rate, they will present you with fresh approaches to many of your problems.

We have attempted to make this book more useful than other books of its kind by including a brief cross-index at the end. Although the titles in the table of contents are descriptive, we feel that some cross-indexing will be helpful when you are searching for a particular technique. Does it really help? Should it be more extensive? Only you, the working engineer will know that, so we solicit your comment.

Another important area for feedback is in accuracy of the designs. Errors of which we were aware have been corrected, if more exist we would like to know about them. We also hope that you will keep a critical eye on the "Design Awards" section of *EDN* and let us know when printing errors occur there.

Far from being "professional plagiarism," the use of this book can save you many hours of frustration, so use it as a time saver whenever you are faced with a new problem. It can save you from many reinventions of the wheel, but there will be times when nothing in here helps you. When that happens and you come up with that neat and simple solution, send it along to us; we'll help you share it with thousands of other designers.

Bill Furlow
Associate Editor
EDN Magazine

SECTION I

Analog, Linear and Communications Designs

The publisher cannot assume responsibility for circuits shown nor represent that they are free from patent infringements.

Vertical sync separator has no integrating network

Walter G. Jung,
Forest Hill, Md.

Standard approaches to separating the vertical synchronizing pulse interval in an EIA TV composite sync waveform involve an integrating network to remove the rapid horizontal rate changes and a level detector to sense the longer duration vertical rate pulses. This scheme by its very nature must contribute a delay due to the integration.

An interesting method to sense the vertical rate information digitally is shown in the figure. This system uses the controlled clocking sequence of a J-K flip-flop to detect the presence of the vertical sync interval.

Referring to the timing diagram, the flip-flop is fed complementary TV sync signals at its J and K inputs (waveforms J and K). It is clocked by a pulse delayed slightly longer than the horizontal sync interval (10 μsec) and with a narrow width (1 μsec). During the normal scan time, the clocking sequence of the J-K flip-flop reads the time interval of 11 μsec after the leading edge as a LOW; thus, the first clock pulse after this change clocks the flip-flop Q output to a ONE. It remains at this level for six clock periods and reverts to the ZERO state again after the vertical sync interval has passed.

The system offers a much shorter delay time, as the leading edge error will only be slightly greater than the time required to "bracket" the horizontal pulse width.

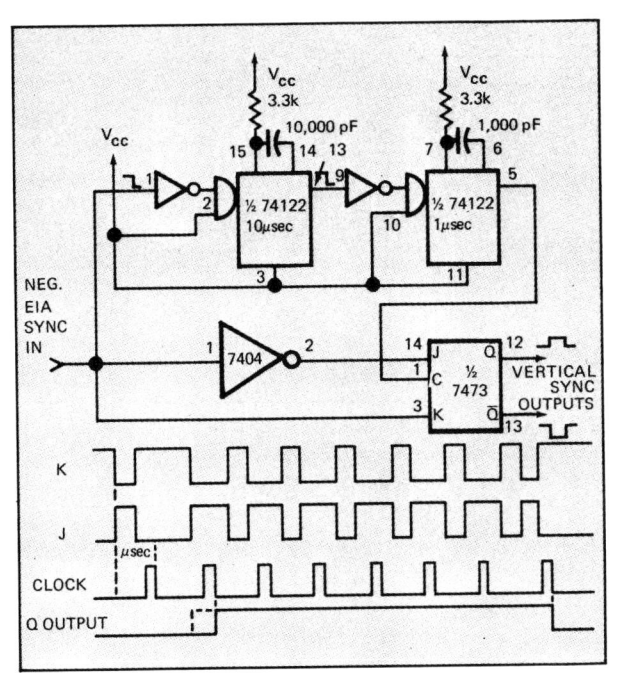

The clocking sequence of the J-K flip-flop determines the presence of the vertical sync interval.

Precision clipper operates from millivolts to volts

R.S. Burwen,
Analog Devices, Norwood, MA

This precision clipper circuit limits very sharply the excursion of a dc input voltage to a level equal to a dc reference input. When the reference is 0, the circuit can be used as a half-wave rectifier for sinusoidal inputs at frequencies up to 100 kHz. The circuit uses feedback to overcome the breakover characteristics of diode D_2 thereby providing sharp clipping at signal levels from mV to V.

When the input voltage is more pos. than the reference, the output of A_1 is neg. and Diode D_2 is cut off. For input voltages more neg. than the reference, the amplifier output goes sufficiently pos. to cause D_2 to conduct and hold the output voltage at the reference level. For best performance the circuit requires an amplifier capable of large differential input voltages without drawing input current, a high slewing rate and fast overload recovery to minimize the time delay in the conduction of D_2 while the amplifier output is swinging from neg. overload. The AD513J used in the application is stabilized in a manner that provides high slewing rate using capacitors C_1 and C_2. Because C_2 has little effect until diode D_2 begins to conduct, it does not provide feedback which would reduce the slewing rate, as does C_1.

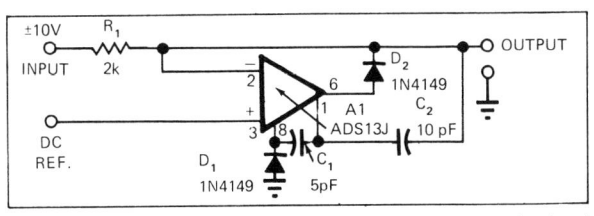

The dc input voltage is limited by the clipper circuit to the level of the dc reference input.

When the dc reference is at ground potential, a diode, D_1, can be added to reduce the neg. output swing thereby lessening the time delay before D_2 begins to conduct.

The circuit is useful to millivolts levels at dc and low frequencies and at 10 kHz from 70 mV rms to 7V rms when D_1 is not used. With D_1, the circuit is useful at levels down to 0.3V rms at 100 kHz. By changing R_1 to two series resistors and tapping the output off the junction the circuit can be used as a function generator which provides unity gain for inputs more pos. than the reference and less than unity gain for inputs more neg. than the reference. The polarity can be changed by reversing D_1 and D_2.

Digital ICs serve as audio filters

by Robert J. Battes
Nova-Tech
Redondo Beach, Calif.

IN MOST conventional marker beacon receivers, coded audio frequencies (400, 1300 or 3000 Hz) associated with particular indicator lamps are selected by LC tuned circuits. Since the tuned circuits operate at audio frequencies, they are large and heavy. Other components sometimes used to obtain the required audio selectivity, such as RC notch filters, are cumbersome and often unreliable since tight component tolerances and typical long-term instability leave much to be desired.

In the alternate system shown in Fig. 1, the more conventional filter circuitry is replaced entirely by digital ICs. The arrangement shown is actually a simple, low-frequency counter. Since the counter capability required for effective audio-frequency selectivity is low, the time-base frequency can be selected such that it may vary as widely as 25% from its initial nominal value. Reliable long-term operation is thus assured and, of course, no periodic tuning is required.

Operation of the counting circuit is conventional, with counting and display periods of approximately 4 ms each. When the gate-control line is positive, the gate is closed, and the previous count is displayed. At the instant the gate-control line goes low, the counter flip-flops are reset, and the gate (which also serves as part of the monostable multi that shapes the input-signal waveform) is enabled, allowing the counter to accumulate one count for each input cycle that occurs while the gate is open.

After the gate closes, the accumulated count is displayed. Since the duration of the count cycle (which is determined by the time-base period) is constant, or nearly so, then the count that is accumulated and subsequently displayed is proportional to the frequency being counted. With the values given, each of three frequencies of interest will light an associated lamp.

A tabulation of the lamp-display output states as a function of input frequency over the range of 100-4000 Hz, the normal marker beacon receiver audio-frequency bandwidth, is shown in Fig. 2. The tabulation is for illustration only. It does not consider the normal ±1-count ambiguity. Any degree of audio selectivity can be provided at the cost of added circuit complexity. The principal aim here is to provide a minimum reliable realization of the required function.

The indicator lamps are disabled during the counting period to prevent display-lamp flicker. The system is extremely immune to noise-induced false-count displays since a fresh count is made and displayed every 8 milliseconds.

The use of inexpensive plastic-package RTL, like the MC-700P series, is quite suitable if extreme temperature range is not a problem. Five chips are required to implement the design as shown.

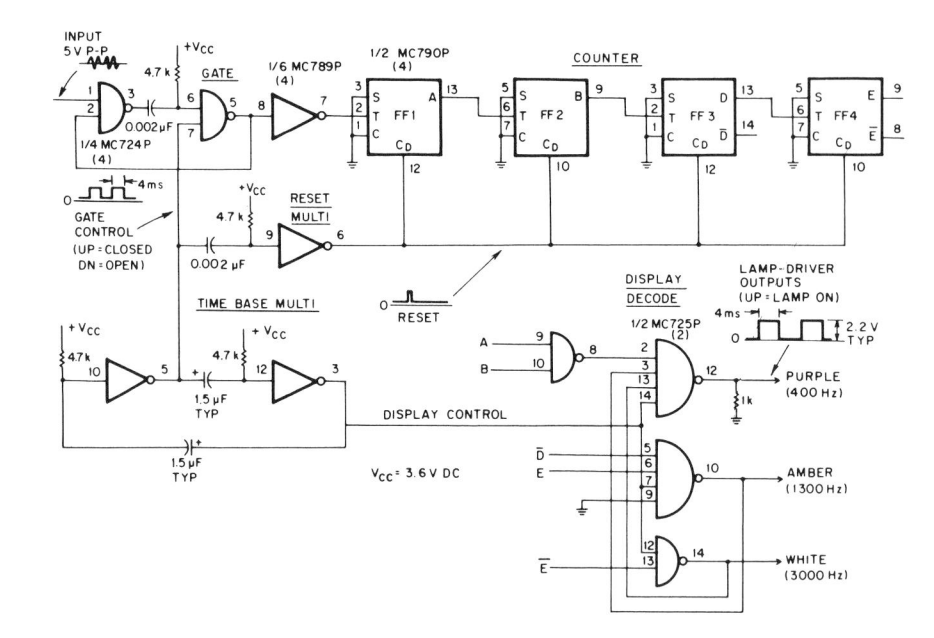

Fig. 1. A digital-IC frequency counter selects the proper coded audio signals for driving the proper display lamps.

Input Freq (Hz)	Cycles in typical 4-ms count	Lamp activated	Nominal lamp modulation
100-250	0	None	None
250-1000	1, 2 or 3	Purple	400 Hz
1000-2000	4-7 inclusive	Amber	1300 Hz
2000-4000	8-15 inclusive	White	3000 Hz

Fig. 2. Typical lamp states as a function of input frequency.

An error-stabilized analog divider

by Gary M. Fitton
E.M.R.
Princeton, N.J.

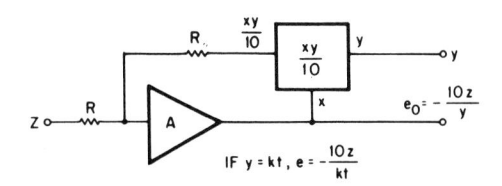

Fig. 1. This is the usual method of analog division.

FIGURE 1 illustrates a circuit for analog division. In this method an analog multiplier is placed in the feedback loop of an operational amplifier. However, this method is limited by the drift of the multiplier. Fig. 2 shows a circuit for cancelling error due to multiplier drift.

In the new circuit, the multiplier output (XY/10) is sampled at $t = 0$. This signal is inverted and fed into the op-amp's summing point, effectively cancelling the error at $t = 0$. Fig. 2 illustrates this method for a wide-band error-stabilized analog divider.

The system generates a waveform of the form $e = Z/kt$ where Z varies 40:1 and K is variable from 1.7 V/μs to 0.2 V/μs. System waveshapes are illustrated in Fig. 3 for a prf of 15 kHz. The multiplier selected, a Hybrid System 105, has an inverted output which is used to simplify the error-cancellation circuit. In order to maintain loop stability at the high speed desired, the op-amp unity-gain frequency must be equal to or less than the 3-dB frequency of the multiplier. A limiter loop was also included to prevent overdrive of the op amp and multiplier.

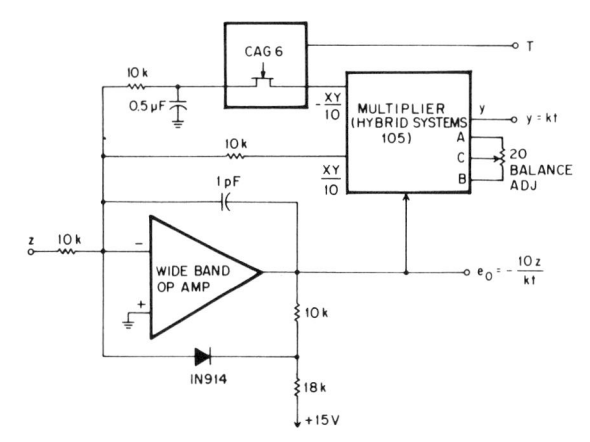

Fig. 2. This circuit shows a method of eliminating multiplier drift.

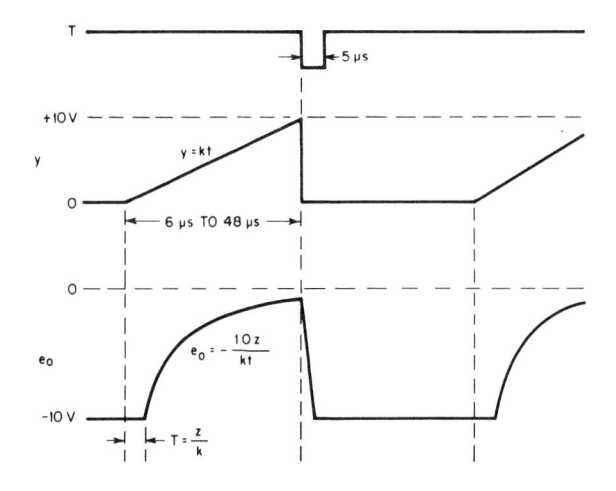

Fig. 3. The waveshapes illustrated are for a hyperbolic broadband generator.

Integrate and dump circuit uses no power supply

Ralph Riordan
NASA—Goddard Space Flight Center,
Greenbelt, MD

Since this integrate and dump circuit works without using any power supply to bias the translator, the circuitry is greatly simplified. It also improves the operation of the circuit since there is never any chance of integrating a bias voltage. The transistor is used merely as a switch. The output waveform of the integrator is like those seen in textbooks—it is not distorted in any way. The input signal on the "dump" line is TTL compatible, and the integrator will dump when the dump signal is a logic ONE (+5V). With the components shown the dump time is about 1 μsec for full discharge. R and C determine the time constant of the integrator.

Since this circuit uses no power supply, it can only drive a high-impedance load. This problem can be easily solved by using a FET input op amp. I used an Analog Devices 40J which sells for $11.

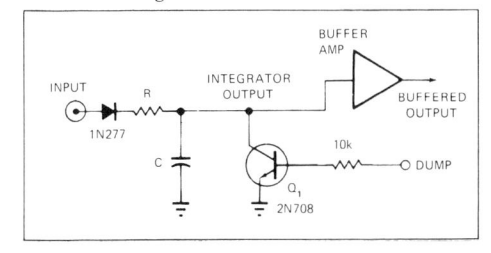

Phase-Locked Frequency Discriminator

By Hart Anway
General Electric Co.
Utica, N. Y.

This circuit may be used to phase lock a low frequency oscillator to some desired frequency. It employs a flip-flop, a filter, and a dc level shift to provide the required center frequency voltage control to the oscillator.

In the circuit, a trigger generated from the master frequency enters at B, while a trigger from the phase locked frequency enters at A. Operation depends on the fact that when the frequency at A is higher than the frequency at B, the average Q_2 collector voltage is positive, while for B frequency higher than A frequency, the average Q_2 collector voltage is negative. In either case, an error-voltage sensing circuit corrects the frequency at A, thus tending to maintain Q_2 collector at 0 V average.

In operation, the trigger at B causes Q_2 collector to become negative. A trigger at A causes Q_2 collector to become positive. Thus, if more B triggers are received than A triggers, (B frequency greater than A frequency) Q_2 collector will be negative more often than it is positive, as shown in Fig. 2a.

The pulse signal appearing at the Q_2 collector is filtered by a resistor in series with a large capacitor, and the resulting dc is applied to Q_3 base. Q_3 acts as a dc shift and emitter follower. The error voltage at Q_3 output is used to correct the oscillator frequency. When correction is complete, Q_2 collector voltage appears as in Fig. 2b, and both frequency and phase lock are achieved.

The oscillator used in conjunction with this circuit may be a voltage-tuned multivibrator or a unijunction sawtooth generator. As shown, the circuit was used to sync a unijunction sawtooth generator to operate at 10X the frequency at B. B frequency had a center value of 16 Hz and varied about this value by ±25 percent. A 1/10 counter followed the unijunction and its output was used to generate the trigger input to A.

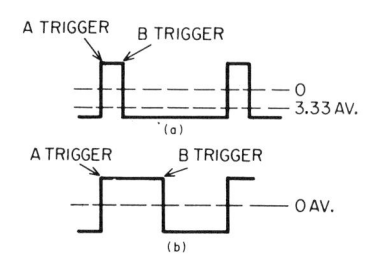

Fig. 2. Waveforms at Q_2 collector for: B frequency greater than A (Fig. 2a) and equal, phase-locked frequencies (Fig. 2b).

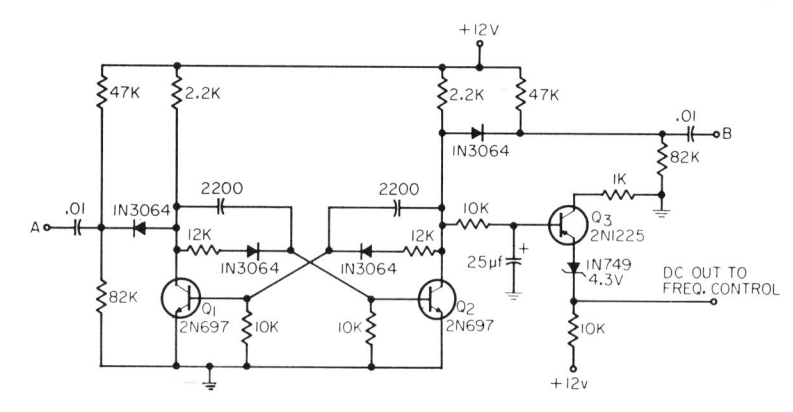

Fig. 1. Frequency detector phase-locks signals at A and B.

Exclusive-OR gates simplify modem designs

Peter Alfke,
Fairchild Semiconductor, Mt. View, Calif.

The inherent self-clocking property of binary phase modulation makes it the most popular technique for transmitting digital data over a single line. Exclusive-OR (NOR) integrated circuits and a retriggerable monostable will simplify design of both the transmitter and the receiver.

In transmitters with a 50% duty cycle clock, a simple Exclusive-OR tie between the clock and the data generates the output signal. Without a symmetrical clock, the output signal can be generated by a toggling flip-flop and a double frequency clock source. In fast systems, data propagation delay could cause spikes on the output; these can be eliminated by another flip-flop operated by the same double frequency clock.

The receiver must regenerate the clock and the data stream. A 9601 adjusted to 3/4 of the data-bit time is connected in the non-retriggerable mode. Any incoming level change will trigger the 9601. One Exclusive-OR and an Exclusive-NOR connected as an inverting delay element will perform this function. The output of the monostable can be used as a clock. The level of the incoming line at the end of the pulse (the rising edge of Q) defines data, retrieved by an edge triggered flip-flop.

This system remains synchronized as long as the monostable pulse width is between 50% and 100% of the data-bit time.

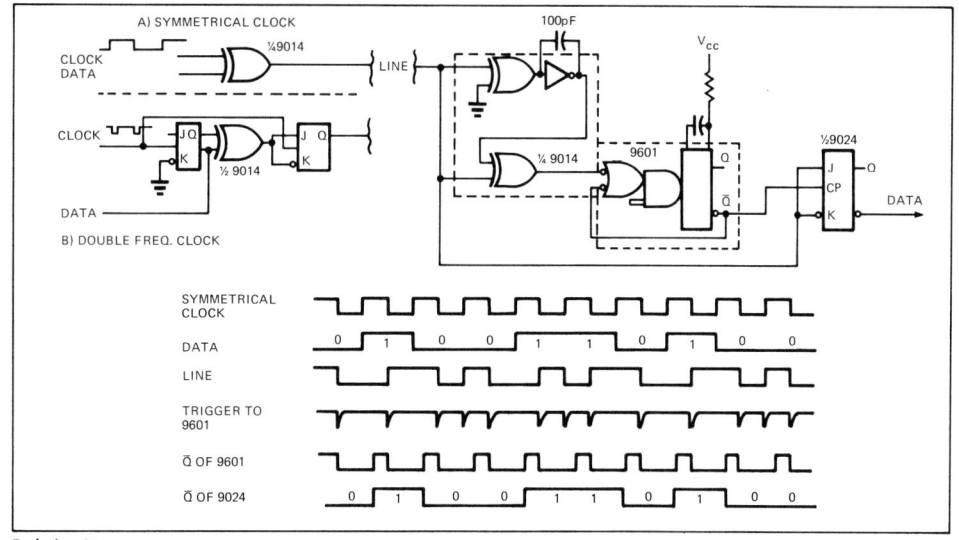

Exclusive-OR/NOR gates and a retriggerable monostable multivibrator greatly simplify designs of both data transmitters and receivers. Circuit timing functions are shown with the schematic.

Circuit selects larger of two analog signals

Werner Gruenebaum
Loral Electronic Systems, New York, NY

Using a differential current source and a few more components, it is possible to design a circuit requiring no switching to select the larger of two analog voltage inputs and provide a precision output of that voltage.

Considering the first op amp, A_1 in the schematic, which is similar to a common bilateral current source, but with V_B added:

$$e_0 = (e - V_A)\frac{(R_1 + R_3)}{R_1} + V_A \quad (1)$$

$$e_0 = e + \left(I + \frac{e - V_B}{R_2}\right) R_4 \quad (2)$$

Since **Eq. 1** must equal **Eq. 2**, this results in

$$I = \frac{(e - V_A)\frac{(R_1 + R_3)}{R_1} + V_A - e}{R_4} - \frac{(e - V_B)}{R_2}$$

$$= \frac{V_B}{R_2} - \frac{V_A R_3}{R_1 R_4} + e\left(\frac{R_3}{R_1 R_4} - \frac{1}{R_2}\right)$$

If $R_1 = R_2$ and $R_3 = R_4$.

$I = \dfrac{V_B - V_A}{R_1}$, independent of the load through which I is conducted (provided the amplifier remains in the linear region).

The current I will flow through D_2 only if it is positive, namely, $V_B > V_A$. Otherwise, it is conducted through D_1.

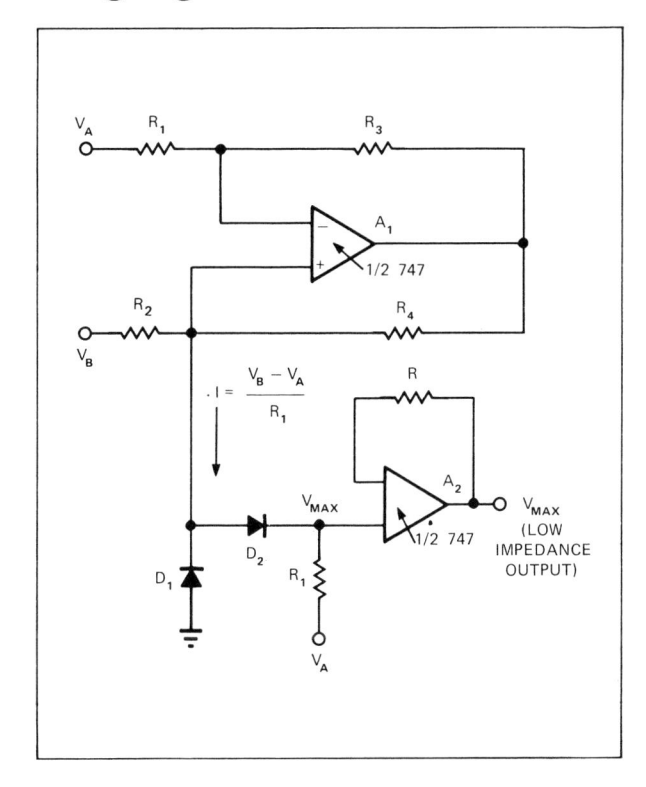

Therefore, the positive input to A_2 (V_{MAX}) is equal to:

$$V_{MAX} = V_A \text{ for } V_A > V_B$$

$$= V_A + \left(\frac{V_B - V_A}{R_1}\right) R_1 = V_B \text{ for } V_B > V_A.$$

Amplifier A_2 (a voltage follower) is added if a low-impedance output is required.

Analog, Linear and Communications Designs 7

Current-to-voltage converter for transducer use

R.S. Burwen,
Analog Devices, Norwood, MA

This current-to-voltage converter circuit is useful with current output transducers such as silicon photocells. Because the source current is fed into the low-impedance summing junction of the FET operational amplifier the frequency response is relatively insensitive to the capacitance of the source. Circuit noise is that produced by the 200 kΩ feedback resistor, which for a given voltage output is substantially less than the noise that would be obtained if the source were terminated in a low resistance to achieve the same frequency response.

For widest frequency response the circuit values should be chosen for the particular source current and capacitance. In this circuit it has been assumed that the source capacitance, C_S, is 22 pF and that the amplifier has been stabilized by a small 5 pF capacitor, C_2. Capacitor C_1, across the feedback resistor, eliminates ringing in the 500 kHz region. With the 22 pF source the frequency response is down 3 dB at 600 kHz. Increased gain can be attained by increasing R_1 although at a sacrifice in bandwidth.

By adding an input coupling capacitor to reduce the dc gain, this circuit can also be used with an inductive source such as a magnetic tape head. The feedback via R_1 dampens the resonance that would normally occur if the head were terminated in such a high resistance as 200 kΩ.

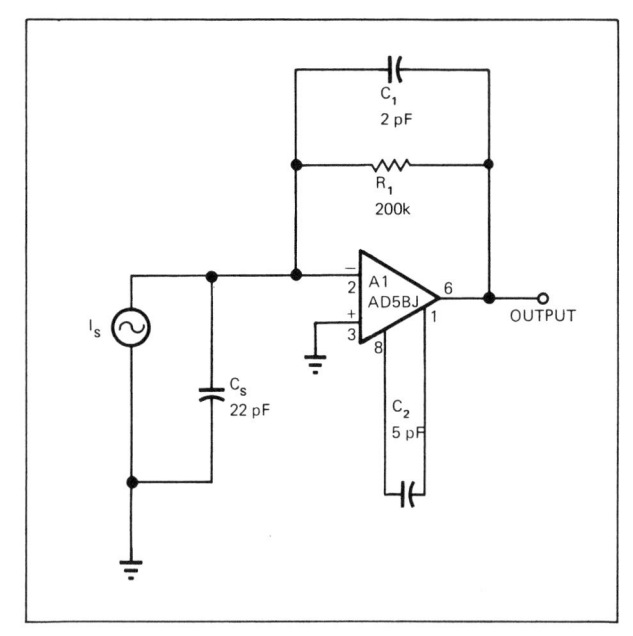

Frequency response of the converter is not greatly affected by source capacitance, because the source current is fed to the low-impedance summing junction of the op amp.

Set-reset latch uses optical couplers

Robert N. Dotson,
Motorola Inc., Phoenix, AZ

This optically coupled set-reset latch provides almost complete isolation between each input and the output as well as between the inputs. Momentary application of about 2V at 14 mA to the SET terminals will allow up to 150 mA to flow between E and E' with a 1.6V drop. This current will continue to flow, even after the SET LED is turned off, until approximately 2V at 15 mA is applied to the RESET terminals, or until the current from E to E' is reduced to less than 1 mA. Note that when the circuit is off, a maximum of about +30V or −9V is all that can be applied from E to E'.

Basically, the MPS6518 and the SET coupler's Darlington make up a discrete approximation of an SCR. The two resistors are used to reduce the gain of the transistors so that the circuit will not automatically latch up. RESET is accomplished by shunting the base of the SET coupler with the RESET coupler, or by reducing the current from E to E' until the betas of the SCR transistors drop low enough to unlatch the circuit.

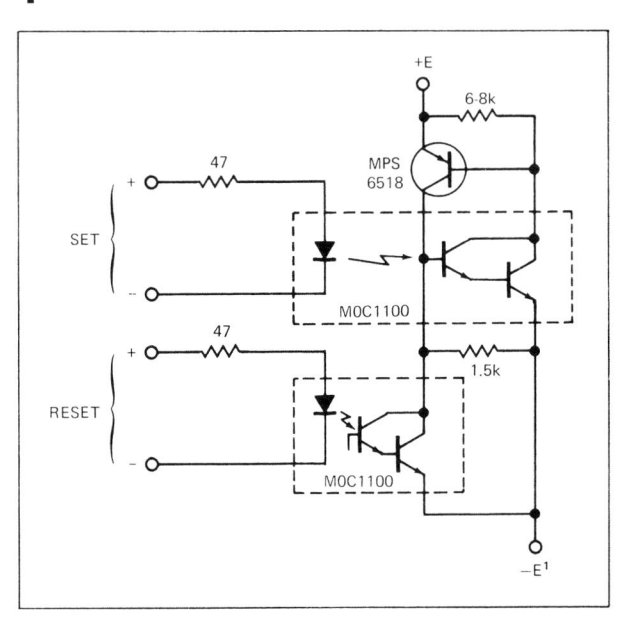

Almost complete isolation between the inputs and output of the set-reset latch is provided by the optical couplers.

Product Detector Uses FET Tetrode

By H. Olson
Palo Alto, Calif.

THE CHARACTERISTICS of the dual-gate FET makes it ideal for use in a solid-state version of the common pentagrid product-detector circuit. Since each gate has independent control of the drain current, the beat frequency oscillator voltage can be injected into one gate and the signal into the other. As in the pentagrid circuit, the output current is then the product of the two voltages.

In the circuit of Fig. 1, an emitter-follower is used to isolate the BFO, and provide the proper bias to G_2.

The dynamic range of the detector, shown in Fig. 2, is greater than that found in most typical tube-type product-detectors used in modern single-sideband receivers.

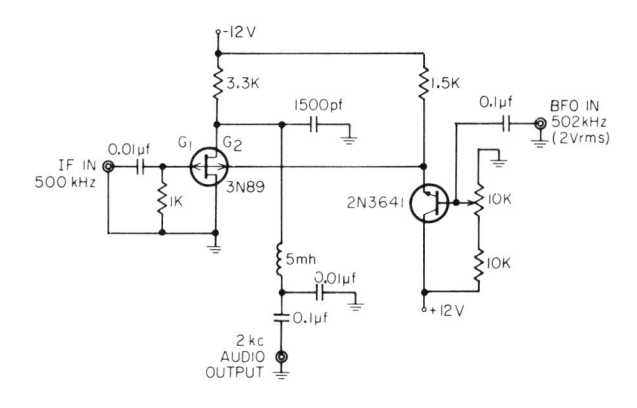

Fig. 1. Pentagrid-type operation of dual gate FET makes it suitable for product-detector circuit.

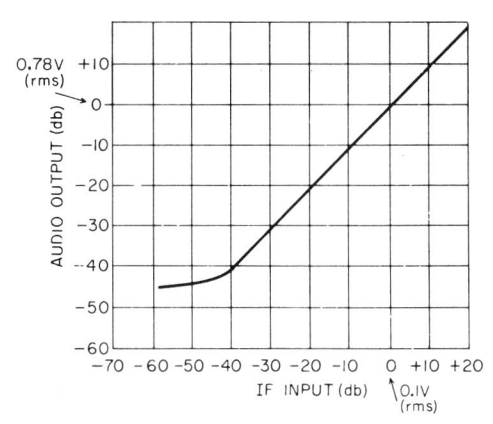

Fig. 2. Output of product detector is linear over a wide dynamic range.

Low-Noise Preamplifier Uses FET

By William F. Blodgett
Bunker Ramo Corp.
Canoga Park, Calif.

IF THE SOURCE impedance is high, a FET gives less noise than a bipolar transistor, even at frequencies as low as 10 Hz. Thus, for an amplifier to be used with a high-impedance transducer, a FET provides the best input circuit.

The preamplifier shown here, was designed for use with a 330 pF hydrophone over the frequency range 10 Hz to 10 kHz.

Positive feedback through C_1 reduces the shunting effect of the source impedance. The hydrophone has an impedance of approximately 48 MΩ at 10 Hz. The voltage gain of the FET stage is 20 dB. This stage gain reduces the effective noise contributed by subsequent transistors. Resistor R_1 provides 40-dB negative feedback, giving a closed loop gain of 40 dB for the complete preamplifier. This resistor also provides dc feedback which stabilizes the bias conditions over a wide range of operating temperature.

Measured voltage gain of this amplifier varies from 40.2 dB to 40.5 dB over the stated frequency range. Measured equivalent-input-noise voltage is −113 dBV at 10 Hz and −157 dBV at 10 kHz.

Voltage gain can be increased to 60 dB by reducing R_2 to 90.9 ohms, with little effect on noise. But this will reduce the 3-dB bandwidth.

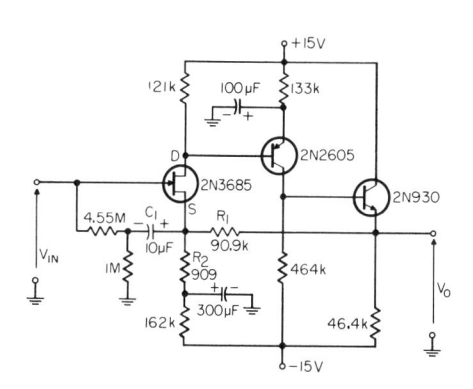

Low-noise preamplifier for low-capacitance transducer.

Comparator with noninteracting adjustments

by Gene Tobey
Burr-Brown Research Corp.
Tuscon, Ariz.

When a comparator is used in industrial applications, such as relay driving, it is often desirable to add some hysteresis to the operating characteristic. This is done to prevent "chatter" caused by noise when the input-signal level is near the trip point. Conventional circuits for adjusting hysteresis, however, cause interaction between the hysteresis adjustment and the trip-point adjustment. The circuit shown in **Fig. 1** avoids this drawback and allows the user to adjust trip point and hysteresis independently.

Amplifier A_1 provides the basic comparator action and clips the output waveform at a negative voltage determined by the feedback limiter (R_4, R_5 and CR_2). Diode CR_1 limits the positive output excursion of A_1 to approximately 0.6V.

Amplifier A_2 provides polarity inversion and precision rectification of the output of A_1. The output of A_2 is fed back to A_1 through R_8 (positive feedback), thus providing the hysteresis effect. The amount of hysteresis is determined by the ratio R_3/R_8.

$$\Delta V = V_L (R_3/R_8)$$

where V_L is the positive output level, given by the equation,

$$V_L = 15 \left(\frac{R_5}{R_4}\right)\left(\frac{R_7}{R_6}\right)$$

Trip point is determined by the setting of R_1, or alternatively, it may be programmed by a voltage applied to R_2. As the trip point is varied, the amount of hysteresis (ΔV) remains constant.

Fig. 2 shows the input-output characteristic of the comparator. As shown, the hysteresis is the difference between the trip and reset levels:

$$V_S = V_R - \Delta V$$

The two output levels are zero (within a fraction of a millivolt) and the positive level V_L. With the resistor values shown, the circuit has output levels of zero and 5V.

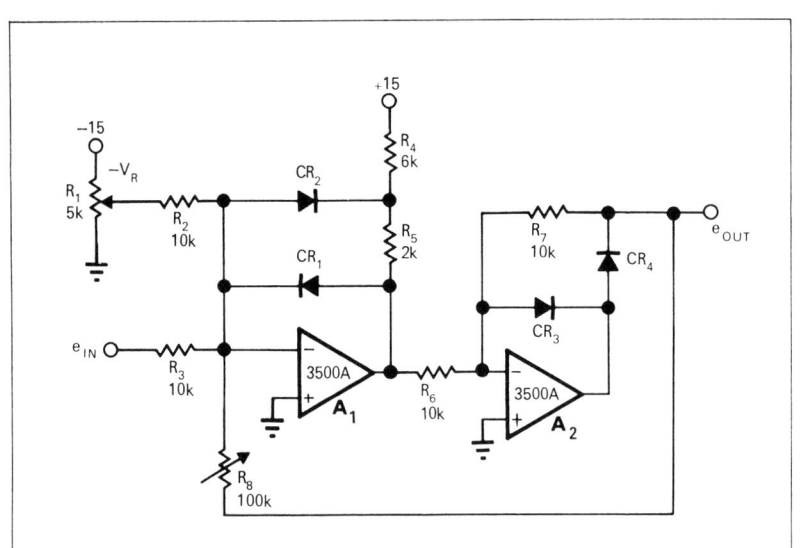

Fig. 1 – **This improved** comparator has independently adjustable trip point and hysteresis. R_1 determines trip point and R_8 determines hysteresis.

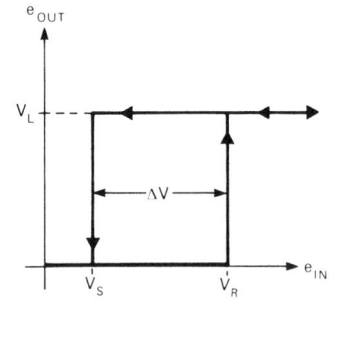

Fig. 2 – **Input-output** characteristic shows the effect of hysteresis (ΔV).

High-voltage source follower

by Stephen A. Jensen
Radiant Industries, Inc. North Hollywood, Calif.

This circuit allows conventional voltage-follower modules or ICs to operate from high-voltage dc supplies and to handle large signal-voltage swings.

The same technique can be used with any unity-gain amplifier that operates on ±15V supplies. The GPS amplifier specified in the schematic is a precision follower. This type was needed to provide sufficient accuracy in the original application which involved interfacing a tachometer signal to the ±100V input amplifier of an analog computer. However the circuit technique can also be used with ICs such as the LM110 or the µA741 strapped for unity gain.

With this circuit, the positive supply lead to the amplifier is always held at 15V above the output lead. Similarly, the negative supply lead is held at 15V below the output lead. Thus the amplifier module or IC always sees the correct supply voltages. Large common-mode voltages are handled by the transistors in the supply lines rather than by the amplifier itself.

If, for example, a positive 75V signal is applied at the input, the power-supply terminals of the basic amplifier will see voltages of ±90V and +60V, respectively, with respect to ground. Thus the correct potential of 30V is applied across the supply terminals.

With this bootstrapping technique, the complete circuit retains the voltage and gain stability of the low-voltage amplifier, while handling much higher input voltages.

The allowable input-voltage range depends solely on the V_{CEO} ratings of transistors Q_1 and Q_2. Thus, with 1000V transistors, the circuit would be able to handle input signals of ±475V.

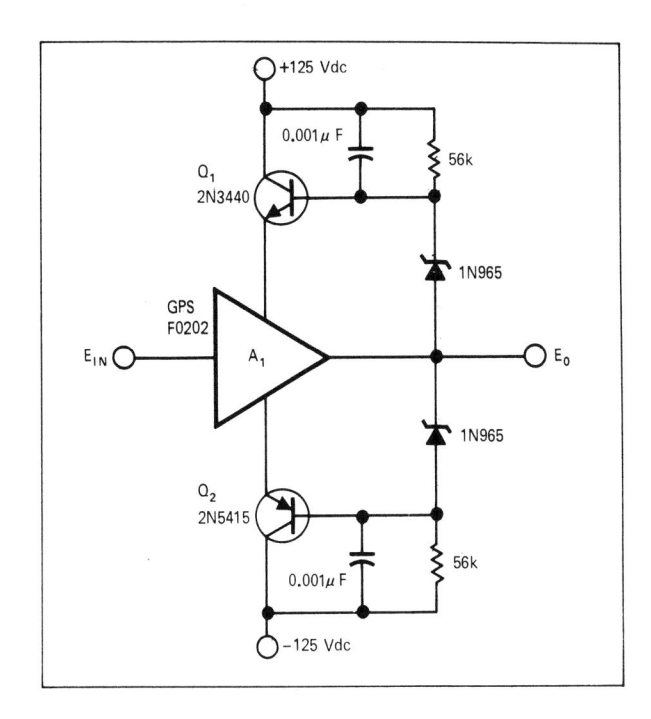

Bootstrapping technique allows low-voltage unity-gain amplifiers to operate from large supply voltages and to handle large input signals.

With the component types specified in the schematic, the input capability is ±100V pk-pk for frequencies from dc to 10 kHz. The output capability is 5 mA pk at ±100V. The input impedance is the same as for the amplifier A_1 (i.e. $10^{12}\Omega$ for the type specified). Similarly the output impedance is less than 10Ω.

Sample-Hold Circuit

By Bert Pearl
Hallicrafters Co.
Chicago, Ill.

MOST SAMPLE-and-hold circuits are quite complex. Some simpler ones have been published in the last year. However, in these circuits because charge and discharge of the hold capacitor can occur simultaneously, a resistor is placed in one of these paths to limit the current. The resistor drastically limits the speed with which the output responds to new input levels. For short sampling periods, in the order of 1 µsec, the slow speed precludes the use of these circuits.

The circuit shown here does not have this limitation. Charge and discharge do not occur simultaneously, and no current limiting resistor is needed. There are two inputs (see figure); the signal whose amplitude is being sampled, and a positive sampling-pulse train. Between pulses the sampling train is at ground potential, holding Q_2 in saturation. It's collector voltage approaches +12 V, cutting off Q_3, whose emitter is then at +12 V. This reverse-biases low-leakage diode D_1, preventing any discharge of C_1 between pulses. At this time, Q_1 is also turned off, since there is no positive signal. Its emitter is at −6 V, reverse-biasing low leakage diode D_2. This prevents charging of the hold capacitor between pulses.

During sampling periods, if the new signal pulse is more positive than the previous one, C_1 is charged rapidly to the new level through emitter follower Q_1 and diode D_2. The collector voltage of Q_1 drops below +12 V because of the charging current. This drop, applied through diode D_3, drives Q_2 into saturation. As before, Q_2 in saturation cuts off Q_3, and back-biases D_1 so that discharge cannot occur during the charging period.

If the new signal pulse is less than the previous one, Q_1 remains cut off. Also Q_2 remains cut off, and emitter follower Q_3 conducts heavily, providing a low-impedance discharge path for C_1 through D_1. This discharge continues until C_1 has discharged to the amplitude of the new pulse, less the diode drops in Q_1 and D_2. At this time Q_1 turns on, stopping the discharge.

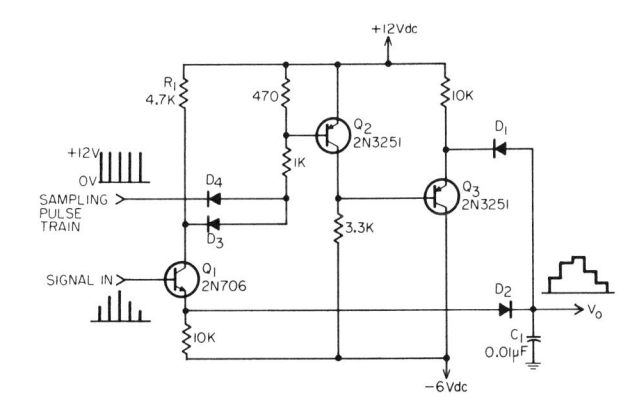

Sample-and-hold circuit in which a new signal level is sensed only when a sampling pulse arrives at the other input.

Diodes provide switching and isolation for absquare circuit

William R. McWhirter, Jr.
Naval Ship Research and Development
Center, Annapolis, MD

The absolute-value-squaring (absquare) circuit provides the designer with a peculiar and useful nonlinear function. The circuit voltage-transfer characteristic is $V_{out} = +V_{in}/V_{in}/10$. Using a few diodes, resistors, an op amp and a general purpose multiplier, a circuit designer can closely approximate a fairly complex function which has been available only on the more modern analog and hybrid computers. The outstanding feature of this circuit is that it uses common, inexpensive components comprising a relatively simple circuit with the tradeoff being the loss of computer-level accuracy.

In circuit operation, diodes D_1 and D_2 act as switches to assure that only absolute values of V_{in}, $/V_{in}/$, are seen at the X terminal of the multiplier. Op amp A_1 acts as the inverter for negative V_{in} values. The amplifier input and feedback resistors should be, at least, good commercial grade types. The Y terminal is permitted to see the actual value of V_{in}; so that the appropriate polarity will be observed at the output of multiplier/divider module. Diode D_3, along with D_1 and D_2, provides circuit isolation so that improper signals cannot stray back to the input signal source, V_{in}, while maintaining desired transfer characteristic integrity.

The Hybrid Systems Corp. multiplier/divider module (Model 107C) was chosen because its output characteristic is $+XY/10$. However, any type multiplier/divider circuit may be selected as long as the proper (and desired) output polarity is observed.

As was stated previously, this circuit can easily be applied if analog and hybrid computer accuracy is not necessary. However, if external "null/gain" and "balance" trimmer potentiometers are used, then ±2% full scale, 4-quadrant accuracy can be achieved. Adjusting these trimmers can help to compensate for the distortion in the voltage-transfer characteristic caused by the diodes forward-voltage drop.

The absquare circuit is particularly useful in simulation and control circuits for continuous, real-time, mechanical systems. One of its more common applications is simulating hydraulic servo valve multiple-flow-gain; as the circuit transfer characteristic closely approximates the nonlinear relationship that can exist between servo valve main-stage spool displacement and valve main-stage output flow. The circuit can provide convenient intentionally nonlinear control for linear-type systems.

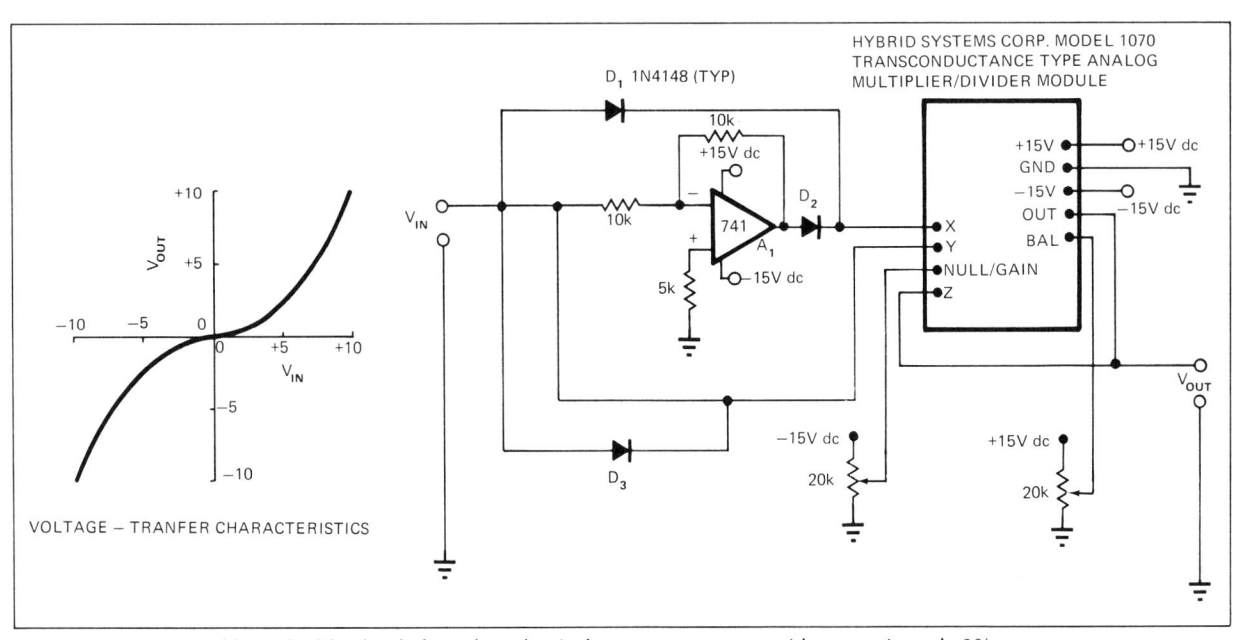

Absquare circuit provides valuable simulation of mechanical system responses with approximately 2% accuracy.

Improved absolute-value circuit

By Miles A. Smither
General Electric
Houston, Tex.

OTHER CIRCUIT designers have described[1,2,3] simple circuits whose output is equal to the absolute value of the input voltage. But all these circuits have the disadvantage that they need several pairs of resistors that are matched to close tolerances. Fortunately, it is possible to design improved absolute-value circuits that achieve high accuracy, yet which need only a single pair of matched resistors.

Let's look first at the conventional circuit shown in Fig. 1. Both operational amplifiers function as inverters. With a negative input voltage, the right-hand amplifier multiplies e by -1, while the left-hand amplifier contributes nothing to the output. With a positive input voltage, both amplifiers invert the signal and the output is $2e_x - e_x = e_x$. The output is therefore positive for either positive or negative input signals — thus realizing the required absolute value $|e_x|$. For proper circuit operation, the following must be true:

$$R_1/R_3 = 1.000$$
$$R_1/R_0 = 1.000$$
$$R_2/R_0 = 0.500$$

The accuracy of these ratios determines the accuracy of the absolute value $|e_x|$. But, for a precision circuit, it is difficult and expensive to match resistors to the required close tolerances.

Figure 2 shows an improved circuit which requires only one pair of matched resistors (R_1 and R_5). With positive input signals, the output of A_1 attempts to go negative while the output of A_2 goes positive. Diode D_2 is therefore back-biased. Diode D_1 prevents A_1 from saturating in the negative direction; thus it speeds circuit operation when the polarity of the input signal changes.

With negative input signals, the output of A_2 attempts to go negative while the output of A_1 goes positive. Diode D_4 therefore becomes back-biased. Diode D_3 serves the same purpose for negative inputs as D_1 serves for positive inputs. Though type 1N914 diodes are indicated in the schematic, any general-purpose silicon type is suitable.

Note that, in the improved circuit, A_1 acts as an inverter while A_2 is a voltage follower. This contrasts with the conventional circuit in which both amplifiers are inverters.

The circuit of Fig. 2 has been tested using 709-style op amps. With these amplifiers compensated so that their upper 3-dB frequency is 10 kilohertz, the absolute-value circuit works well with input frequencies up to 1 kilohertz. This bandwidth was adequate for the original application (an A/D converter system).

Bandwidth of the amplifiers should be several times greater than the highest sinusoidal frequency of interest — this is necessary because high-frequency components generated within the absolute-value circuit must be accommodated to preserve the circuit's accuracy. Ideally, the bandwidth of the op amps should be at least ten times greater than the bandwidth of the input signal.

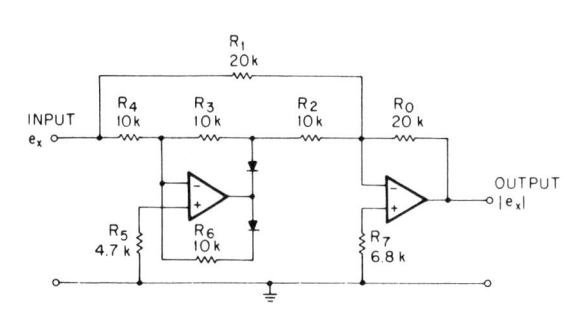

Fig. 1. This widely used absolute-value circuit has the disadvantage that many of the resistors must be accurately matched.

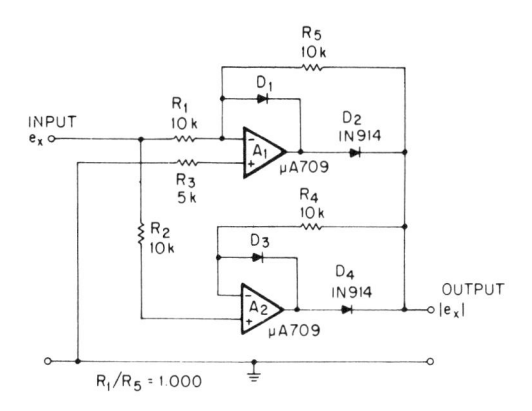

Fig. 2. In this improved circuit only one pair of resistors (R_1 and R_5) is critical.

References

1. "Handbook of Operational Amplifier Applications," *Burr-Brown Research Corp*, 1963, p.73.
2. J. N. Giles, "Linear Integrated Circuits Applications Handbook," *Fairchild Semiconductor*, 1967, p.150.
3. "Applications Manual for Operational Amplifiers," *Philbrick/Nexus Research*, 2nd edition 1966, p.59.

Fixed bias extends zener range

by Peter Kestenbaum
General Instrument
Hicksville, N.Y.

ADDING A SINGLE constant-current diode to a conventional zener regulator allows the zener to serve as a reference over a very wide range of input voltage. The constant-current diode fixes the zener's bias current, eliminating the usual problem that stems from the fact that the zener's bias current is a function of input voltage.

If a series-pass transistor is used with the zener circuit, the regulator becomes an unsophisticated dc supply for inputs of varying dc or rectified ac with large ripple.

In the circuit shown, a 1N-5302 constant-current diode provides the small bias necessary to sustain V_Z. Other diodes are available for constant currents in the 4- or 5-mA range. With a power transistor, these allow the design of hefty supplies for dc into the ampere range without a Darlington arrangement. They offer great ripple reduction.

For a particular supply, one selects the zener according to V_{OUT} and the no-load dissipation. The constant-current diode must supply

$$I = \frac{I_{out}}{\beta} + I_{zener\ bias}$$

while the transistor is chosen for the required dissipation and high beta. The constant-current diodes require about a 3-V drop to stabilize. This fixes $V_{in\ min}$ to $V_Z + 3$, which is the only real restraint on the circuit.

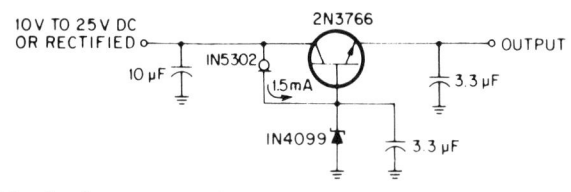

This simple zener regulator covers a very wide range of input voltages — even with large ripple content. It delivers at least 40 mA at a nominal 6 V.

Gating circuit detects positive or negative peaks

Clark S. Penrod and **W. T. Adams**
Applied Research Labs., Austin, Texas

The following circuit provides a method of detecting and sampling the positive or negative peaks of a voltage waveform. The circuit is amplitude restricted only in that the input voltage should not exceed the maximum for the comparator or analog switch. Frequency sensitivity is determined by the Q of the phase lag network. Duration of the sample time is controlled by the monostable. The output of the circuit is a pulse of the same amplitude as the input but only as wide as the monostable "ON" time.

The circuit works as follows. An input voltage is compared to a delayed version of itself. Whenever the voltage is increasing, it will be larger than the delayed version. When the signal begins to decrease, the delayed version will become larger, since it follows, in a time sense, the input. When this happens, the output of the comparator changes state and triggers a monostable. The monostable output drives an analog switch which samples the input for the duration of the monostable. The circuit shown will sample the positive peaks. Negative peaks may be sampled by driving a monostable with the output at point B.

When the attenuator is unnecessary, a more accurate representation of the peak voltage may be obtained by sampling at point A instead of the input. If this is done, care should be taken not to load the output of the phase lag network.

This circuit has been constructed at our facility with the simple RC phase lag network and FET switch shown in **Fig.1**, and was used to detect peaks in the waveform of a sonar video signal.

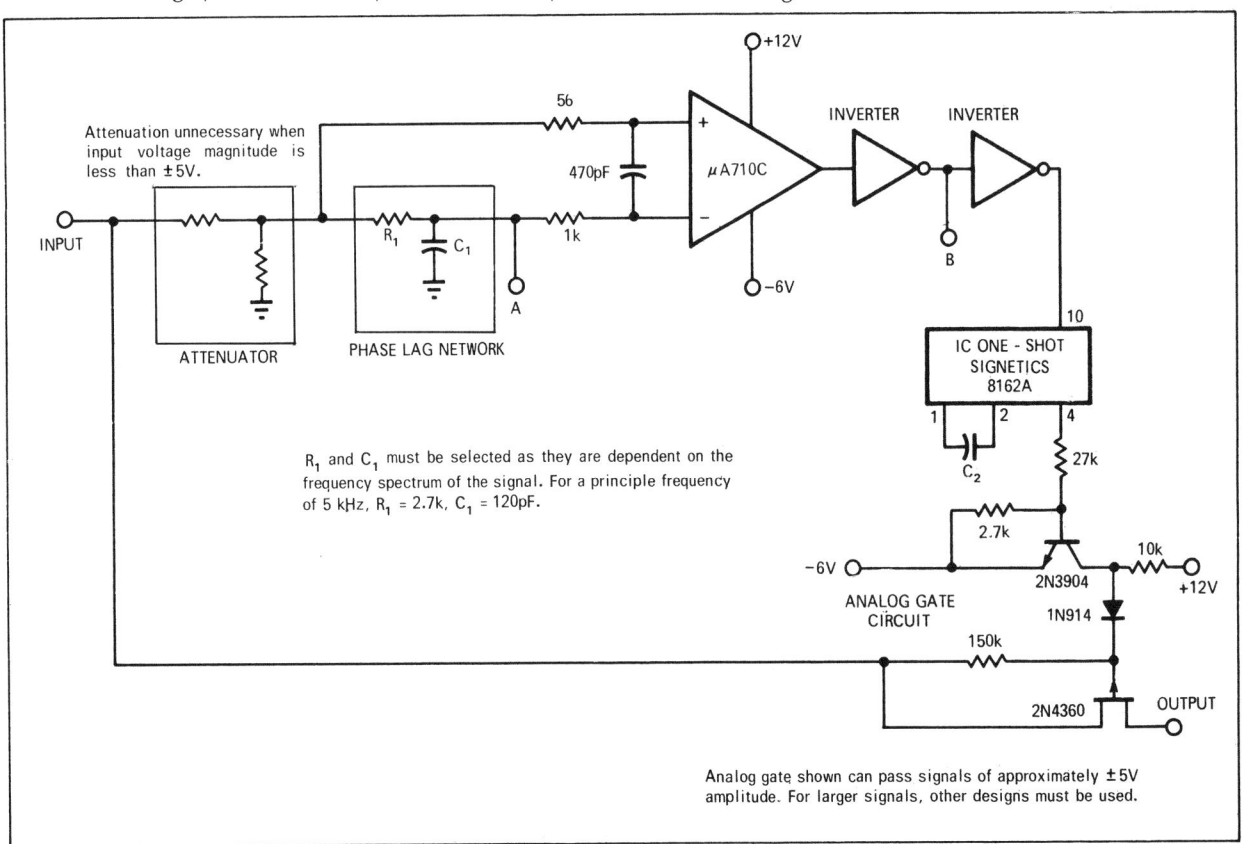

Fig.1—Waveform peaks are detected with this circuit by comparison of the waveform with a delayed version of itself. Detection of a positive peak is triggered when the original input falls below the voltage level of the delayed input.

True RMS measurements using IC multipliers

Karl Huehne and **Don Aldridge**
Motorola Semiconductor Products Div. Phoenix, AZ

Mathematically, the RMS value of a function is obtained by squaring the function, averaging it over a time period T and then taking the square root:

$$V_{RMS} = \sqrt{\frac{1}{T} \int_o^t V^2 dt}$$

In a practical sense this same technique can also be used to find the RMS value of a waveform. Using two MC1594 multipliers and a pair of op amps, an RMS detector can be constructed. The first multiplier is used to square the input waveform. Since the output of the multiplier is a current, an op amp is customarily used to convert this output to a voltage. The same op amp may also be used to perform the averaging function by placing a capacitor in the feedback path. The second op amp is used with a multiplier as the feedback element to produce the square root configuration.

This method eliminates the thermal-response time that is prevalent in most RMS measuring circuits.

The input-voltage range for this circuit is from 2 to 10V pk. For other ranges, input scaling can be used. Since the input is dc coupled, the output voltage includes the dc components of the input waveform.

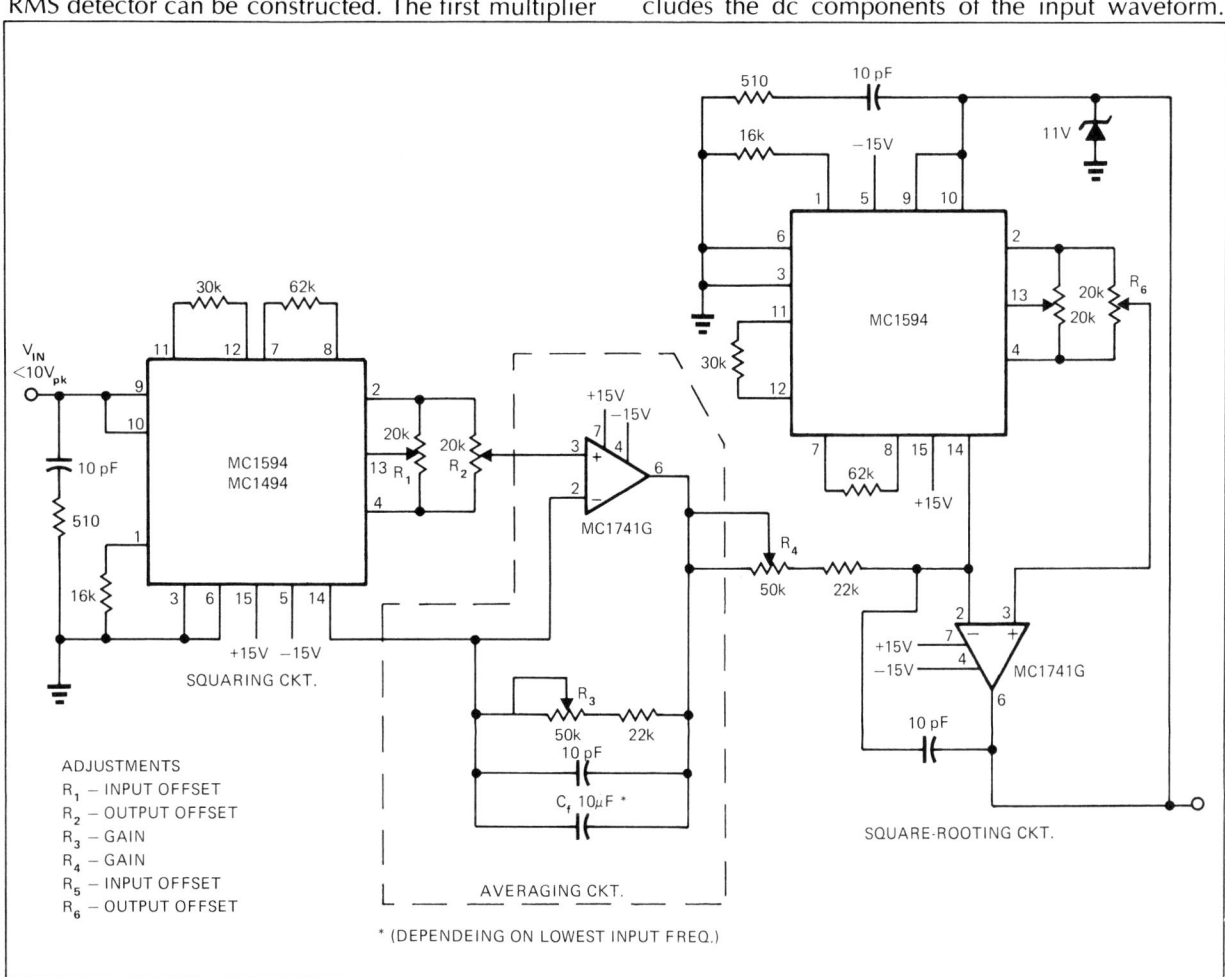

This RMS circuit follows math procedures of squaring instantaneous input values, averaging over time and taking the square root.

Fast-Rise Current Switch

By Thomas W. Collins
IBM Corp.
San Jose, Calif.

This circuit demonstrates the use of the law of conservation of electrokinetic momentum (i.e., constant flux linkages) in driving highly inductive loads which require fast-rise currents. Using this principle, the current is not limited to the slow L/R risetime constant. Instead, the current will rise as fast as it can be switched into the load.

The law of conservation of electrokinetic momentum states that the product ($L \cdot i$) of an electromagnetic system cannot change except as the system is operated on by an external voltage; in which case the change of ($L \cdot i$) is equal to the product ($v \cdot t$) of time and voltage.

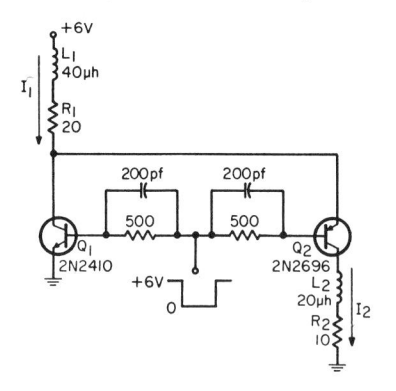

Fig. 1. Fast-rise current switching circuit.

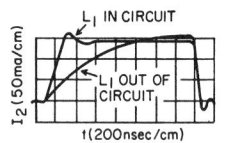

Fig. 2. Load current traces.

The law and its refinement are reduced to practice in the circuit of Fig. 1.

The circuit is in its stand-by condition with Q_1 on and Q_2 off. The standby current I_1 is flowing through L_1, R_1 and Q_1. When a negative input pulse is applied at t_0, Q_1 turns off, and Q_2 turns on. A current i_2 now flows through L_1, R_1, Q_2, L_2 and R_2. At this time ($t > t_0$), the equation for $i_2(t)$ is:

$$i_2(t) = \frac{V}{R_T}\left[1 - e^{-\frac{R_T}{L_T}t}\right] + \frac{L_1}{L_T}I_1 e^{-\frac{R_T}{L_T}t}$$

where
$$R_T = R_1 + R_2$$
$$L_T = L_1 + L_2$$

It can be seen that at t_0+, the current (i_2) immediately jumps to:

$$I_2(t_0+) = \left(\frac{L_1}{L_1 + L_2}\right)I_1$$

where $I_1 =$ steady state standby current $= +V/R_1$.
Three conditions are available at t_0+:

(a) $I_{2_{0+}} > I_{2_\infty}$,

where I_{2_∞} is the steady state value of i_2 as t approaches infinity

$$I_2 = V/R_1 t R_2$$

(b) $I_{2_{0+}} > I_{2_\infty}$
(c) $I_{2_{0+}} = I_{2_\infty}$

Condition (c) is the refinement of the law. If this condition is fulfilled, the load current (i_2) will rise as fast as the transistors can switch, and it immediately assumes its steady state value without any L/R transient. This condition can be fulfilled if:

$$L_1 = L_2(R_1/R_2)$$

This analysis neglects the voltage drops across the "on" transistor. The only limitation to the circuit is the high voltage which appears across Q_1 when Q_1 goes off. This is a function of the switching time of the transistors. In order to protect Q_1, it must be turned "off" so that the maximum collector voltage of Q_1 is not exceeded. Therefore,

$$L_2 di_2/dt < V_{CBX_{Q1}}$$

Figure 2 shows oscilloscope traces of the actual load currents in the circuit. When L_1 is out of the circuit, the time constant $= L_2/R_1 + R_2$, and risetime $= 840$ nsec. With L_1 in the circuit, the law of constant flux linkages now applies; current risetime improves remarkably (time constant is effectively 0; rise time $= 100$ nsec).

This circuit is applicable as a current driver for digital computer memory arrays, or it may be used for driving relay coils for faster operation.

Analog switch and IC amp controls 80V rms

Dominick J. Musto
Austin Electronics

This circuit is invaluable wherever speed, accuracy and reliability are needed to switch high level analog signals into a circuit operation. It has been successfully used in digital-to-synchro converters whose application required a peak output signal of 80V rms. By adjusting the gain and increasing the negative-cutoff voltage applied to the FET gate, the output can easily reach 90V rms. Its most unique feature is the manner in which the high-level output is obtained while incorporating a feedback network to stabilize the output against changes in circuit parameters.

FET analog switching at the input of op amp A_1 uses the "virtual ground point" of the inverting input to simplify the switching procedure. A $-12V$ dc applied to the gate of the FET is sufficient to maintain channel cutoff for ac signal inputs between $\pm 10V$. With a peak signal of $+10V$, the $-12V$ at the gate is sufficient to maintain channel cutoff ($-2V$ between gate and drain). Voltage between gate and source is virtually constant at $-12V$. During channel cutoff, R_2 provides a ground path for the op amp inverting input, thereby minimizing voltage offset and preventing noise pickup on the input lead.

A ground potential applied to the gate of the FET turns on the channel which has a low R_d of 50Ω. With the FET ON, the input signal is amplified by op amp A_1 and applied to the unity gain, 0° phase shift, power amplifier A_2. The output of A_2 feeds the primary of autotransformer T_1. The transformation ratio of T_1 is chosen at 5:1, i.e., $\frac{V_o}{V_i} = 5$, so as to not overdrive the power amplifier.

The output V_o is dependent upon the feedback ratio of the op amp and in this case is regulated at 19.8, i.e., Loop Gain = $\frac{R_F}{R_S}$ = 19.8.

The autotransformer connection provides a closed loop for dc feedback, thereby maintaining the quiescent dc potential at the output of A_2 very near to ground potential. The RC combination, R_1 and C_1, compensates for phase shift, and in this circuit application it is chosen to resonate with T_1 at 400 Hz to increase the load impedance at A_2. Voltage spikes present from input switching are shunted to ground by C_2.

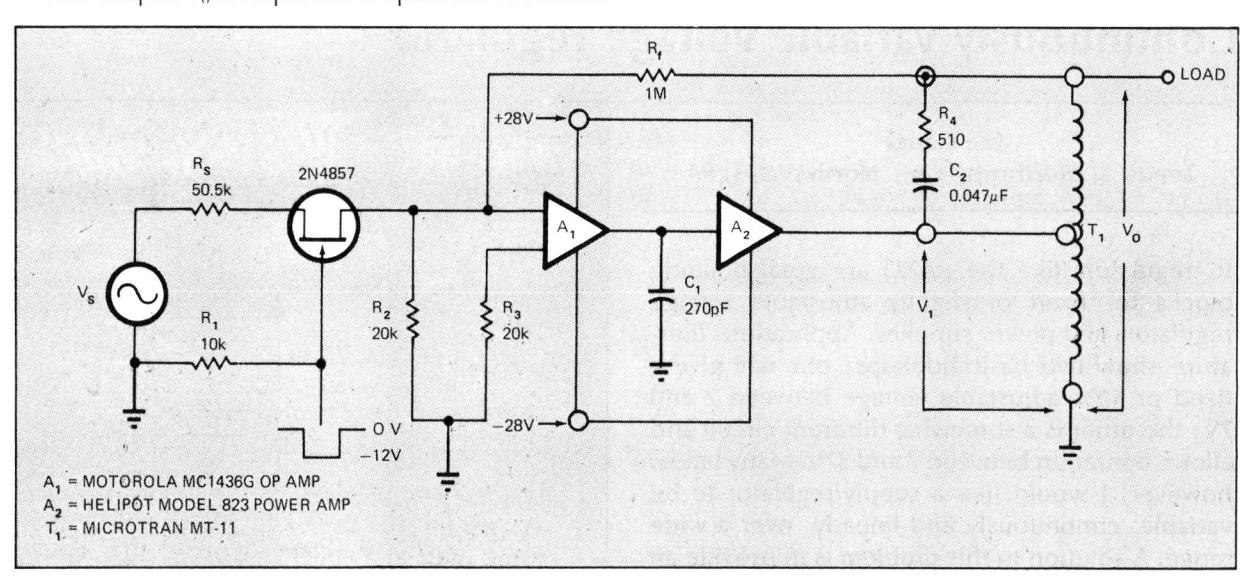

Analog arithmetic unit offers good accuracy

Charles F. Wojslaw,
National Semiconductor, Santa Clara, Calif.

In automated or computer-controlled test systems, it is often necessary to process several sequential analog signals prior to their being digitized and stored in memory. The analog circuit shown here offers the accuracy, speed and flexibility to perform sample-and-hold, divide, subtract, or amplify functions. The circuit can achieve a ±1mV + 5μV/msec accuracy. If it is designed with J-FET analog switches, the circuit provides the necessary speed and reliability to operate in a computer test system.

The circuit can have gain by closing S_4 which configures A_1 as a non-inverting amplifier with gain equal to $(R_2 + R_1)/R_1$.

For the subtract function, the timed switch S_3 is opened during V_1 test time to store V_{t1} test time, again storing V_{t1} across C_1, and then closing S_1 to transfer the voltage. The output voltage, $V/_{out}$, is equal to K (V_{t1}).

Simple division for V_{t1} is achieved by closing S_2 and opening S_3. The output voltage V_o is equal to $K \frac{C_1}{C_1 + C_2} \cdot V_{t1}$.

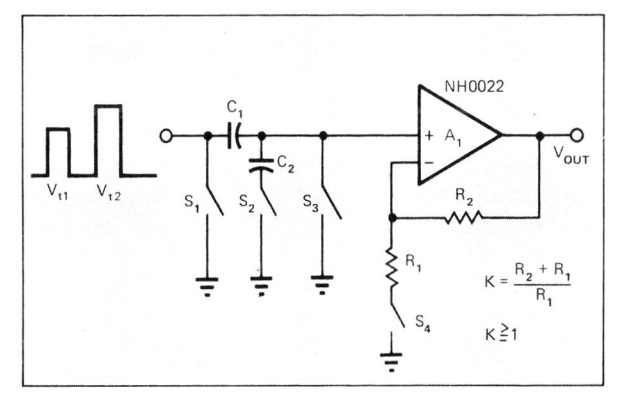

Analog arithmetic unit can sample-and-hold, divide, subtract or multiply. Using FET analog switches in place of mechanical switches shown in the circuit can provide processing speeds compatible with computer test systems.

Analog monitor has threshold and hysteresis controls

Irving Krell
US Engineering Co., Los Angeles, Calif.

Unlike the usual window detector which provides a positive or zero output voltage, this circuit yields negative, zero, or positive output voltages.

Trimming potentiometer R_1 and R_2 adjust the upper and lower threshold, and R_3 and R_4 the upper and lower hysteresis. For example, if e_{in} is to have an acceptable range of +4.5V to +5.5V, when e_{in} exceeds +55V, the A_1 output goes negative. This causes Q_1 and Q_2 to saturate, and the output of Q_2 goes from +5V to zero. The TTL logic now tells that e_{in} has exceeded +5.5V. Upper hysteresis assures that the A_1 output will remain negative until the input e_{in} has dropped to, say, +5.3V, depending upon the setting of R_3. Similarly, when e_{in} drops to less than +4.5V, the A_2 output goes positive, saturating Q_3. Now the TTL logic tells that e_{in} is less than +4.5V. Lower hysteresis, adjustable by R_4, makes the A_2 output remain positive until the input e_{in} has risen to, say, +4.7V.

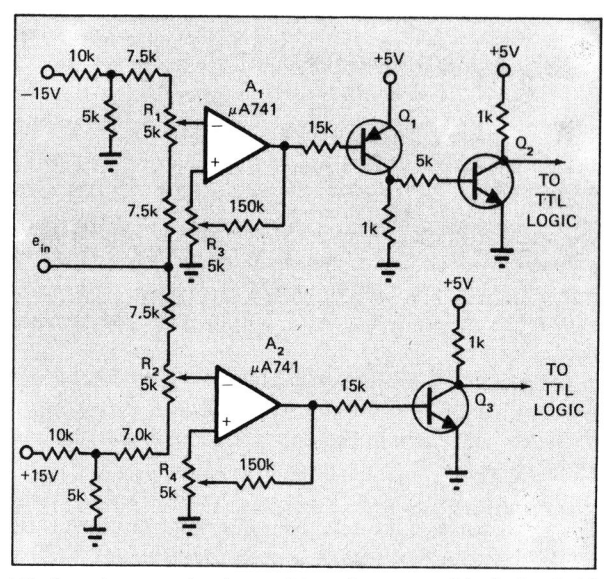

Window detector circuit provides adjustment of both threshold and hysteresis levels. Outputs indicate high, low or acceptable input levels, rather than the normal in-tolerance or out-of-tolerance indication.

IC op amps make inexpensive instrumentation amplifiers

Helge Mortensen, National Semiconductor Corp.
Santa Clara, Calif.

Instrumentation amplifiers using IC op amps are normally dependent on closely matched resistors for good CMRR performance. Further, the adjustment of gain and CMRR interact unless a rather sophisticated design approach is taken, and building an instrumentation amplifier with high input impedance usually requires at least three op amps. By using a 741 op amp with feedback to the offset adjustment as shown in **Fig. 1**, rather than to the non-inverting input, many of the above problems can be eliminated.

Some new limitations are encountered though. For example, while most instrumentation amplifiers can be programmed for unity gain, this set-up cannot. To program for unity gain, R_2 would be 2 kΩ, which means that pin 5 would have a potential of 2/3 of $-V_{cc}$, which would turn off the internal transistor Q_6. Unity gain will also overdrive the input stage. From experimental results, this configuration works best when gain is kept >50. The basic instrumentation amplifier using a 741, as shown in **Fig. 1**, provides 5V output for 100 mV input. The 100 kΩ resistor (R_2) connected from the output terminal to the offset null provides a feed-back path to the internal 1 kΩ resistor. In order to balance the set up, another 100 kΩ resistor (R_1) is added from ground to pin 1.

The gain is given by: $R_2/2 \times 1k$, and the output offset can be varied over the full output range by referencing R_1 to voltages other than zero. In addition, an output sense point is provided so that a unity gain power booster may be added in the feed back loop, as depicted in **Fig. 2**.

The gain stability for the amplifier is dependent on the temperature coefficient of the diffused resistors ($\approx 0.25\%$/°C). By shunting the 1 k's with smaller resistors this effect is minimized with increased gain.

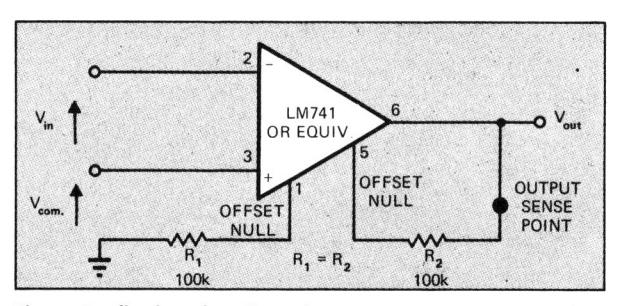

Fig. 1 — Feedback to the offset adjustment of an IC op amp, rather than the normal feedback to the input, converts the device to an inexpensive instrumentation amplifier.

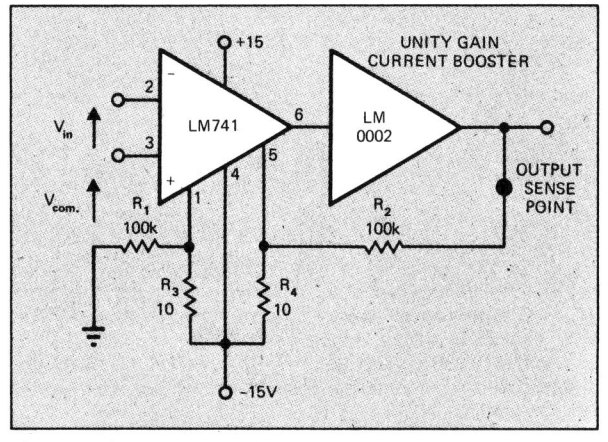

Fig. 2 — Unity gain power booster added to the basic circuit of **Fig. 1** provides more flexible operation at only a modest increase in cost, and is still less expensive than more complex designs.

Voltage-controlled current source

by Peter T. Skelly
University of Washington
Seattle, Wash.

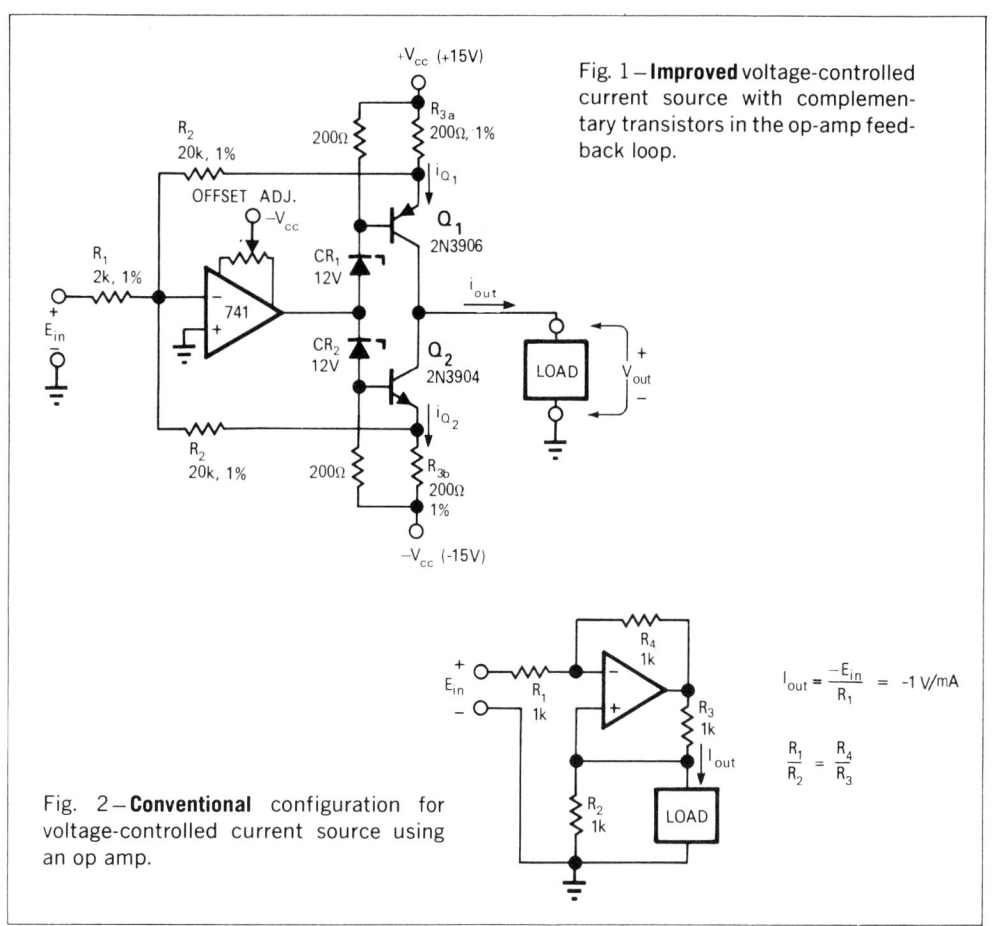

As shown in **Fig. 1**, a voltage-controlled current source can be constructed using complementary transistors in the feedback loop of an operational amplifier. This type of circuit is superior to the more usual op-amp current-source configuration (**Fig. 2**) because the improved circuit always has near-zero common-mode voltage at the input to the op amp. With the conventional circuit of **Fig. 2**, the common-mode input voltage is approximately equal to the voltage drop across the load.

The circuit of **Fig. 1** was designed for use in an integrator with a ground-referenced integrating capacitor. With the component values indicated, it produces 1 mA/V. The general transfer equation is:

$$i_{out} = \frac{2V_{in} R_2}{R_1 R_3}$$

Resistors R_{3a} and R_{3b} sense the current through Q_1 and Q_2 so that a voltage proportional to the difference ($i_{Q1} - i_{Q2}$) is fed back to the input of the op amp for comparison with the input voltage. The output current I_{out} also equals this current difference ($i_{Q1} - i_{Q2}$) if we neglect the base currents of Q_1 and Q_2.

The zener-diode voltages determine the quiescent-current level. Operation at very low quiescent currents is possible without crossover distortion because of the current-sensing feedback.

Lowest output-offset currents can be achieved by using transistors with large matched betas and by nulling the offset voltage of the op amp. The op amp should be compensated for operation at a gain of R_2/R_1. Frequency response of this type of circuit is limited by op-amp bandwidth and slew rate and by the frequency response of the transistors. For the circuit shown, frequency response is limited to 1 MHz by the performance of the specified op amp.

Fig. 1 – **Improved** voltage-controlled current source with complementary transistors in the op-amp feedback loop.

Fig. 2 – **Conventional** configuration for voltage-controlled current source using an op amp.

Linear signal compressor has wide dynamic range

Richard Karwoski
Raytheon Co. Sudbury, Mass.

Here is an unusual and flexible way to compress or limit signals linearly and over a wide dynamic range.

Specific applications for this type of circuit occur in high-quality sound recording or other situations where small-amplitude audio signals must be compressed or limited to prevent overdrive. Most conventional limiting systems have drawbacks. Either they use nonlinear devices which distort the signals or they provide an inexact compression region which usually involves some sort of calibration procedure and always involves trial-and-error adjustments.

With the circuit shown in **Fig. 1**, amplifier A_4 produces a dc control voltage. The audio input signal is amplified by A_3 and rectified by CR_1 to provide the dc input for A_4.

The main signal path is from input to output via A_1. The local feedback around A_1 allows compression and gain adjustments. With R_f shorted, the control circuitry has no effect and the system becomes simply a linear voltage follower. The gain equation:

$$e_o = e_i \left[1 + R_f \left(\frac{R_s + R_n}{R_s R_n} \right) \right] \quad (1)$$

reduces to $e_o = e_i$

The system shown in **Fig. 1** is adjusted to provide unity gain with an input of 100 mV, regardless of the setting of R_f. No additional gain adjustments are needed even if the limiting is readjusted.

The limiting mechanism is based on the following equation for the configuration shown:

$$\frac{X}{1 + X} = Y \quad (2)$$

For small values of X, the equation degenerates to Y = X, which is equivalent to one-to-one compression. For large values of X, the output is asymptotically limiting.

Note the similarity between Eq. 2 and the well-know gain formula of Eq. 3.

$$\frac{Ae_i}{1 + A\beta} = e_o \quad (3)$$

In the limiter circuit, the feedback element β is an analog multiplier, the conductance of which depends on the control voltage from A_4, which in turn is proportional to the input amplitude.

The transfer function for the complete circuit is as follows:

$$\frac{e_o}{e_i} = \frac{\left[1 + R_f \left(\frac{R_s + R_n}{R_s R_n} \right) \right]}{1 + 410 \, (R_f/R_s) e_i \, pk} \quad (4)$$

A suitable value of R_n can be selected so that the 100-mV unity gain point remains the same for any setting of R_f. Then, with respect to this initial setting, one can adjust compression without affecting gain. **Fig. 2** shows various compression characteristics that can be obtained with the component values shown in **Fig. 1**.

Any suitable analog multiplier can be used in this circuit. A Teledyne Philbrick Type 4452 was used in the prototype version and it provided the required performance.

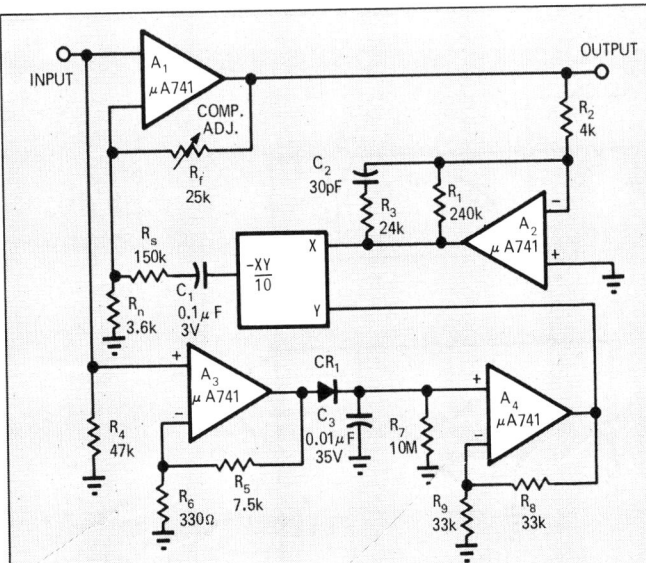

Fig. 1—**Signal-compression** circuit uses analog multiplier as gain-control element. Input signal is rectified and applied to multiplier which determines feedback to A_1.

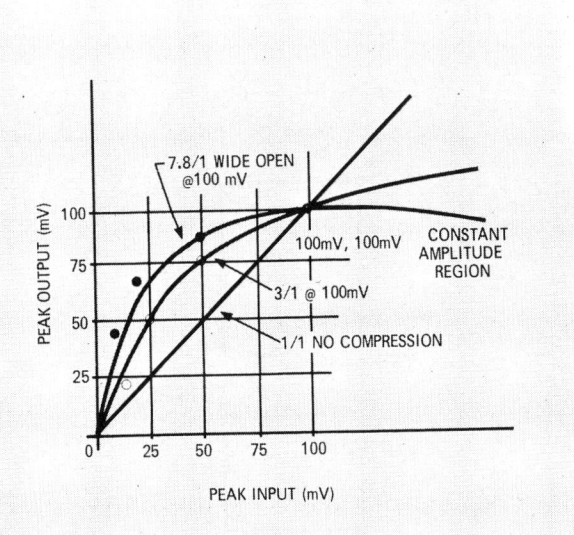

Fig. 2—**Various input-output characteristics** for different settings of R_f in the circuit of **Fig. 1**. Circuit is initially adjusted so that it provides unity gain with a peak input of 100 mV.

IC functions as sampling amplifier or tone-burst gate

By Gerald S. Givens
LTV Electrosystems
Greenville, Tex.

A LOW COST IC amplifier can be gated by applying digital pulses to its AGC line. Depending on the relative frequencies of the signal and the gate pulses, the circuit can be used either as a sampling amplifier or as a tone-burst gate. Only one IC gate and a few resistors are needed in addition to the basic IC amplifier.

With the components specified in the schematic, the circuit can provide up to 500,000 samples of the input signal per second. The minimum width for the sample pulse is 0.8 µs. If the load is reduced from 56 kΩ to 10 kΩ, then the minimum usable sampling-pulse width increases to 2 µs.

A type CA3001 amplifier (70-Ω output impedance) could probably be used instead of the indicated CA3000 (7-kΩ output impedance) to provide improved load driving capability. In the original application, however, the CA3000 was chosen because it has a larger AGC range (90-dB typical, compared with 60-dB for the CA3001).

The possible sampling frequency of 500,000 per second allows the circuit to be used for sampling signal frequencies up to 250 kHz. Alternatively, the width of the gating pulse can be increased so that it is wide compared with the period of the signal frequency. The output will then consist of a tone burst with duration equal to the length of the gate pulse.

The IC amplifier gives an output of +5 Vdc when full AGC is applied. In the amplifying mode, dc output doesn't exceed +4 V with a signal level of 2 V pk-pk. Thus the outputs of several amplifiers can be combined, using diodes biased to clip signals at a level of 4.3 V. This type of circuit would lend itself to multiplexers or tone-burst sequencers.

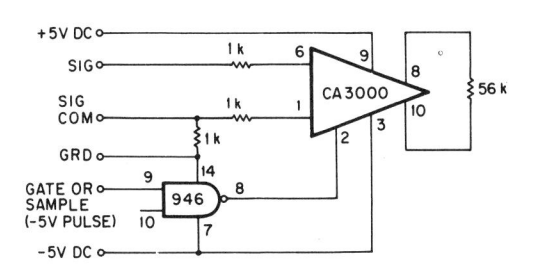

Depending on the signal frequency and on the width of the gate pulse, this simple circuit is either a sampling amplifier or a tone-burst gate.

Use opto-isolation in your next video amp

Dave Weigand
Gulf & Western Research, Swartmore, PA

Often a low-frequency (AGC) gain control or multiplier function is required in a video amplifier. The standard technique is a JFET/resistor voltage-divider circuit similar to the one shown in **Fig. 1**. This works fine for low-level video inputs up to 100 mV or so, but not for high dynamic range video inputs of up to 10V. The FET resistance becomes a function of the input signal voltage level, as well as the AGC gate voltage.

The circuit in **Fig. 2** (Hark, the old Raysistor) replaces the JFET with a photocell/LED combination. Due to the linear resistance of the photocell, the video amplifier gain is not effected by input signal dynamic voltage swing, or by frequencies to 20 MHz. Only direct AGC input-voltage controls the amplifier gain.

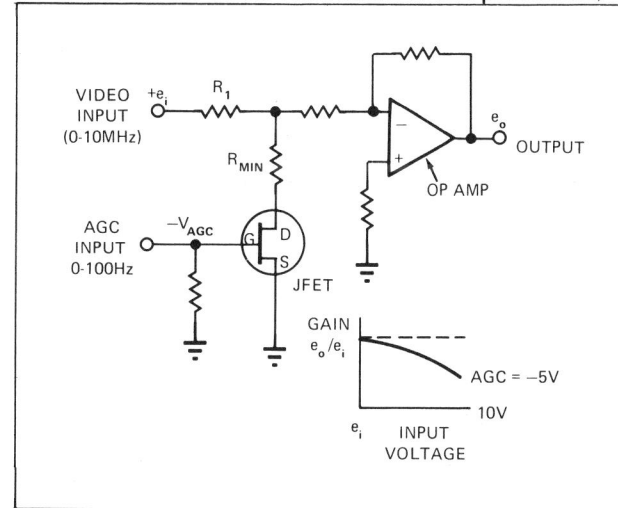

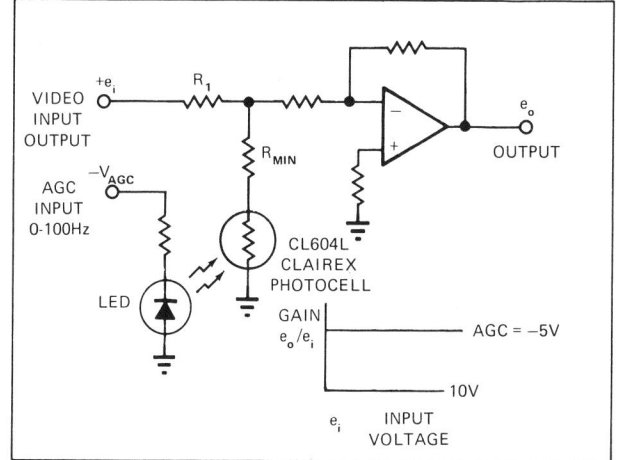

Fig. 2—An LED, used within its linear range of light output vs current input, and a photocell eliminate the interaction of input signal and AGC voltage.

Fig. 1—Conventional JFET-AGC circuit is susceptible to variations of R_{ON} for the FET caused by the applied drain-to-source (signal) voltage.

Analog, Linear and Communications Designs

Op amp in active filter can also provide gain

Arthur D. Delagrange
Sykesville, MD

A common single pole-pair Butterworth filter (top circuit) has unity gain in the passband. Frequently a designer requires some additional gain when part of the signal is filtered out, since the amplitude of the remaining signal is normally smaller. The op amp can also provide the gain in most instances.

A modification of the circuit for this purpose is shown (bottom circuit). The filter characteristic is the same except for the gain factor shown. This can be easily shown by taking the Thevenin equivalent of the part of each circuit to the right of the dotted line. It will be the same for both cases, so the voltage at the op amp noninverting input will be the same for both circuits.

This method also works for the low-pass version of the filter, but the lower voltage divider will consist of two capacitors. The method also applies to other filter characteristics, such as Chebyshev.

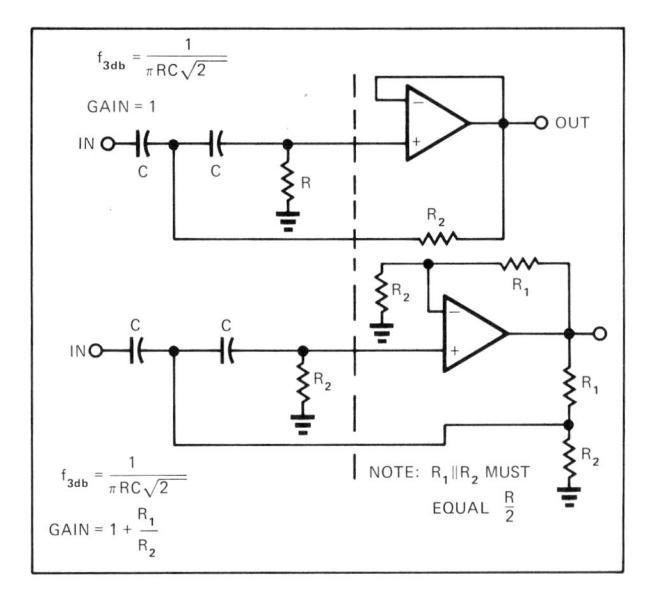

A unity gain Butterworth filter (top) can be modified (bottom) to provide gain to compensate for input-signal attenuation during the process of filtering.

Transistor Bridge Detector

By Denis P. Dorsey
RCA Sarnoff Research Center
Princeton, N. J.

THIS CIRCUIT develops a dc error voltage proportional to the phase difference of two applied signals. It can be used to correct oscillator outputs and as a pulse-width discriminator.

It is shown with a 60-cps reference, but can be designed to operate at other frequencies. Three possible phase relationships can arise when comparing the input drive to the reference emitter supply:

☐ the input drive can be in phase with the reference (zero phase difference).
☐ the input drive may lead the reference (negative phase error).
☐ the input drive may lag the reference (positive phase error).

Transistors Q_1 and Q_2 form a bridge-balancing network, with emitters directly coupled to the secondary of a transformer which provides the reference signal. When the 60-cps emitter voltage is negative, npn transistor Q_2 allows current to flow in its arm of the bridge. If the emitter voltage is positive, pnp transistor Q_1 allows current to flow in the pnp portion. However, in the absence of input drive, bridge current will not flow regardless of emitter reference polarity.

When input drive pulses are applied, Q_1 is saturated. At the same time, inverter amplifier Q_3 saturates Q_2. If the emitter reference voltage is passing through zero volts at the instant both transistors are saturated, the applied input and reference signals must be in phase and no error voltage is developed. If the transistors are saturated when the reference voltage is zero, point A receives no signal and therefore the output error voltage, which is the result of filtering the signal of point A, is also zero.

If the drive pulses phase-lead the emitter reference voltage, the collector of Q_2 is pulsed to the instantaneous emitter reference source. This results in negative pulses being applied to point A, with amplitude directly proportional to the amount of phase lead. The signal at point A is then heavily filtered producing a negative output error voltage that is a direct indication of the leading phase error.

If the drive pulses phase-lag the reference, the collector of Q_1 is pulsed to the instantaneous emitter reference source. Point A now receives positive pulses, with amplitude directly proportional to the amount of phase lag. Again, the signal at point A is heavily filtered establishing a positive output error voltage that is a direct indication of the lagging phase error. The error voltage also is proportional to the phase and pulse duration of the signal at point A. Therefore, a narrow pulse will not establish the same error voltage as a long pulse. Under these conditions, the bridge detector can be used as a pulse-width discriminator, where the output error is a direct indication of the phase relation and pulse duration.

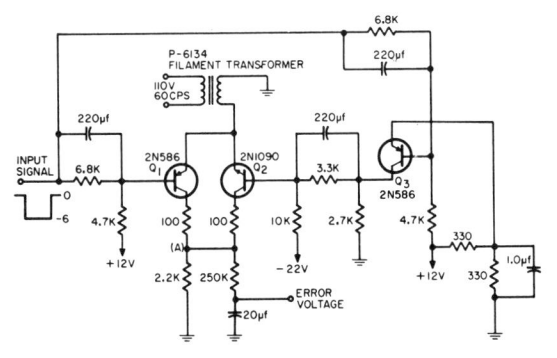

Transistor bridge phase detector.

Op-amp comparator with latching function

by Tom Cate
Burr-Brown Research
Tucson, Arizona

BY ADDING just a few components to an operational amplifier comparator, one can obtain a latching function. The circuit compares an input signal against a reference voltage, and when the reference voltage is exceeded the output switches and the comparator latches up. This circuit is the electronic equivalent of a latching relay. Resetting is accomplished manually or electronically. Input impedance and trip point are independent of the input level E_1.

In the figure, the input signal and reference level are summed together through R_1 and R_2. If E_o is positive, then Q_1 will be saturated and D_5 will block the 60 mV of Q_1 saturation voltage from the summing junction of the operational amplifier. When the input level goes more positive than $-E_R$, the output of the op amp will swing negative to about $-1.6V$ and Q_1 will switch off. With Q_1 off, approximately $+2.5$ mA will flow into the summing junction through R_3 and R_4.

The latching operation is regenerative and analogous to the operation of a flip-flop. The amplifier will remain latched up with its output negative because

$$\frac{+15V}{R_3 + R_4} > \left| \frac{E_1}{R_1} + \frac{E_2}{R_2} \right|$$

with any combination of input voltages.

Resetting may be accomplished by several means, one of which is to short the collector of Q_1 to the common. A manual pushbutton can do the job, or an extra transistor Q_2 can be used if a logic signal is available.

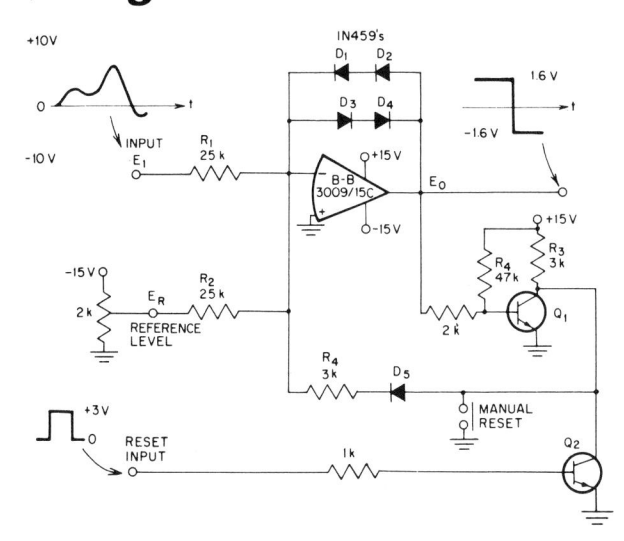

This op-amp comparator latches up when the input signal exceeds the reference.

Accuracy of the circuit depends almost entirely on the match between R_1 and R_2. The latching-circuit sensitivity can be varied somewhat by varying the bias on Q_1 through the resistor R_4. The choice of op amp is not critical and most IC or inexpensive discrete units will perform well.

Phase-locked loop stereo decoder is aligned easily

Bruce Korth
Motorola Semiconductor Products, Phoenix, AZ

Stereo decoders normally incorporate LC filters both to provide 19 kHz selectivity and as part of a doubling circuit to generate the required 38 kHz subcarrier. One of the problems with these decoders has been system alignment, since it is necessary to adjust the coils very accurately to obtain maximum stereo separation. Even after obtaining the correct alignment, the problem of aging of the coils and temperature and humidity variations still exists.

A solution to the problem is the phase locked loop stereo decoder shown. The MC1310 requires only one adjustment, which is controlled by a potentiometer. The correct procedure for alignment is to open the input, pin 2, and monitor pin 10 with a frequency counter. The potentiometer is adjusted until a 19.00 kHz frequency reading is obtained.

For those having limited test equipment, a very simple procedure may be followed which will result in separation of within a few dB of optimum. This method consists of simply listening to a stereo broadcast and adjusting the potentiometer until the stereo pilot lamp turns on. Then the center of the lock-in range is found by rotating the potentiometer back and forth until the center of the lamp ON range is found. This method produces separation of 40 dB (typical) with total harmonic distortion typically 0.3%.

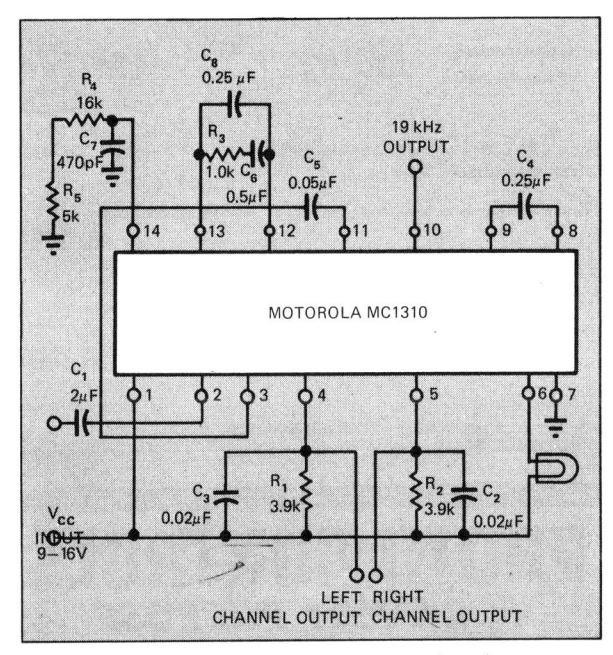

All adjustments for the stereo decoder are made with R_5.

Reduce integrator transients with synchronized gate signals

Entry by Roland J. Turner
RCA Missile and Surface Radar Div. Moorestown, N. J.

When new signals are gated into a high-Q tank circuit serving as a bandpass integrator, ac gating transients may cause adverse ringing of the integrator. This is especially true if the series signal gate is closed when the ac input signal is at its peak level. When this happens the integrator will have a long settling time and will not properly represent the integrated value of the input until after this transient dies out—which may take considerable time, because by definition the bandpass integrator has a high Q and a long memory.

The circuit shown in **Fig. 1** reduces ringing significantly by the following process:

1. The ac input signal to be integrated is amplified in two broadband differential amplifier stages, A_1 and A_2.
2. The amplified signals are differentiated and selected by diodes D_1 and D_2.
3. The input gate pulse, P_1, and the differentiated pulses then drive "and" logic which generates an output pulse, P_S, coincident with the zero crossings of the ac input signal. The leading edge of P_S will always occur at the first signal zero crossing after P_1 initiates the gating action.

The leading edge of the output pulse, P_S will always occur at a zero crossing of the ac input signal and will not cause ringing of the bandpass integrator. If pulse P_S is used to drive a balanced diode bridge, both the gating pedestal and the ac signal transient are eliminated, and the high-Q bandpass integrator will have a fast settling time and may be operated at a faster repetition rate.

The timing waveforms of this circuit are shown in **Fig. 2**. The NAND gates in the logic portion of the circuit are Texas Instruments Incorporated or Fairchild IC packages.

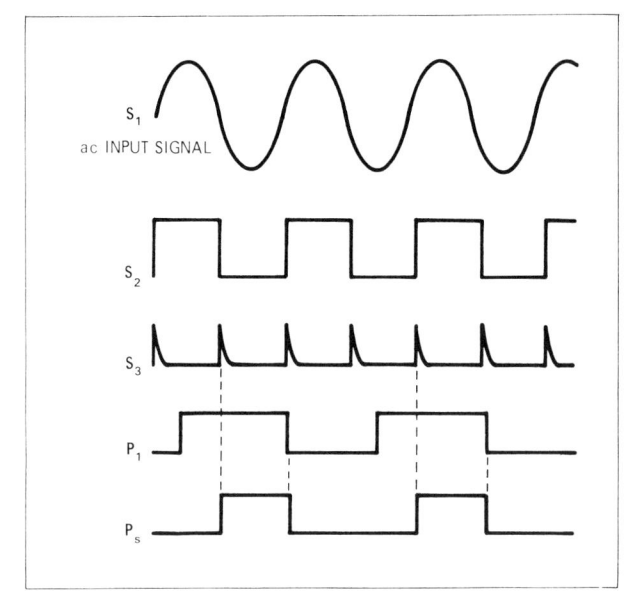

Fig. 2 – **Input signals** (S_1) generate zero crossing pulses (S_2). Normal enable gate pulses (P_1) are then withheld from integrator until zero crossing occurs (S_2), at which time the zero-crossing gate (P_S) enables the integrator tank circuit.

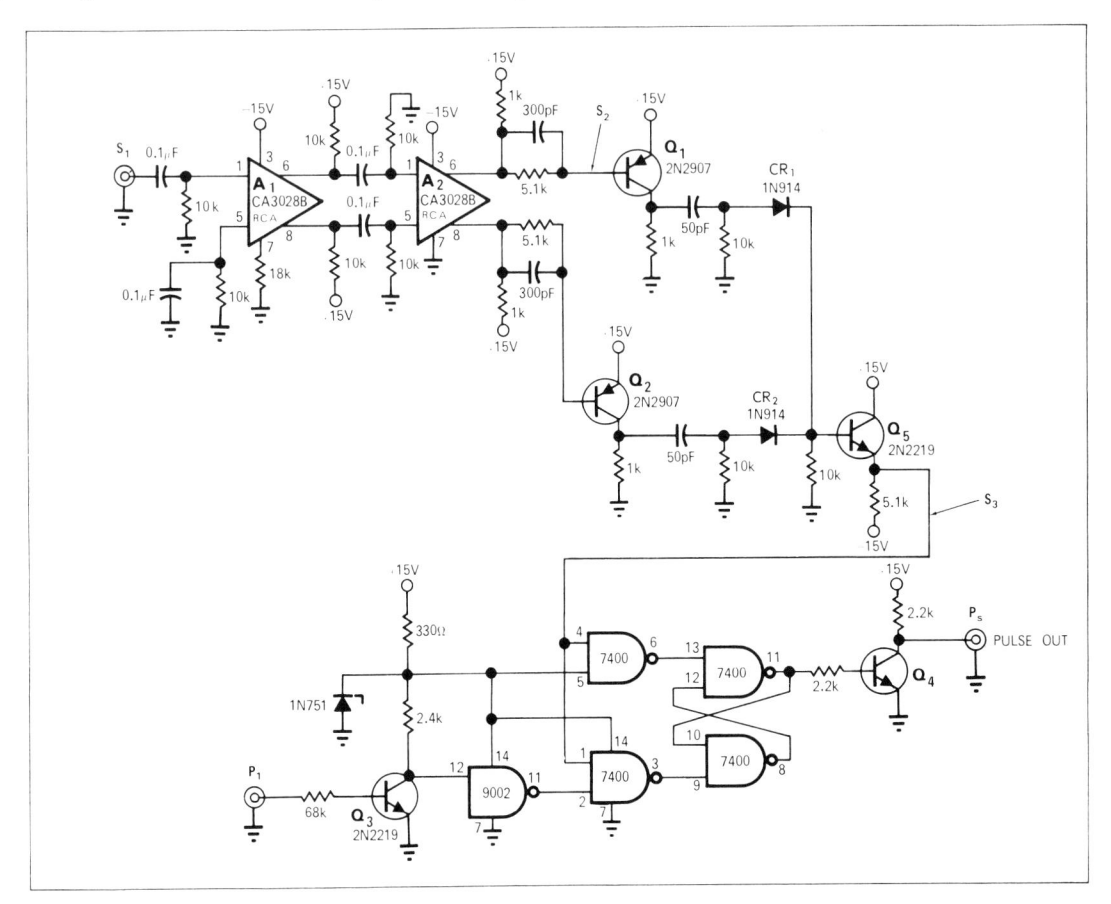

Fig. 1 – **Ringing and long settling times** – characteristic of shock-excited high Q tanks – are eliminated in bandpass integrators by this circuit. Inputs are gated into the tank only at input zero crossings, avoiding shock excitation of the high-Q tank.

Constant-Voltage Current Sink

By Gerardo Rogoff
Technical Measurement Corp.
North Haven, Conn.

WHENEVER CLAMPED LOGIC CIRCUITS are used, the need arises for a power supply with "negative" output current, or, in other words, a constant-voltage sink. Many methods are commonly used to get that current sink, namely:

■ The use of a bleed resistor R_B (Fig. 1a) calculated to take, at V_{CL}, the maximum clamp current plus the holding current of the power supply. When the clamp current goes to a minimum the power supply must provide $I_{CLmax} - I_{CLmin} + I_{hold}$. The disadvantages here are: the high power dissipated in R_B and the need of a separate power supply for the clamp. The regulation is as good as that of the power supply.

■ The clamp supply is replaced by a zener diode (Fig. 1b) that will develop V_{CL} when I_{CL} flows through it. Capacitor C helps to filter the transients that might appear. The disadvantage is that the regulation is poor. The impedance is that of the zener.

■ Two power supplies are connected in series (Fig. 1c). PS_2 provides for V_{CL}, and $PS_1 + PS_2$ give V_{CC}. $I_{S2} = I_{CC} - I_{CL}$. The main disadvantage is the need for two power supplies, one of them (PS_1) being floating.

The new method proposed here (Fig. 1d) consists of a dc amplifier driving an emitter-follower power transistor of the proper polarity (pnp for positive clamping voltage).

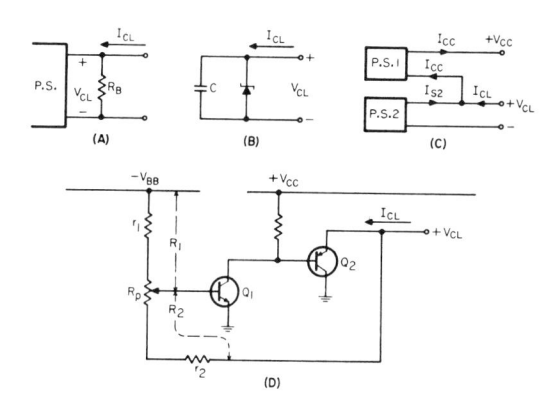

Constant-voltage current sinks: bleed resistor, 1a; zener diode, 1b; two supplies, 1c; and recommended dc-amplifier, emitter-follower circuit, 1d.

The input of the dc amplifier is the comparison between the clamping voltage being controlled and a regulated voltage of opposite polarity (the base-bias voltage of the system would do). With some simplifications, the output impedance Z_6 is

$$Z_0 = \frac{h_{ie}}{\beta_1 \beta_2} \frac{R_1}{R_1 + R_2}$$

where the parameters apply, of course, to the operating point, determined by R_3 (h_{ie} is the most nonlinear in this case).

Modified Emitter Follower Has No Offset

By Anthony C. Caggiano
Safety R & D Corp.
Ridge, N. Y.

FOR MOST emitter-follower applications, such as impedance matching of ac signals, dc offset is unimportant, so the conventional circuit is adequate. The simple modification shown here gives a more versatile emitter follower, suitable for dc as well as ac applications. This modification eliminates the inherent difference between output and input dc levels caused by V_{BE} of the transistor. A further advantage of the modified circuit is that it's less sensitive to temperature variations.

Diode D_1 is selected to match the V_{BE} of the transistor. Thus a silicon diode is used with a silicon transistor, and germanium with germanium. In order to achieve a high input impedance and to correct for errors due to base current and leakage current, the base of the transistor is biased by a resistive divider. (For low input impedances, a simple resistor to ground may be adequate.) The ratio of the divider network is adjusted to give zero volts at the base.

Making the value of R_3 equal to the parallel combination of R_1 and R_2, i.e.

$$R_3 = \frac{R_1 R_2}{R_1 + R_2}$$

minimizes drift due to variations in power-supply voltage.

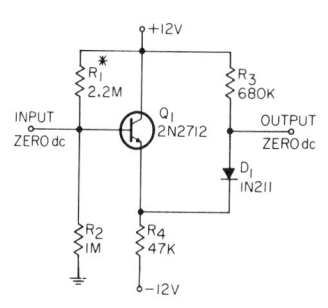

Modified emitter follower has equal dc levels at input and output. Offset due to V_{BE} is eliminated by diode D_1.

Differentiator circuit produces a relative change

Fred W. Etcheverry
Southwest Regional Laboratory, Los Alamitos, CA

In simple differentiator circuits, a voltage is produced proportional to the absolute change of an input signal.

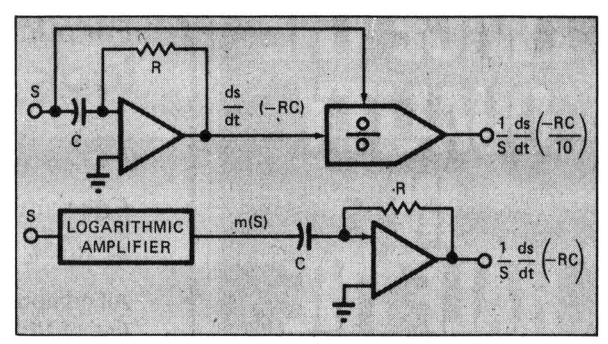

Relative change from a differentiator circuit can be obtained with either of these two circuits. The top circuit, however, does not have the accuracy within the dynamic range of signals with interesting relative change. By using a logarithmic amplifier, as in the bottom circuit, this problem can be solved.

However, it is often desirable to develop a voltage which is proportional to relative change. Such a relative-change circuit can be realized by feeding the input signal to a differentiator and dividing the output of the differentiator by the input signal, as shown (top circuit). This circuit, however, does not have the accuracy within the dynamic range characteristic of signals with interesting relative change.

With the bottom circuit, the input signal is fed to a logarithmic amplifier. The output of the logarithmic amplifier, ln(s), is then fed to a differentiator. The differentiator's output, 1/s (ds/dt) (-RC), is then proportional to relative change

Logarithmic amplifiers can be purchased in modules or constructed from IC op amps by placing a transistor in the feedback loop. If a high degree of stability and accuracy is desired, transistors, such as Fairchild's μ726, and IC op amps, such as Fairchild's μA727, both fabricated on temperature regulated chips, can be used.

Circuit yields linear pulse-width modulation

James D. Doss and Charles W. McCabe,
Los Alamos Scientific Laboratory, Los Alamos, N.M.

This circuit provides very linear pulse-width modulation, provided that the R_2C_1 time constant (see **Fig.**) is very large compared to the pulse width desired. This is generally an easy requirement, since R_2 may be quite large.

Operation of the circuit is as follows: Without an input trigger, the operational amplifier output is approximately −15V due to the positive dc assumed to be 15V. This positive reference would generally be added to the modulation signal in a previous op-amp stage. When the trigger input pulse (amp) train is applied, the op-amp output is driven to about +15V by each trigger. This same +15V signal is coupled through C_1 and D_1 to the non-inverting op-amp input to sustain the positive output.

At the same instant, the charging of C_1 by a constant current will begin. This will result in a very linear negative ramp on the transistor collector and, therefore, on the non-inverting op-amp input. When this ramp has dropped to the level of the instantaneous modulation voltage on the inverting input, the op-amp output will switch to −15V, ending the pulse. Thus, the output pulse is linearly related to the instantaneous applied modulation voltage. The constant current that charges C_1, resulting in the linear ramp, is approximately:

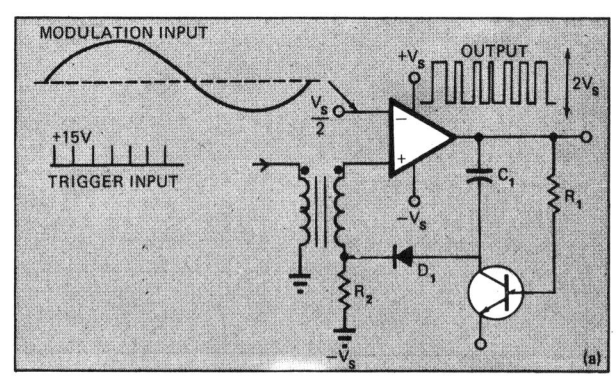

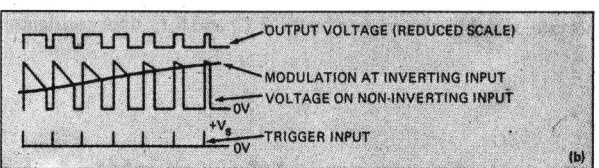

Single op amp pulse-width modulation circuit (a) generates the linear voltage waveforms shown in (b).

$$\frac{2\beta V_s}{R_1}$$

where β is the transistor dc current gain and $2V_s$ is the approximate potential across R_1 when the op-amp output is driven to $+V_s$.

This circuit has been used with satisfactory results in several applications.

Current boosters for IC op amps

by Jiri Dostal
Research Institute for Mathematical Machines
Prague, Czechoslovakia

ASIDE FROM the matter of cost, there's an important functional limitation to increasing the current output from IC op amps: thermal feedback to the input circuitry causes voltage offsets. The best way to overcome the problem is to use a physically separate current booster.

The figure shows (A) a 100-mA booster with zero quiescent current. It uses the full output current of a preceding op amp (typically 5 or 10 mA). If the output current level is less than V_{BE}/R_e, the entire output current is supplied by the amplifier itself. Currents above V_{BE}/R_e are supplied by the complementary emitter followers.

We can extend the output to the 1-A range by cascading a similar booster (B in the figures), by adding complementary transistors (C) or by combining both (D). And if we want to limit short-circuit currents, we can use the popular circuits (E,F).

Diode types are not critical and the transistors are chosen to satisfy power and bandwidth requirements.

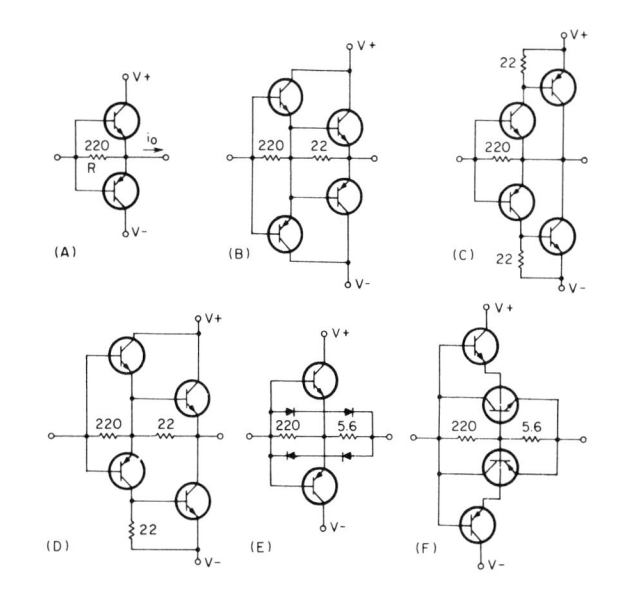

Various schemes for boosting current from IC op amps to 100 mA (A) and to 1 A (B,C,D) and for limiting short-circuit current (E,F).

One Transistor, 50-db Dynamic Range Compression Amplifier

By Robert W. Cotterman
ITT Kellogg
Ft. Wayne, Ind.

NON-SATURATING AMPLIFICATION of widely ranging video signals often is difficult to achieve in transistor circuits. The amplifier shown provides a minimum output signal level of 1.0 v with an input of 20 mv, but does not saturate with a 5 to 6 v input. In the actual application, the signal is later processed for risetime and duration information.

The circuit provides a minimum gain of 1 and a maximum gain of 15. The diodes across the collector load resistors change the stage gain and impedance by shunting the resistors, one by one, as signal input is increased. The 100-K pot is initially adjusted with zero signal to provide about 50 μa collector current. Q_1 must be a high beta at low current transistor such as a 2N336A. Capacitor C_E is selected for proper frequency response. The voltage transfer function for the circuit is shown in the accompanying curve.

In actual application, two such circuits are cascaded. The second amplifier provides a maximum gain of 5 and minimum gain of 1. Overall input-output characteristics are 20 mv to 5 v input with 1.5 to 8 v output. By selecting load resistors and diodes, the gain characteristic can be made to follow many curves.

For Class A amplification, the ac-coupled dynamic load is similar except that two diodes are connected in opposite polarity in parallel with the various load resistors and the biasing circuit is altered.

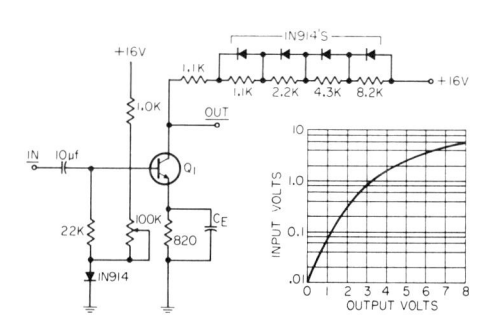

Wide dynamic range compression amplifier with voltage transfer function.

Analog, Linear and Communications Designs 27

Optical sensor has built-in hysteresis

David C. Hoffman
Industrial Nucleonics Corp., Columbus, OH

It is often difficult for an optical sensor with a digital output to operate at zero speed as well as high speeds and still have suitable rise times to drive digital logic. This is especially true when the logic device is at the other end of a transmission line. This sensor has built-in optical hysteresis so that it does provide digital outputs at low, and even zero-operating speeds.

When the optical path from D_1 to Q_1 is completely obscured, Q_1, and also Q_2 and Q_3, are OFF. The output is pulled LOW by R_6. As more light passes between D_1 and Q_1, Q_2 and Q_3 begin to turn ON. The rising collector of Q_3 adds more current through R_2 to D_1. This gives Q_1 more light and drives Q_3 into saturation.

As the light dims, Q_1 begins to turn OFF, and as the output falls, the extra current is cut off turning Q_3 OFF. Because of this hysteresis, there is no constant light level where the circuit will oscillate. The circuit has been tested at 4-kHz operation and is speed-limited at about 20 kHz. □

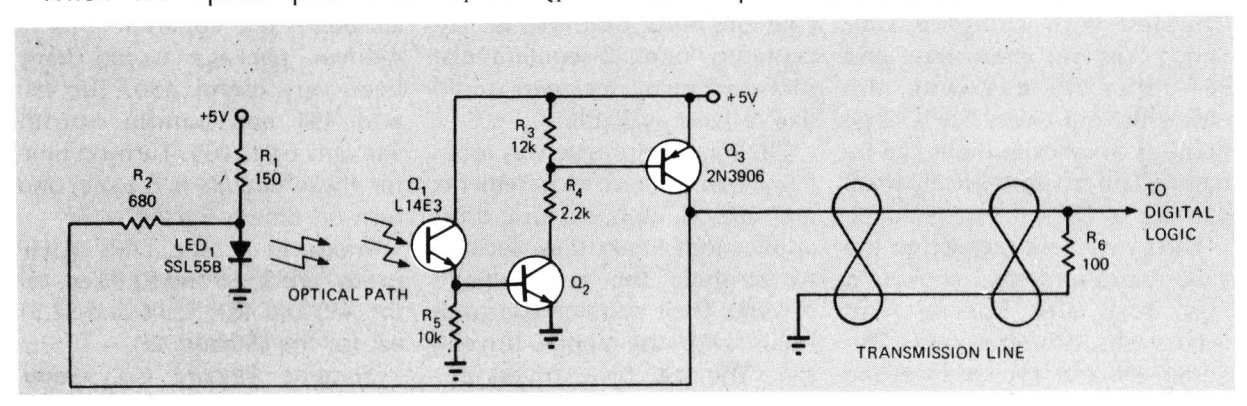

Lamp test becomes circuit test

Edwin R. DeLoach
Astro-Dynamics Electronics, New Orleans, LA

Lamp test circuits commonly use diodes connected as illustrated in (a). This arrangement provides a testing means to locate faulty lamps or other indicators. However, it does not check the circuit for faulty or loose connections, broken wiring or faulty sensor switches.

This can be accomplished with no additional components merely by placing the test diodes in the circuit, as shown in (b).

The circuit functions as a monitor status indicator when the test switch is in the nontest position. Then when a switch completes the circuit, the lamp is energized, indicating status.

Circuit testing is provided when the polarity on the system is reversed. Thus, every switch is "shorted" by the action of the diodes. This provides lamp testing and continuity check of the entire circuit. Note that switches 2 and 3 are connected differently than switch 1. This configuration necessitates current flow through the switch contacts, and provides indications of dirty switch contacts or open wipers as well.

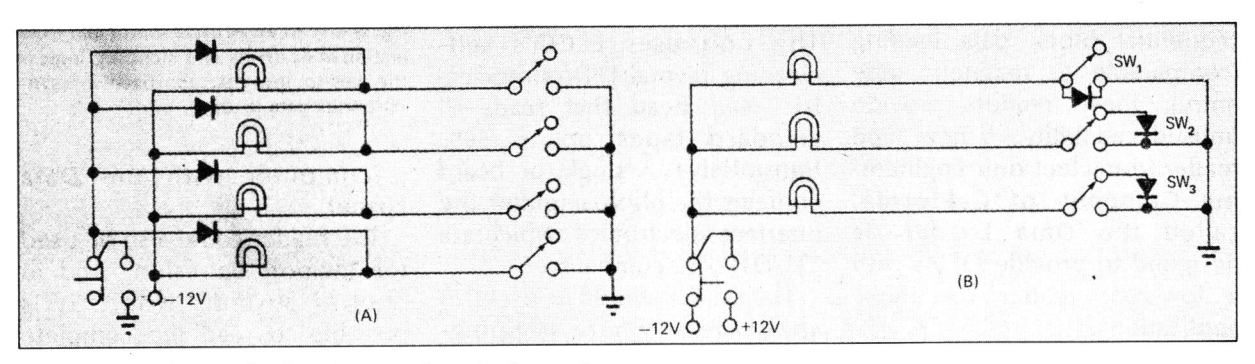

Circuit continuity can be tested, as well as the lamps, by connecting the diodes of the conventional lamp testing circuit (A) as shown in (B).

Multi-Aperture Solar Cell Amplifier

By Glenn R. Hearn
Sanders Associates
Bedford, Mass.

THIS CIRCUIT is presently being used as a source of strobe pulses in a high speed commercial card reader. The input is connected to a long solar cell which is masked by an aperture plate having equally spaced windows to allow light to shine on the cell at the required strobe times. With a long, 10-aperture cell and one circuit, this scheme will generate 10 strobe pulses. Eight circuits and eight solar cells are required to generate the 80 strobe pulses needed when reading a conventional punched card.

A common-base input circuit is used so that the solar cell operates in its linear region. The base of Q_1 is biased at $-V_f$ of D_1 so that the emitter of Q_1 is essentially at 0 V, thereby minimizing cell leakage current. The maximum input current which the circuit can accept is determined by

$$I_{L\,MAX} + NI_A + \frac{V}{R_1} \leq \frac{V}{R_3} + I_{B2}$$

where $I_{L\,MAX}$ equals the cell leakage current, N equals the number of apertures uncovered and I_A equals the cell current generated each time an aperture is uncovered.

Figure 2 shows waveforms at various points of the circuit. The waveform of the signal at the base of Q_3 consists of a series of voltage steps, where the changing portion is a linear voltage which changes by $I_A R_3$ in the time required to uncover an aperture. The flat part of the waveform between steps represents the time that it takes the card to travel between apertures.

In the remainder of the circuit, Q_3 is an emitter follower which has a high input impedance and is capable of driving the $R_5 C_2$ differentiator. Q_4 is an inverter which amplifies the differentiated analog signal and drives the digital pulse shaping output stage, Q_5.

The purpose of C_1 and C_2 is to limit the frequency response of the amplifier so that noise, ever present in most electro-mechanical systems, will not generate digital output strobe pulses.

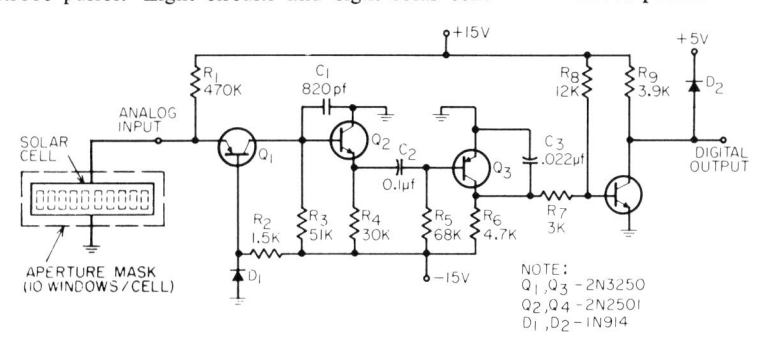

Fig. 1. Solar-cell amplifier for reading punched cards.

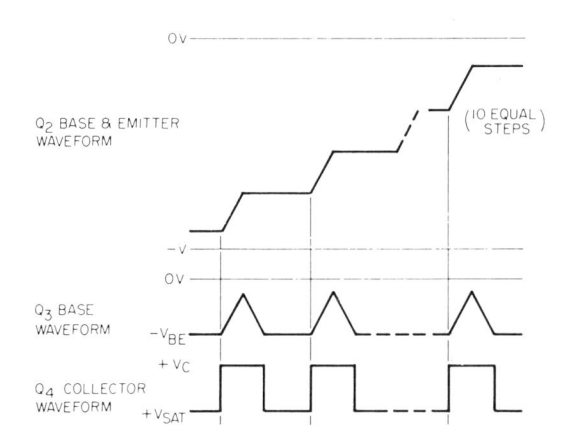

Fig. 2. Waveforms at critical points in the circuit.

Coaxial Cable Driver Circuit

By Ben Strunk
General Precision, Inc.
Little Falls, N.J.

HERE IS A CIRCUIT that can drive digital information through long lengths of coaxial cable. Pulses of 30-nsec rise and fall time have been sent through 1155 ft of 50-ohm co-ax (RG/188U), and through 650 ft of 93-ohm co-ax (RG/62U). Since risetime (τ_r) can be related to frequency (f) by the equation $f = 0.352/\tau_r$, a risetime of 30 nsec means that a frequency of 10 to 20 mc is coupled into the supply and ground leads. R_4, C_2, and C_4 comprise a decoupling circuit which eliminates this. These capacitors, C_2 and C_4, must be good quality, high-frequency units. Since tantalytic and electrolytic capacitors become inductive in this frequency range, they must not be used here. Sprague monolithic capacitors were found to be suitable. Also, great care must be used in laying out this circuit—the power and ground leads as well as input and output leads, must be as short as possible.

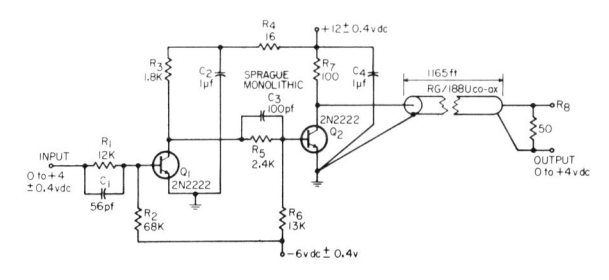

Coaxial cable driver circuit.

For use with a different characteristic-impedance co-ax, merely replace R_8 with the characteristic impedance of the co-ax selected. The output level, of course, will change. This can be easily calculated.

There is no noticeable deterioration in the wave shape (input to co-ax vs. output) with a ±20 percent variation in R_8.

High-speed circuit detects phase-difference

John C. Freeborn, Honeywell, Inc.,
West Covina, Calif.

This circuit detects a small relative phase shift between two ac signals and generates an output which indicates the magnitude and direction of the shift. It was designed for a metal detector, but other applications are possible.

Two dual D flip-flops, SN 7474s, are interconnected as shown in **Fig. 1**, so that two seperate outputs are generated as a result of the phase comparison between the input signals. Characteristics of this dual D flip-flop are: On a positive going clock pulse to input C the level at input D is transferred to Q; on a negative going (clear) pulse to input C1 the Q goes to "ZERO". As long as the clear input (C1) is maintained at "ZERO," clock pulses are not effective.

At the beginning of a cycle, all four flip-flops are in the reset (a = "ZERO") state. Two ac signals, S_1 and S_2, are applied to the two input points of the SN 7474S. If S_1 leads S_2 in time, an output pulse will appear across V_1 and $\bar{V}_1$, the duration of which will be equal to the phase difference. If S_2 leads S_1, an output pulse will appear at V_2 and $\bar{V}_2$. In practice, useful pulses as narrow as 20 nsec have been generated. No dead spot or zero phase shift condition has been observed, indicating that the circuit's ability to detect phase differences is limited by noise and drift rather than by response of the circuit.

The circuit compares the inputs at the positive-going clock-pulse threshold level (about +2.0V) and requires a reset prior to the next expected comparison. In the circuit shown, both S_1 and S_2 go negative, and this half of the signal is used through D_1 to reset the flip-flops.

The detailed operation of the circuit can be seen from the timing diagram and truth table of **Fig. 2**. These show signals vs time for the case in which S_1 leads S_2. Square signals are used for clarity. Significant times are labeled t_0, t_1, t_2, t_3. These time symbols are used to separate various circuit states in the truth table. The subscripts refer to flip-flop number assignment (1 through 4) made in **Fig. 1**.

Note the addition of a dc bias to the input signals will influence the apparent dc detection level and thus provide a means for balancing or zeroing.

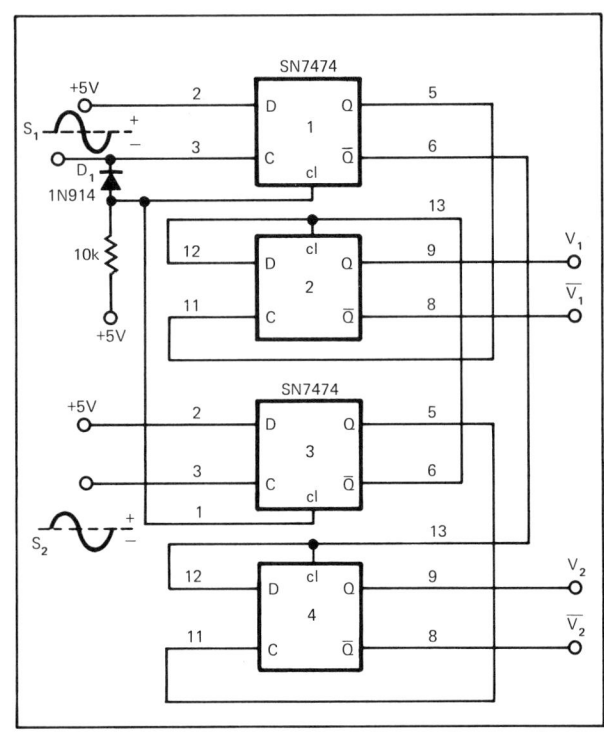

Fig. 1—Phase shift detector requires only 2 ICs and 2 discrete components. When input signals at S_1 and S_2 are not in phase, a pulse will be present at the output of the leading phase.

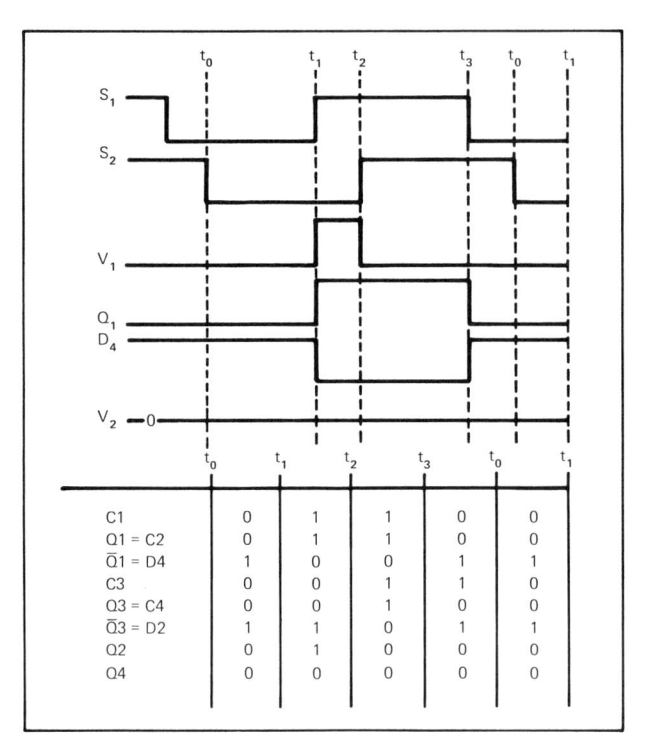

Fig. 2— Timing waveforms and truth table depict circuit operation when S_1 leads S_2 in phase. All flip-flops are reset on the negative half-cycle; thus, phase detection occurs at the input signal rate.

	t_0	t_1	t_2	t_3	t_0	t_1
C1		0	1	1	0	0
Q1 = C2		0	1	1	0	0
$\bar{Q}$1 = D4		1	0	0	1	1
C3		0	0	1	1	0
Q3 = C4		0	0	1	0	0
$\bar{Q}$3 = D2		1	1	0	1	1
Q2		0	1	0	0	0
Q4		0	0	0	0	0

Linear modulator has excellent temperature stability

By R. G. Kelly
Texas Instruments
Dallas, Texas

THIS CIRCUIT provides a precise degree of linear amplitude modulation. With the specified transistors, and with low tempco resistors, modulation errors due to temperature variation can be neglected.

In the schematic, components have been chosen for audio-frequency operation. Carrier frequency f_c is 400 Hz and modulation frequency f_m is 1 Hz. However, the same basic circuit can be used for frequencies from zero up to several megahertz. Note that there are no capacitors. Bandwidth is determined primarily by the characteristics of the ICs. Use of closed-loop op amps and chopper transistors gives stable gain, unaffected by temperature.

The modulation signal, at the output of op amp A_1, rides on a positive dc potential as shown in the schematic. Potentiometer R_1 adjusts the dc level. At A_1 output, the signal never goes below ground potential except when over modulation occurs. Chopper transistors Q_1 and Q_2 alternately sample A_1's output waveform at the carrier-frequency rate f_c.

Op amp A_2 inverts the incoming signal, and A_3 provides a noninverted signal. The output signals of both amplifiers are summed at one input terminal of A_4. When A_2 conducts, A_3 is off, and vice versa. Thus the modulated-carrier wave at the output of A_4 is a series of alternating positive and negative pulses, with amplitude determined by the modulating signal. Full 100-percent modulation occurs when the peak of the modulating wave is equal to the dc quiescent level at the output of A_1.

Any temperature drift is likely to be caused by the resistors rather than by the op amps and transistors. Op-amp gains are stabilized by feedback, and V_{sat} for the switching transistors changes less than 0.2 µV/°C. So, low-tempco resistors should be used for optimum temperature performance.

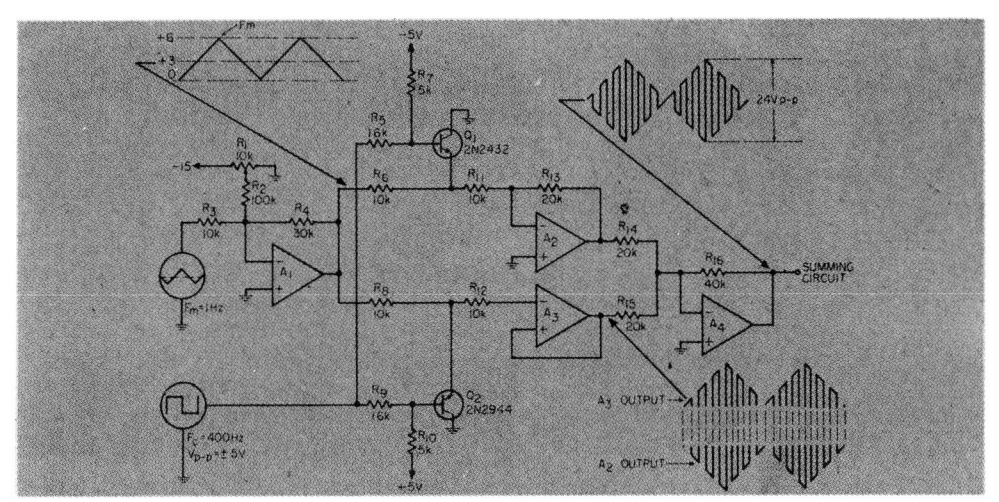

Circuit provides accurate amplitude modulation that isn't temperature dependent. Pot R₁ adjusts quiescent level of the modulating signal.

LED modulator

by Greg Schmidt
San Diego State College
San Diego, Calif.

A COMMON difficulty in high-speed pulse modulation of Gallium Arsenide Phosphide LEDs is providing a low driving-point impedance. This is required for fast turn-on. In addition, dynamic current limiting is needed to prevent excessive current overdrive.

This circut overcomes these two difficulties. Transistor Q_1 is an amplifier that supplies a dc level as well as modulation information to emitter-follower output stage Q_3. Output current is sensed and limited by Q_2. Current limiting occurs when the voltage drop across R_7 forward biases CR_1 and draws base current from Q_2. For the values shown, current limiting occurs when LED current is about 30 mA. Since the LED is dark for voltages below 1.5 volts, turn-on time for full brightness is 12 ns.

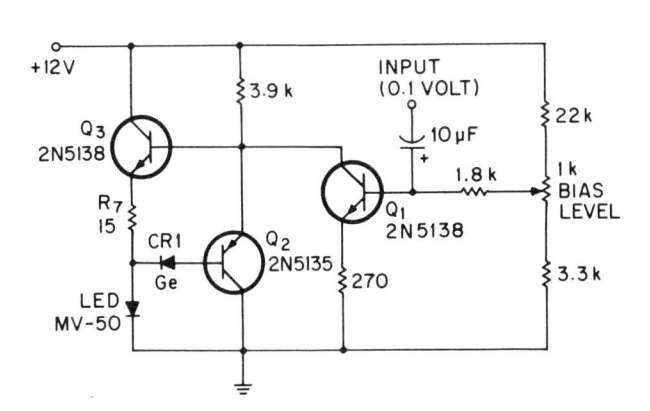

Fig. 1. This circuit provides a low-impedance, current-limiting, high-speed driver for a LED modulator.

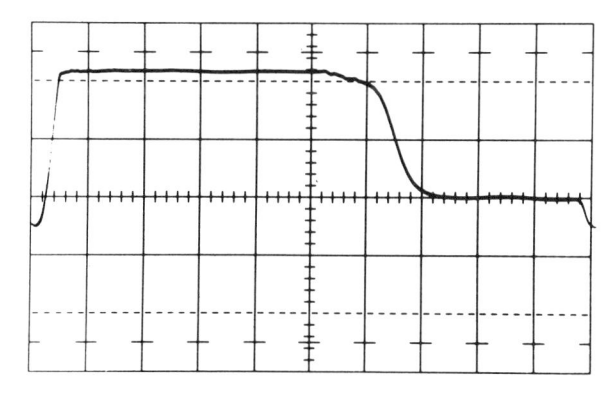

Fig. 2. LED voltage at a 1 MHz rep rate. Vertical scale is 0.5 V/div and horizontal scale is 0.1 µs/div. Zero volts dc is 1 division below the center line.

Analog, Linear and Communications Designs

JFET switched resistor controls gain of op amps

Kenneth C. Bower, ESL Inc.
Sunnyvale, California

Often it is desirable to remotely change the gain of a dc amplifier by logic-level signals without causing a dc shift at the output. The circuit shown in **Fig. 1** switches the gain of the op amp in discrete steps, using a TTL control input. When a logic ONE is applied at the switch input of the control circuit, JFET Q_3 is turned OFF, and R_3 is removed from the circuit. The gain of the op amp is then, simply: $A = (R_1 + R_2) \div (R_2)$.

When the logic level input is changed to a logic ZERO, R_3 is in series with the ON resistance of the JFET and these parallel R_2. The gain then becomes: $A = (R_1 + R_f) \div (R_f)$, where R_f is the parallel value of $R_3 + R_d$ and R_2. R_d will not exceed 40Ω for a 2N4869 JFET. If R_3 is kept in the range of 4KΩ, then R_d will not noticeably effect the circuit, and gain calculations are made, as above, on the basis of R_1, R_2 and R_3. This method can be used with several JFET switches paralleled to obtain as many discrete step changes in amplifier gain as may be required for your design.

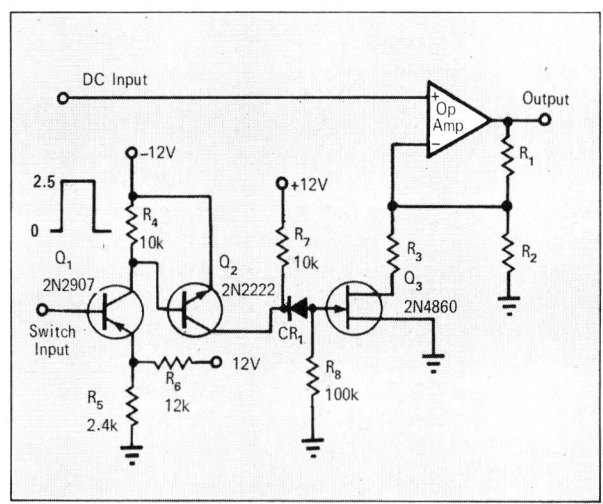

Fig. 1—TTL control of op amp gain is achieved by effectively varying the value of R_2. JFET Q_3 switches R_3 in and out of the circuit in parallel with R_2 to achieve differing feedback ratios.

6-bit D/A converter uses inexpensive components

Don Aldridge and **Karl Huehne,**
Motorola Inc. Semiconductor Products, Phoenix, AZ

A complete 6-bit D/A converter can be built, as shown in the illustration, for a total IC cost of less than $5.00 in 100-up quantities. This is possible because the MC1406 D/A converter requires a reference voltage and an operational amplifier for many of its applications. Since the operational amplifier is normally used as a current-to-voltage converter and its output need only go positive, the MC1723 voltage regulator can provide both of these functions in one package, with the added bonus of up to 150 mA of output current. The primary departure is that, instead of powering the MC1723 from a single positive supply, it uses a negative bias as well. This allows use of its output amplifier as a classic current-to-voltage converter with the non-inverting input grounded.

Although the reference voltage of the MC1723 is then developed with respect to that negative voltage, it appears as a common-mode signal to the ladder current-drive amplifier in the D/A converter. Since ±15V and +5V are normally available in a combination digital/analog system, only the −5V need be developed. A resistive divider is sufficient inasmuch as the operating range for the −5V is between −2V and −8V and between −4V and −6V for 7-bit accuracy.

Fine gain adjust can be provided by making either R_1 or R_2 variable. Full scale output is approximately 10V but may be increased to as much as 32V by increasing R_2 and raising the +15V supply proportionately to a maximum of 35V. Resistor R_3 provides temperature compensation for input bias current as well as impedance for the 50 pF Miller-effect compensation capacitor.

Response time is significantly improved by R_4, which provides a passive pull-down for the output. Replacing the 100 pf capacitor with a 24 pf unit between pins 2 and 6 results in a less than 1 μsec setting time for full-scale changes. The current limiting provision of the MC1723 is not shown but could be easily incorporated.

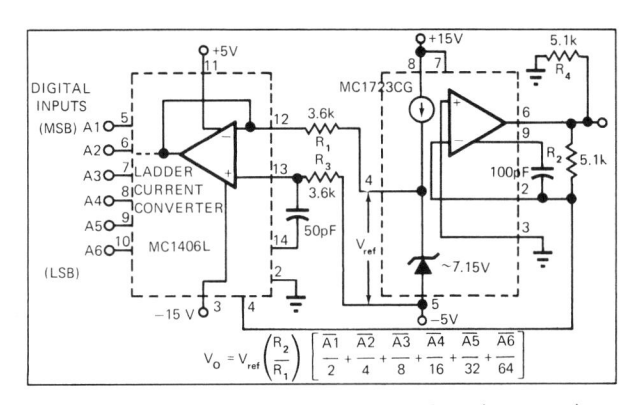

$$V_O = V_{ref}\left(\frac{R_2}{R_1}\right)\left[\frac{\overline{A1}}{2} + \frac{\overline{A2}}{4} + \frac{\overline{A3}}{8} + \frac{\overline{A4}}{16} + \frac{\overline{A5}}{32} + \frac{\overline{A6}}{64}\right]$$

The MC1723 serves as both an op amp and a reference voltage source for the MC1406 D/A converter.

Linear circuit integrates pulse trains

Andres J. Baracz,
Sparta, N. J.

The circuit shown performs linear integration in a very simple manner. The circuit consists of a complimentary transistor which can be turned on and off by means of short pulses of alternate polarity, or by spikes applied to the base of transistor Q_1. A differentiated square pulse will produce a positive spike corresponding to its leading edge, and a negative spike, corresponding to its trailing edge. The positive spike will turn the transistors on, charging the capacitor via the constant current device, a negative spike will turn the transistors off. Thus, the charge on the capacitor C_3 will be controlled by the time interval between positive and negative spikes, or by the duration of the square pulses.

The charging of the capacitor via the constant current device will generate a linear voltage ramp at the terminals of the capacitor C_3. The final magnitude of this voltage will be directly proportional to the time integral of Q. Any subsequent pulse will produce an additional voltage increment proportional to its duration.

A train of pulses will generate a staircase function, with increments proportional to the duration of the pulses. Square pulses in the nsec range will actuate the circuit.

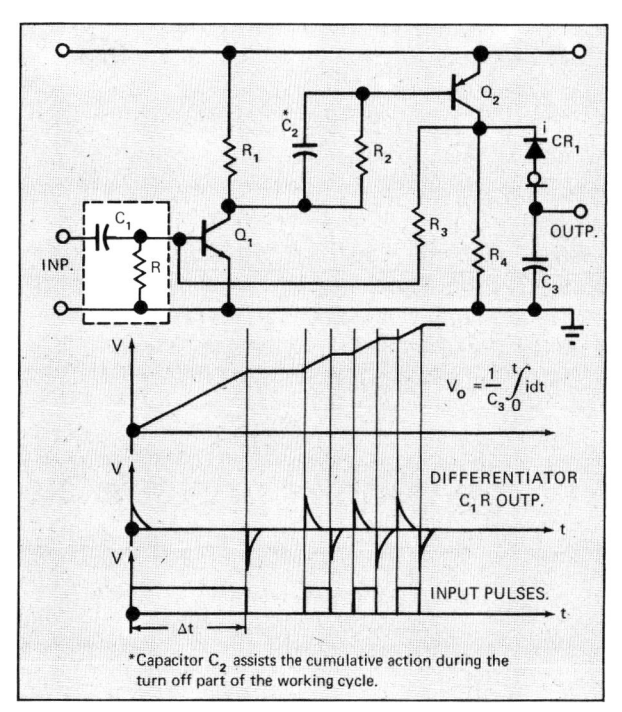

Pulses in the nsec range will actuate this linear integrator. It's output is a linear function of input current and pulse duration. Capacitor C_2 assists the cumulative action during the turn off part of the cycle.

Low-Cost UJT Raster Generator

By Fred Stevens
Cooke Engineering Co.
Alexandria, Va.

THIS CIRCUIT ANSWERED the need for a simple, low-cost, stable raster generator for use in a breadboard transistorized flying-spot scanner. It could be adapted for similar use in closed-circuit television cameras and monitors.

Unijunction transistor Q_1 is a relaxation oscillator at approximately 10 kHz, the frequency chosen as the horizontal sweep rate. Positive pulses at base 1 of the UJT serve two purposes: they are used as the sync pulses for horizontal sweep and as the drive to the vertical waveform generating circuitry.

The pulses are amplified by Q_2, which operates into saturation, and clamped by CR_1 to 8 v. Thus, uniformity of pulse shape is assured. They are then ac-coupled into Q_3, which serves as a source of constant-current pulses for a UJT staircase generator stage. The staircase waveform across capacitor C_3 is applied to the vertical input of the scope and the sweep rate is adjusted so that each step on the waveform is slightly longer than one sweep of the trace. Potentiometer R_7 adjusts the current per pulse into Q_3 and therefore the voltage per step across the capacitor.

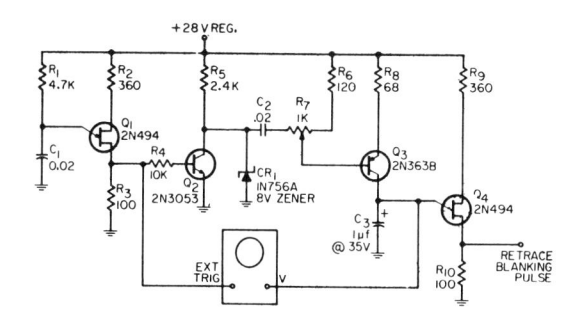

Unijunction transistor serves as relaxation oscillator at horizontal sweep frequency.

Since the UJT fires and discharges C_3 at a voltage equal to the intrinsic standoff ratio of Q_4 times the supply voltage, adjusting R_7 effectively determines the number of steps on the staircase. Because the number of steps per cycle is the number of lines per field, and also a stable integral multiple of the pulse repetition rate of Q_1, interlaced scanning is easily and reliably achieved. Since the field retrace time is the time during which C_3 discharges through Q_4 into R_{10}, the positive pulse available across R_{10} is conveniently used for retrace blanking.

Isolated current source

Ralph Tenny
Texas Instruments Inc., Dallas, TX

I needed a floating current source to power a strain gage bridge, and the battery operated circuit shown here gave me adequate stability for a system with ±0.2% accuracy. As shown, the voltage compliance is equal to $E_1 - 1.4V$, and the temperature stability is +0.7 µA/°C from 0-50°C, with a current output level of 20 mA.

Q_1 and Q_2 control the current in the external loop with R_1 sensing the current. If R_1 is chosen for 0.5Ω/mA, the sensitivity is adequate without dissipating excess power. R_2, R_3 and R_4 provide an offset voltage to set the output current. R_5 and D_1 insure that the circuit will start regardless of the initial offset of the op amp.

Four alkaline pen cells provide ±3V for the 741, and E_1 should be chosen to give adequate voltage compliance for the intended load at maximum load current and lowest acceptable end-point voltage for E_1. The end-point voltage may be defined by rise in noise output of E_1 as it is depleted.

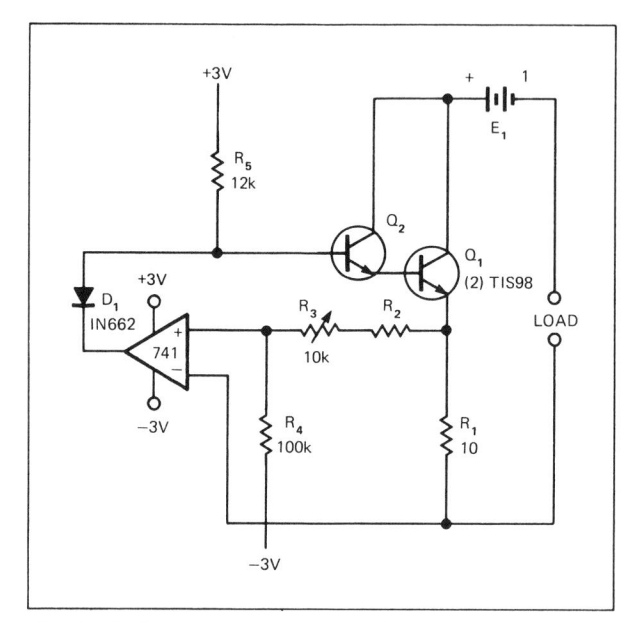

This simple, battery operated op-amp circuit provides good current and temperature stability over a wide range.

Squaring circuit makes efficient frequency doubler

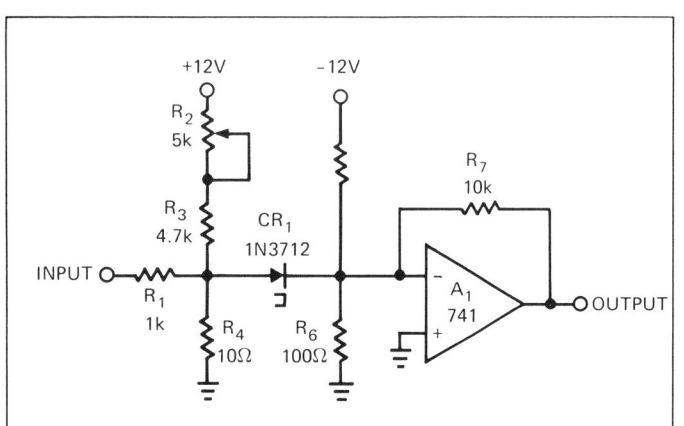

Fig. 1—Simple tunnel-diode circuit provides efficient frequency doubling without the use of tuned circuits.

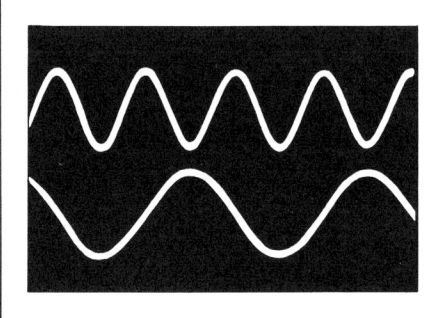

Fig. 2—Input (bottom) and output (top) waveforms at 1 kHz. Voltage scales are 1V/cm for the bottom trace and 0.1V/cm for the top trace.

by **Russell Kincaid**
Sanders Associates
Nashua, N. H.

Using the tunnel-diode squaring circuit shown in **Fig. 1**, it is possible to double the frequency of input signals, without the use of tuned circuits.

With an input signal A sin ωt, the output will be A cos 2 ωt. The fundamental and other harmonics will be at least 30 dB below the level of the frequency-doubled output signal. Satisfactory operation is possible from dc to the upper frequency limit of the amplifier. The scope picture (**Fig. 2**) shows the input and output at 1 kHz.

Potentiometer R_2 in the positive bias supply allows the diode current to be adjusted to the peak of the tunnel-diode bias current, thus preventing offset at the amplifier output. Resistor R_6 shunts the op amp input and prevents oscillation at frequencies above the amplifier response range.

The circuit relies on the parabolic shape of the tunnel-diode characteristic near the peak current, and on the following mathematical relation:

$(\sin x)^2 = 1/2 - 1/2 \cos 2x$

Ideally, the tunnel diode should operate with zero impedance for the source and load. In this circuit, the source is 10Ω and the load is the op amp summing point.

Buffer/filter extracts WWV time codes

Ernest F. Wilson, E.G. and G, Inc.
San Ramon, California

This circuit is capable of extracting the 100 Hz sub-carrier of the modified IRIG-H time code now being transmitted continuously by the National Bureau of Standards radio stations WWV and WWVH. It consists of a dual-op amp with one section operating as a unity gain voltage follower to allow the signal to be extracted from high impedance circuits, such as the second detector of the radio receiver. The other section is a low pass filter to remove all higher frequency components (400, 500, 600, 1000, 1200, and 1500 Hz) without unacceptable distortion of the step modulation of the 100 Hz sub-carrier.

The voltage follower, op amp A in **Fig. 1**, has a very high input impedance, so the circuit from which the signal is tapped sees only the value of R_1. This stage must have sufficient amplitude range to accept the total audio signal, although the 100 Hz component is only a small percentage of the total signal amplitude. It delivers the signal at the low output impedance required by the active filter.

The second section is a 3-pole low-pass Chebyshev active filter with 1 dB ripple and f_o = 130 Hz. This is a reasonable compromise between simplicity and effectiveness.

With an average signal extracted from the second detector of a Specific Products SR-7 receiver, and with a ±15V op amp power supply, a pk-pk mark-output amplitude of about 3V is obtained. The amplitude stability is dependent on the rf signal stability of the signal, modified by the AGG action of the receiver, and the amount of low frequency noise components present.

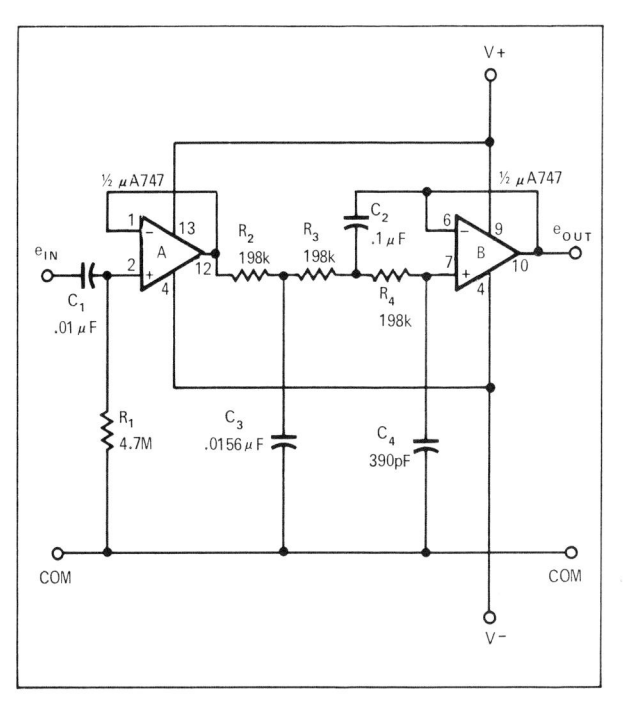

Fig. 1—WWV 100 Hz time signals can be extracted from receiver detector circuits with this simple dual-op amp module. Op amp A is a unity-gain voltage follower, and op amp B is a 3-pole, low-pass 1 dB Chebyshev filter.

Transistor circuit speeds synch time of phase-locked oscillator

J. S. Swartz
IBM Corp., San Jose, Calif.

This circuit can reduce maximum synchronization times of a phase-lock oscillator (PLO) by as much as half. For applications where variable data transmission rates are encountered and the PLO must be synchronized rapidly the circuit offers outstanding improvement in performance.

The circuit includes a compensation network consisting of series capacitors C_1, C_2 and a resistor, R_1, coupled between the two capacitors and to a reference potential, such as ground. The compensation network serves as a filter for the voltage-controlled oscillator (VCO), which provides a signal of nominal frequency subject to error correction for phase locked operation. The stored voltage on C_1, C_2 is proportional to the frequency. If a PLO must be switched from an extreme frequency above or below the nominal center frequency to another data source on the opposite side of the center, the switching time can be reduced by initializing the voltage on C_1, C_2 to the nominal center frequency (OV) before synchronization is attempted. This reduces the maximum time needed to acquire final synchronization.

During operation, an initialization pulse is provided to complementary input transistors Q_1 and Q_2, causing transistors Q_3 and Q_4, connected to the capacitors, to saturate. As a result, the capacitors are forced to ground. The initial frequency deviation of the PLO is thus cut substantially, reducing synchronization time effectively. □

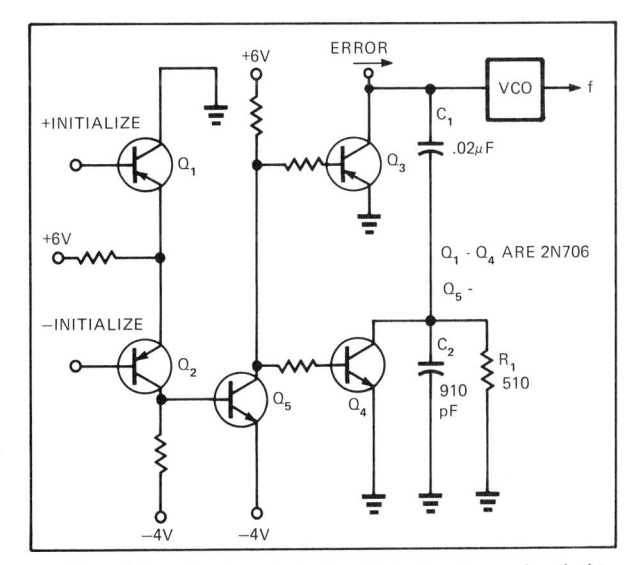

Rapid synching of a phase-locked oscillator is enhanced with this switching circuit that provides potential control of the VCO.

Analog, Linear and Communications Designs

Higher-efficiency chokeless vertical sweep

by Jeremy M. Tucker
Fairchild Semiconductor
Mountain View, Calif.

THE ADVANTAGE of a chokeless vertical deflection circuit for small-screen B&W TV is well known—the elimination of costly copper for choke windings. But the sweep circuit with a choke tends to be about three times more efficient. Its efficiency is typically 25% compared with a bit better than 8% for the chokeless system.

The chokeless circuit of Fig. 1 can be modified to improve the efficiency. In this self-oscillating circuit, feedback from the collector of Q_2 (or from the yoke) to the base of Q_1 and direct connection of Q_2's base to Q_1's collector insure positive feedback. When the supply is turned on, current flows through R_1 and R_2 creating a positive ramp voltage across C_2.

This voltage causes the collector voltage of Q_2 to fall and induces a negative voltage at the base of Q_1, via C_1, thus holding Q_1 off. The emitter voltage of Q_2 also rises as its base voltage rises. This emitter voltage can be fed back to the base of Q_1, tending to charge C_1 such that the base of Q_1 goes positive.

When this happens, Q_1 conducts and turns Q_2 off. The rapid rise of the voltage at Q_2's collector keeps Q_1 conducting for the duration of the retrace period. Retrace time is here controlled by the R_3C_1 time constant which determines the time required for the current to decay to a value such that Q_1 cannot continue to conduct. When Q_2 stops conducting, C_1 starts charging again, repeating the cycle.

Q_2 and Q_3 constitute a totem pole output. As the current in Q_2 rises, the voltage across R_6 rises. The forward-biased diodes D_1 and D_2 couple this voltage back to the base of Q_3, reducing the conduction of Q_3. During retrace, when Q_1 conducts, Q_2 turns off and Q_3 saturates, recharging C_3.

The constant-voltage forward characteristic of the diodes shifts the dc level and couples the voltage across R_6 back to the base of Q_3 without attenuating the ac component.

We can improve the efficiency of this system if we can achieve part of the yoke-current reversal by resonating the yoke with a capacitor and isolating the yoke for the portion of the retrace pulse that exceeds B+. This is done in Fig. 2.

Now let us replace the integrator capacitor C_2 of Fig. 1 with C_4 and C_5 as in Fig. 3, the final circuit. The voltage at the emitter of Q_2 can then be integrated and fed back to provide some linearity correction. Also, C_5 is returned to the yoke to provide overall Miller feedback. C_5 also bootstraps the yoke to the emitter-follower (Q_3) part of the circuit. The size of C_5 affects retrace time. C_9 prevents spurious oscillation.

The final circuit generates a yoke current of 0.6 A pk-pk in a 25-mH, 10-Ω yoke from an 18-V supply. Retrace time is 1.35 ms.

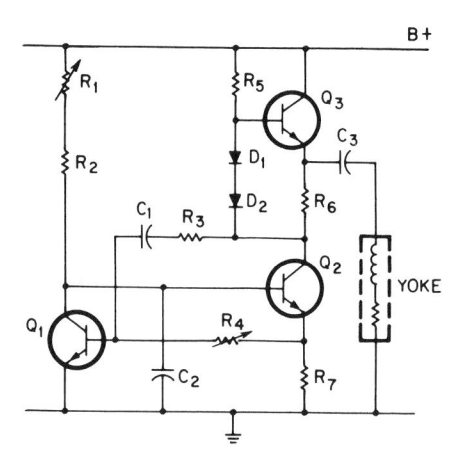

Fig. 1. This self-oscillating vertical-deflection circuit for small-screen black-and-white TV requires no choke. But it is quite inefficient.

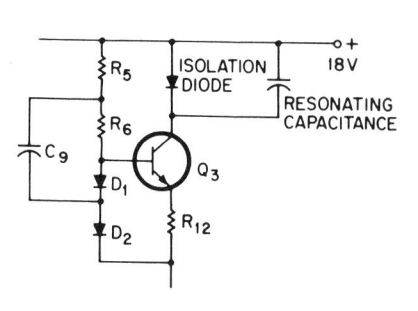

Fig. 2. A resonating capacitor improves the efficiency of the sweep circuit.

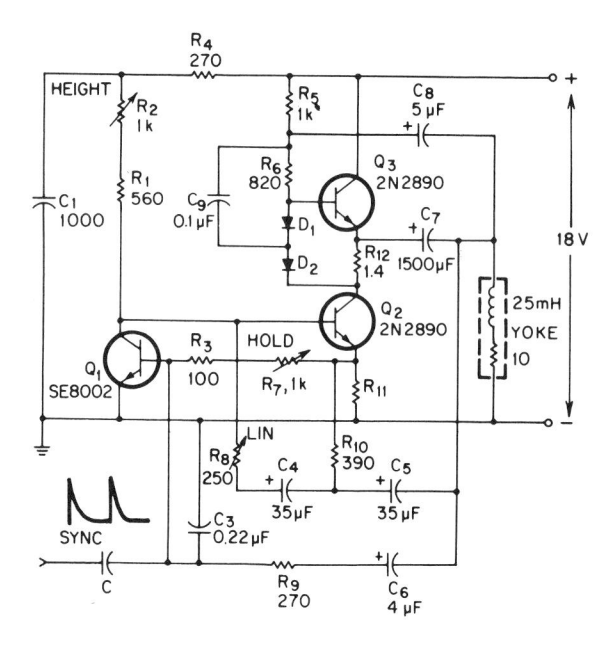

Fig. 3. Further improvements and necessary controls complete the chokeless sweep circuit.

Phase modulator has broad bandwidth.

C. N. Charest,
Philco-Ford, Palo Alto, California

Here is a phase modulator with a bandwidth so large it can be used in multicarrier systems. As currently operated at 23.5 MHz in a communication satellite, the sensitivity is 2.38V RMS per radian, and this can be increased to a maximum of 430 mV per radian by reducing the paddling at the modulation input port. The linearity of sensitivity (with modulation drive) is within ±7 percent of a straight line — out to a modulation index of 0.5 radian.

There is no observable variation in sensitivity over modulation frequencies from 10 Hz to 100 KHz. The thermistor compensates the circuit over +5°C to +35°C to hold sensitivity changes to ±3 percent.

In addition to isolating the phase shifting portion from the modulation source, the 741 op amp sets the −0.25V varactor bias according to its offset. The 741 also serves the important function of driving the varactors from a low-impedance source, making the flat 10 Hz to 100 KHz response possible. The input resistance can also be set as desired.

The principal limitation of carrier frequency is the core material available for the transformers. Circuit gain is independent of changes in R or C, and phase shift is essentially linear with C_1, within the limits discussed above.

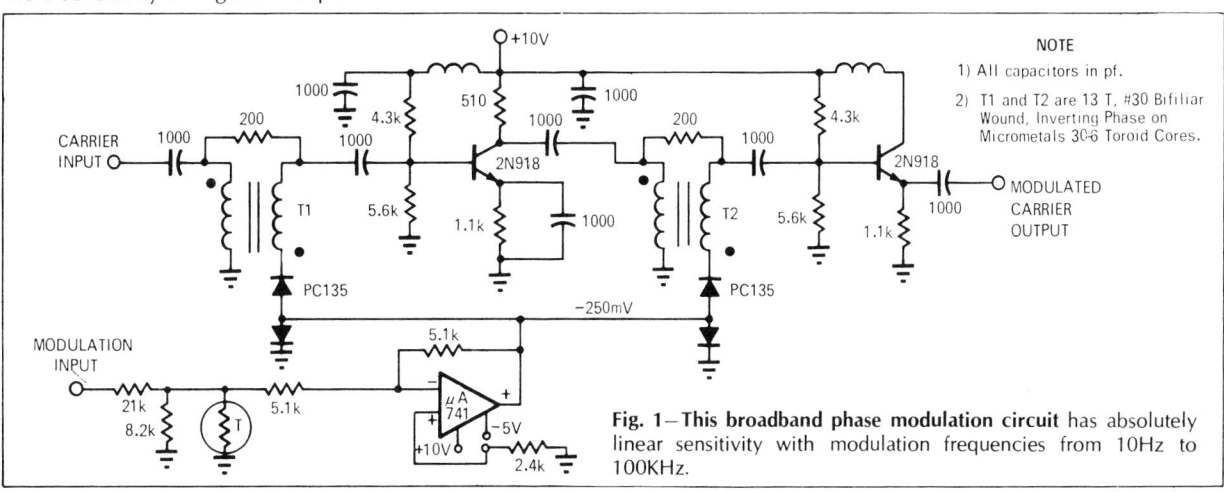

Fig. 1 — This broadband phase modulation circuit has absolutely linear sensitivity with modulation frequencies from 10Hz to 100KHz.

Active bandpass filter with adjustable center frequency and constant bandwidth

By Leslie Robinson
Honeywell
Seattle, Wash.

ACTIVE FILTERS are becoming more and more popular now that low-cost op amps and linear ICs are widely available. The simple bandpass filter, shown here, needs just an op amp plus a few discrete components.

In this circuit, a single resistor controls the center frequency without changing the bandwidth or the gain.

With the component values shown, center frequency can be shifted from 1.6 kHz to 2.4 kHz by changing R_C from 1100 ohms to 406 ohms. The 3-dB bandwidth remains constant at 260 Hz. Center-frequency gain varies only ±0.5 dB from the nominal 26 dB. Bandwidth at the −10 dB points is 775 Hz.

Filters can be designed for a wide range of parameter values. Here is a simple step-by-step procedure for calculating the component values from specified values of bandwidth, gain and range of center-frequency adjustment:

Step 1. Choose values for mid-frequency f (Hz), nominal voltage gain G, and 3-dB bandwidth Δ (Hz).
Step 2. Select a convenient Value C for the capacitors.
Step 3. Calculate the required resistor values from the following equations:

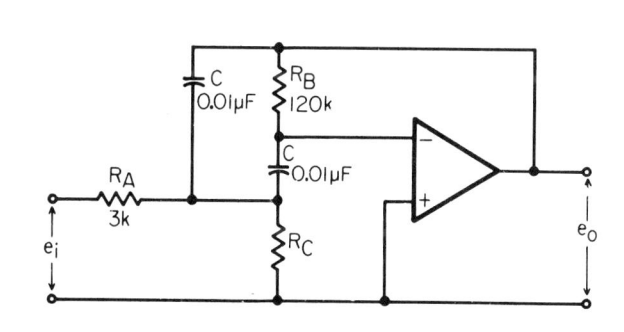

$$R_A = \frac{1}{2\pi \Delta G C} \quad (1)$$

$$R_B = \frac{1}{\pi \Delta C} \quad (2)$$

$$R_C = \sqrt{\frac{1}{2\pi C \left[(2f^2/\Delta) - \Delta G \right]}} \quad (3)$$

Value of R_C determines the center frequency of this band pass filter. Adjustment has no effect on bandwidth.

Circuits, designed by this technique, can be used singly as tunable filters; or they can be used in multiple combinations to form a comb filter in which each section can be individually adjusted to an exact frequency.

A polar clamp

by Richard R. Breazzano
Western Union
Mahwah, N. J.

THIS CIRCUIT provides overvoltage input protection for data communications equipment. The polar clamp normally operates with conventional 600-Hz teleprinter signals of ±6 Vdc at 10 mA. However, it can protect against overvoltage input transients of up to ±120 Vdc at 20 mA.

When a positive voltage input exceeds Q_1's emitter-base breakdown, the junction acts like a zener diode whose voltage rating is somewhat greater than the breakdown value. Transistor Q_2 becomes forward biased and input clamping is completed. With a negative input, Q_1 is forward biased and Q_2's emitter-base zener completes the clamping action. Emitter follower Q_3 maintains proper input impedance while delivering a buffered logic output. Q_1, Q_2 and Q_3 are inexpensive plastic transistors.

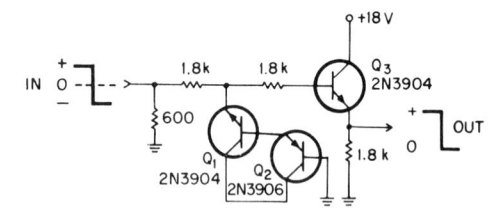

This polar clamp, constructed with an inexpensive plastic transistor, is protected against up to ±120-V at 60-mA transients.

Adjustable Level-Detector

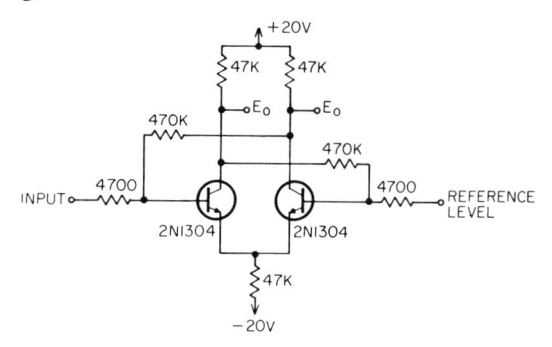

Stable state of this circuit changes when input level crosses the reference level. Output is taken from either collector.

By G. Richwell
Reflectone Electronics
Stamford, Conn.

THIS CIRCUIT is a differential amplifier with positive feedback, which can be used as a level detector. Like a conventional flip-flop it has two stable states and will switch states when the dc input level goes above or below a preset-reference level. The output can be taken from either collector depending on the polarity required. An advantage of this circuit is its inherent dc-stability due to the balanced configuration.

With component values shown, the circuit will change state for an input-voltage differential of 100 mV with respect to the reference. The hysteresis depends on the amount of positive feedback. The reference level may be above or below ground. Output voltage swing at the collector and the range over which the reference may be changed, are determined by the values of the collector and emitter resistors.

Compensated adjustable zener

Ralph Tenny
Texas Instruments Inc., Dallas, TX

This circuit, when constructed with low-cost plastic transistors, has 1Ω or less dynamic impedance from about 1 mA to the maximum allowable dissipation of Q_2. Substitution of different TIS98s for Q_2 has negligible effect on the output, and random substitution of 2N5447s for Q_1 causes only a ±2% change in the output. All these features demonstrate the circuit's suitability in regulator service for battery powered MOS instruments, except for the circuit's linear -0.4%/°C temperature coefficient (20% change 0 - 50°C).

By adding a thermistor (Gulton 35TF1) and R_4, the circuit is compensated to a maximum deviation of ±0.5% over the 0 - 50°C range. Circuit values shown yield this compensation over the output range of 3.5 to 15.5V, with only R_2 changing to set output voltage. Dynamic temperature compensation depends upon low thermal impedance between the thermistor and Q_1, otherwise the time constant of the compensation is equal to that of the thermistor.

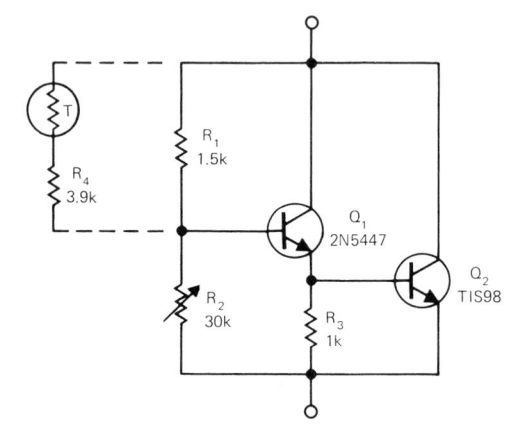

2-transistor adjustable zener has only a 1Ω dynamic impedance. Thermistor compensating network replaces R_1 for ±0.5% stability from 0 - 50°C.

High-speed synchronous detector

by Robert P. Lackey
Texas Instruments
Dallas, Texas

FET OR BIPOLAR-TRANSISTOR choppers cause problems in most synchronous detectors because of gate-to-drain or base-to-collector capacitance. In the FET chopper, for example, the capacitance allows the gate-drive sync signal to enter the drain circuit. This is particularly troublesome at high frequencies and low-input-voltage levels.

A simple solution lies in producing a signal of opposite polarity to the sync signal and feeding it to the drain (in the FET case) through a capacitance that's exactly equal to the gate-to-drain capacitance. This cancels sync feedthrough from gate to drain.

In the circuit shown, the signals at the D_1R_1 and D_3R_3 junctions are equal in magnitude to and 180° out of phase with the sync signals at the gates of Q_1 and Q_2 respectively. If C_1 is made exactly equal to the gate-to-drain capacitance of Q_1, the signals passing through C_1 and C_{GD1} will cancel and eliminate feedthrough. A similar situation holds for Q_2.

These adjustments are best made by adjusting C_1 and C_2 for zero output with the sync signal present and the input voltage absent. The circuit shown has a dynamic range of 80 dB at 100 kHz.

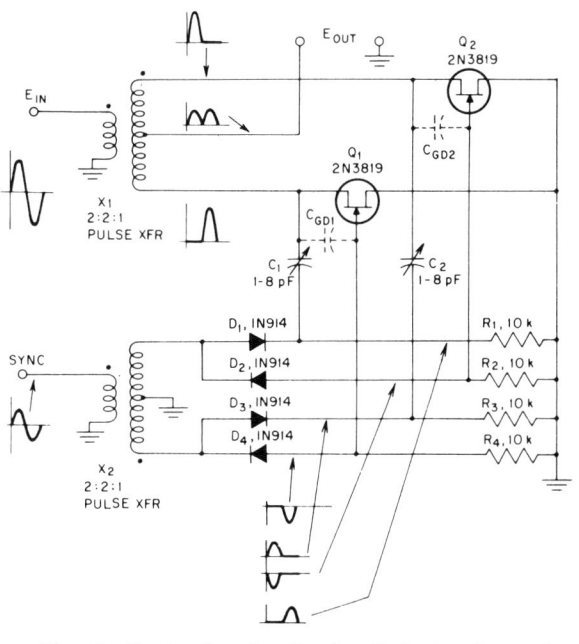

Capacitive feedback of gate signals eliminates forward sync-signal feedthrough, boosting synchronous-detector frequency capability.

Improved Circuit for Constant-Current Source

By Irwin Cohen
Bendix Corp.
Teterboro, N.J.

THE CONVENTIONAL CURRENT source circuit shown in Fig. 1 has the disadvantage that there is considerable variation in output current I_o with variations in supply voltage E. This is because the zener diode has internal resistance. A solution to the problem is to add the resistor R_2 shown dotted in the figure.

Using the Thevenin equivalent circuit, the circuit can be represented as shown in Fig. 2, where R_2 and R_3 have been replaced by a generator and an equivalent emitter resistance R_E. Assuming $I_z \gg I_b$ and $\alpha = 1$, adding voltages around the base-emitter loop gives

$$V_z + I_z R_z - V_{BE} - \frac{ER_3}{R_3 + R_2} - I_o R_E = 0 \quad (1)$$

Where $I_z = \dfrac{E - V_z}{R_z + R_1}$

Therefore

$$I_o = \frac{V_z}{R_E} + \frac{(E - V_z) R_z}{(R_z + R_1)R_E} - \frac{V_{BE}}{R_E} - \frac{ER_3}{R_E(R_3 + R_2)} \quad (2)$$

Then, for changes in load current I_o due to supply voltage changes

$$\frac{dI_o}{dE} = \frac{R_z}{(R_z + R_1)R_E} - \frac{R_3}{(R_3 + R_2)R_E} \quad (3)$$

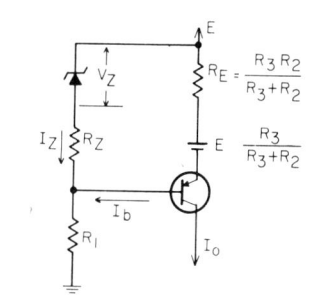

Fig. 2. To simplify the analysis, the circuit of Fig. 1 has been redrawn to show the Thevenin-equivalent generator and impedance in the emitter circuit.

From which, for no change in output current with variations in supply voltage

$$\frac{R_z}{R_z + R_1} = \frac{R_3}{R_3 + R_2} \quad (4)$$

The optimum value for R_2 can be determined from equation 4 if R_1, R_z and R_3 are known. Also it is possible to design current sources in which the output current varies directly or inversely with variations in supply voltage in any required ratio.

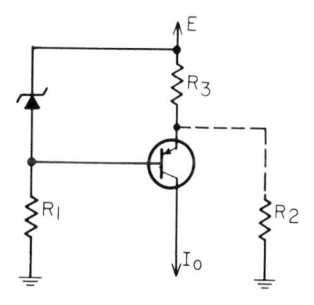

Fig. 1. The performance of this current source circuit can be improved by adding the resistor shown dotted.

PIN-diode turnoff for microwave VCO

By August Barone

AIL Div. of Cutler-Hammer
Deer Park, N.Y.

A PIN DIODE in the feedback loop, provides an excellent turnoff circuit for a microwave VCO. Most conventional turnoff methods, such as applying and removing dc bias or Vcc, cause excessive frequency drift when the oscillator is turned on.

Drift is normally caused by thermal effects in the transistor chip. When dc bias is removed, collector current ceases and the chip cools down. When bias is reapplied, oscillation occurs at a different frequency and only returns to the original frequency after the transistor chip returns to its original temperature.

The circuit shown here causes very little drift, because it does not cut off the transistor. Oscillations are controlled by varying the resistive feedback. With no applied voltage, the PIN diode has a high resistance of about 10 kΩ. When +10 Vdc is applied, the resistance falls to about 10 Ω. This low resistance stops the oscillator, without affecting the collector current of the transistor.

With the component values shown, the VCO has a center frequency of 1500 MHz with a tunable range of 500 MHz. Frequency drift is less than 100 kHz after the VCO is switched off for 15 minutes and then switched back on again.

This circuit is especially useful in applications where the oscillations must be frequently interrupted, such as when several VCOs are used alternately to span a large tunable frequency range.

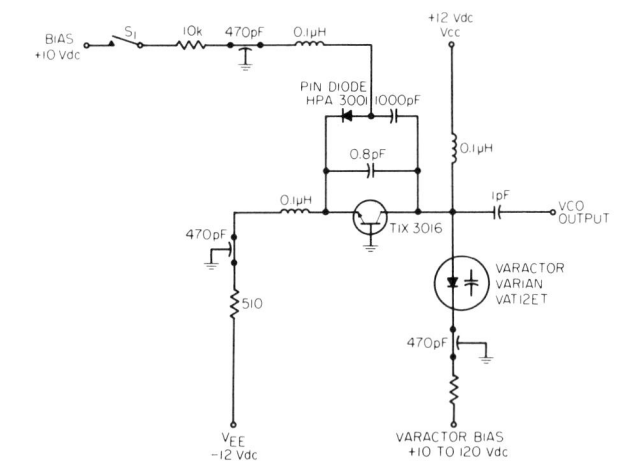

This microwave VCO is switched off by biasing the PIN diode.

Amplitude-switching circuit for large analog signals

G. Panigrahi
Univ. of Illinois, Urbana, IL

This circuit switches the amplitude of a high-voltage analog signal to either of two levels. These levels can be adjusted by R_1 and R_2. Q_1 is connected as an emitter follower and isolates the analog input. The opto-isolators are connected as multiplexed analog gates and by turning on either of the two opto-isolators, the output point is connected to the corresponding amplitude level. Outputs from TTL circuits control the switching, and they are completely isolated from the large analog signal.

The opto-isolators like Monsanto's MCT-2 have typical ratings of BV_{CEO} = 65V, BV_{CBO} = 165V and BV_{ECO} = 14V. When OC_1 is ON and OC_2 is OFF, there is reverse voltage across OC_2, and this should not exceed the BV_{ECO} rating. To take care of the smaller BV_{ECO} rating, a diode is connected in series with OC_2, so that most of the reverse drop is across the diode. When OC_2 is ON and OC_1 is OFF, the voltage across OC_1 should not exceed the BV_{CEO} rating. The BV_{CEO} rating can also be increased by putting another opto-isolator in series. The offset voltage is equal to V_{CE} sat and is only 0.1V. The off resistance is hundreds of MΩ, corresponding to leakage current I_{CEO} of a few nA.

The speed at which the output level can be changed is dependent on the photo-transistor time constants (t_{ON} = 5 μsec, t_{OFF} = 25 μsec in the saturated mode) and the output resistor, R_3. The time in which the output changes from the high level to the low level is slower compared to the other transition and is the time that OC_1 takes to discharge through R_3. This time is 200 μsec and can be improved to about 25 μsec by decreasing R_3. This should be adequate for medium-speed applications.

It should be noted that the alternative to the use of the opto-isolators is to use bipolar analog switches, which can become complex for such large signals. Most MOSFET and JFET analog switches are not rated for such large-amplitude signals. Photo-coupled resistors typically have response times of msecs and higher and cannot meet speed requirements of 200 μsec or less.

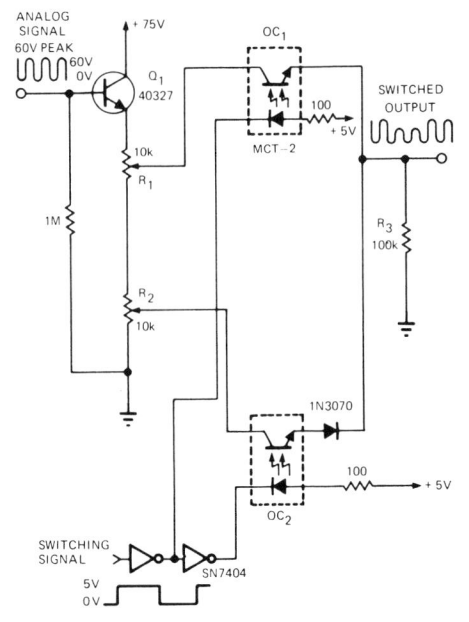

TTL switching of analog signal levels up to 60V is easily accomplished with a couple of opto-isolators.

Multiplexing without FETs

by Marvin K. Vander Kooi
Fairchild Semiconductor
Mountain View, Calif.

A circuit for time-division multiplexing and signal conditioning usually requires FET switches and buffer amplifiers. If FET switches are used, the system will require additional positive and negative supply voltages that are sufficiently larger than the peak signal amplitude to ensure that the FETs turn on and off completely.

The circuit shown here, however, completely eliminates the need for FET switches. This circuit uses the new µA776 op amp, which has an external bias-current-setting feature.

The amplifiers operate from conventional ±15V supplies, and the multiplex inputs (S_1 through S_3) can be driven directly from TTL or DTL devices. Each amplifier is controlled by either supplying or denying its 70 µA master-bias input, thus turning the amplifier ON or OFF, respectively.

For the circuit shown, the isolation between an ON amplifier and a signal fed to the input of another OFF amplifier is 80 dB at 50 kHz. Switching time of the output, from one multiplex to the next, is limited only by the 1 µV/sec slew rate of the amplifier. The SE4021 transistor, indicated in the schematic, has a leakage (I_{cbo}) of only 2nA, which ensures complete amplifier turn-off.

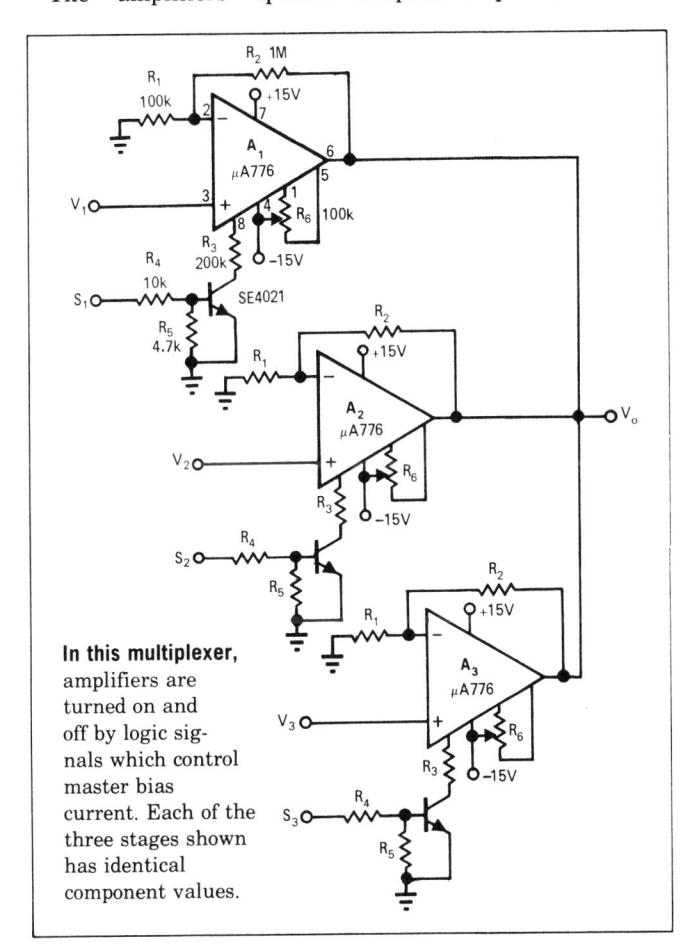

In this multiplexer, amplifiers are turned on and off by logic signals which control master bias current. Each of the three stages shown has identical component values.

Double-Balanced Diode Mixer

By Robert T. West
Fairchild Semiconductor
Mountain View, California

THIS DIODE mixer gives fine performance in the frequency range 2 MHz to 500 MHz, with low conversion loss over the entire bandwidth. Ferrite-bead transformers in the two input ports, ensure good mixer balance at the operating frequencies.

The circuit works as follows: Voltages appearing at points 1 and 2 of transformer T_1, cause diode pairs D_1D_2 or D_3D_4 to conduct, depending on signal polarity. Alternate conduction of these diode pairs causes the ends (3 and 4) of T_2 to switch to ground potential at a rate equal to the frequency at the LO port. Output voltage at the IF port is determined by the instantaneous voltage appearing across T_2 secondary (3 to 4). I-f output also depends on which end of the secondary is at ground potential at a given instant. Thus the output frequencies appearing at the IF port are the frequencies applied to the R and LO ports, plus their sum and difference.

The mixer should be constructed, using good rf layout techniques. Excessive lead length, from transformers to diodes and input connectors, will increase conversion loss by as much as 3 dB at 500 MHz. All ground return leads should be short.

Transformers T_1 and T_2 are identical, except that the center top of T_2 secondary is connected to the IF port, instead of to ground as for T_1. The transformers are wound on a common ferrite bead using 50–Ω bifilar wire, with a third wire twisted around the bifilar pair. This third wire is two to three times longer than the bifilar winding. Thus the three wires form a trifilar winding. The two ends of the long wire are brought together and wound on a separate ferrite bead to form coils L_1 and L_2. These inductors keep the ends of T_1 and T_2 balanced with respect to ground, thus improving mixer balance. The coils have two turns each and the transformers both have four turns. All windings are 36-AWG wire.

The best way to align the circuit is to adjust the positions of the secondary windings of T_1 and T_2 while measuring conversion loss. In this way you can find the condition of maximum coupling between primary and secondary. The mixer derives its operating power from the local oscillator, so the LO power must be around 3-5 mW for best performance. This level can be optimized during alignment to give the best compromise between conversion loss and noise figure.

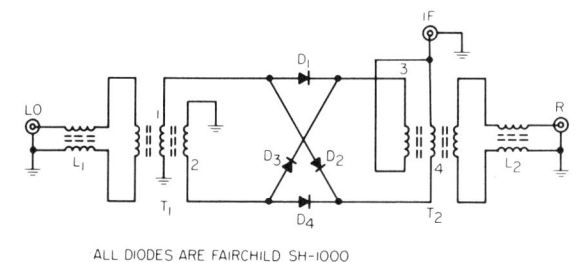

ALL DIODES ARE FAIRCHILD SH-1000

Double-balanced mixer uses hot-carrier diodes. Inductors L_1 and L_2 balance the transformers primaries with respect to ground.

Typical performance figures for an experimental mixer using this circuit, are as follows:
- Frequency Range: 2 MHz to 500 MHz (R and LO), and dc to 500 MHz (i-f).
- Conversion Loss: −7 dB to 500 MHz (i-f = 100 MHz).
- Mixer Balance: −35 dB to 500 MHz (LO power at R) and −30 dB to 500 MHz (LO power at i-f).
- Single-Side-Band Noise Figure: 8 dB to 500 MHz.

All measurements made with 10-dB 50−Ω pad on each port.

Linear period-to-voltage converter with low ripple

by James M. Kasson
Santa Rita Technology
Menlo Park, California

In applications where instantaneous voltages proportional to a signal frequency must be derived at rates comparable to the signal frequency, conventional averaging techniques usually prove inadequate. The circuit shown in Fig. 1 can give accurate and linear period-to-voltage conversion under such conditions.

The circuit consists of a monostable multivibrator (Q_1 and Q_2), a linear ramp generator (Q_4 and Q_5) and a sample-and-hold circuit (Q_4 and A_1). The sample/hold circuit at the output allows accurate tracking of rapid changes in period, because each step is independent of its predecessor.

The voltage across C_5 must be sampled before the capacitor is discharged; therefore a delay must be included in the circuit. In the circuit of Fig. 1, the monostable multivibrator (Q_1 and Q_2) provides the necessary delay. This one-shot activates the sample/hold circuit when it is triggered by an incoming pulse. When the one-shot recycles it discharges C_5.

An input pulse causes the collector of Q_2 to go positive, thus turning on Q_4 and allowing C_5 to charge C_6. Note that the capacitance of C_6 is small compared to C_5. Thus the loading of C_6 on C_5 is not a serious problem. This loading affects circuit performance only when the input period slews rapidly; and the effect can be corrected with a unity-gain buffer.

When the multivibrator recycles, Q_4 is turned off and C_6 remains charged to a voltage proportional to the period of the input signal. Amplifier A_1 acts as a buffer and, because A_1 has unity gain, the voltage on capacitor C_6 appears at the output.

Also, when the multivibrator recycles, a positive-going pulse is produced at the base of Q_5. This transistor then discharges C_5. The capacitor then starts to recharge until another input pulse is received and the cycle begins again.

Figure 2 shows typical waveforms at various points in the circuit, illustrating the sequence of events. In the graph of Fig. 3, output voltage is plotted against input period. The scales are logarithmic and a range of two decades is shown. Note that the response is essentially linear and that the slope is unity.

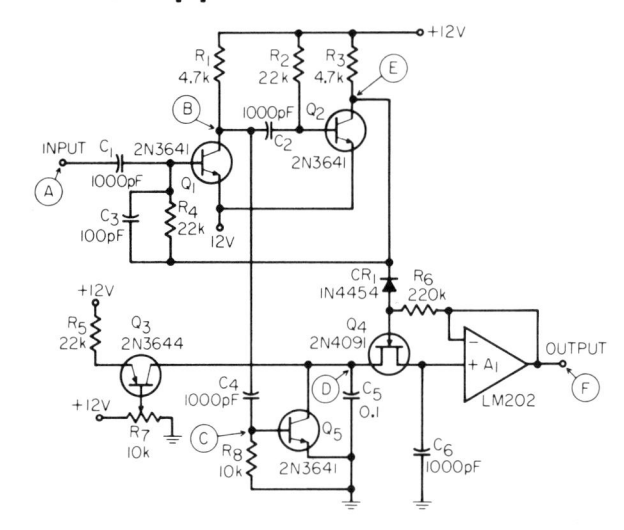

Fig. 1. This period-to-voltage converter allows sampling at rates approaching the input frequency, but without introducing excessive ripple.

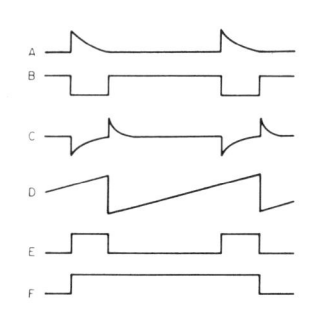

Fig. 2. Timing diagram for the circuit of Fig. 1.

Fig. 3. Measured response shows that output voltage is directly proportional to input period over a range of two decades. The line has a slope of one.

Achieve 0.5 µV/°C drift in IC video amplifiers

Surya Sareen
Aertech Industries, Sunnyvale, CA

DC-coupled, high-speed video amplifiers (Texas Instruments SN5512, Fairchild µA733) are available with a gain of ≅44 dB and a bandwidth ≅80 MHz. These devices have zero signal quiescent output voltage of ≅3.0V dc with a ≅-1.5 mV/°C temperature coefficient. This is troublesome when two amplifiers have to be cascaded and/or operation down to dc is required. A simple circuit which eliminates the above drift without sacrificing the speed or the ac gain is described in the schematic.

The output of the IC is fed to an emitter follower, Q_1, and level is shifted down to V_2

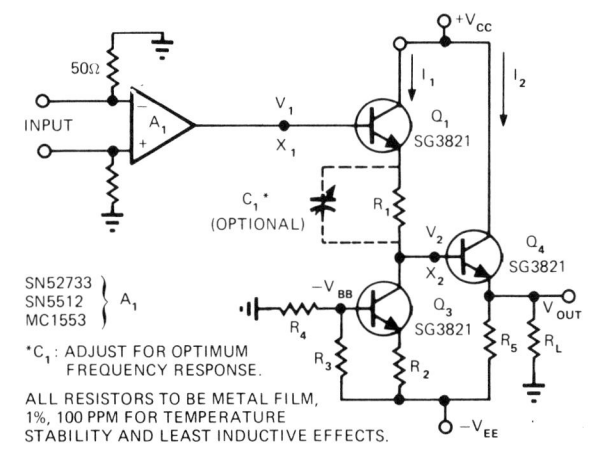

Temperature drift of video amplifiers is reduced to less than 1/2 V/°C by this circuit.

($\cong 0.7$V) by I_1R_1. I_1 is the current set by the constant current sink Q_3. Output of Q_4 is at $\cong 0.0$V as V_2 is set at 0.7V and the V_{BE} of $Q_4 \cong 0.7$V.

Emitter follower Q_4 avoids degradation in the frequency response by isolating the load from the high-impedance, high-capacitance collector node of Q_3.

Writing Kirchoff's voltage law equations for V_{out}, and I_1, we get:

$$V_{out} = V_1 - V_{BE}Q_1 - I_1R_1 - V_{BE}Q_2$$

and

$$I_1 = (-V_{BB} - V_{BE}Q_3) - (-V_{EE})/R_2$$

Assuming,

$V_{BE}Q_1 = V_{BE}Q_3 = V_{BE}Q_4 = V_{BE}$ Tune for any transistor in the SG3821 within ± 0.5 mV.

$\partial V_{BB}/\partial T = \partial V_{CC}/\partial T = \partial V_{EE}/\partial T = 0$ Regulated power supplies.

$\partial R_1/\partial T = \partial R_2/\partial T = 0$ External ± 100 ppm resistors.

$\partial V_{BE}/\partial T = -K$; where K = 2.5 mV/°C

$\partial V_1/\partial T = -K_1$; where K_1 is the temperature coefficient of IC output voltage; a measurable quantity.

and solving for:

$$\partial V_{out}/\partial T = 0; \text{ one gets,}$$
$$\partial I_1/\partial T = (\partial V_{BE}/\partial T)(1/R_2) = 1/R_1(-K_1 + 2K)$$

The above equation simplifies to:

$R_1/R_2 = 2 - K_1/K = K_2$ (Constant)

Thus by merely adjusting the resistor ratio, one can achieve almost zero temperature coefficient and excellent bandwidth. Emitter followers Q_1, Q_4 and very high impedance at the collector of Q_3 compared to R_1 virtually assures unity gain in the level shifting circuit.

This circuit was constructed using Silicon General's SG3821, V_1 = 3.0V dc for SN5512, $I_1 = I_2$ = 2 mA, K = 1.6 mV/°C, K_1 = 1.4 mV/°C, R_1 = 0.8k, R_2 = 0.71k, R_3 = 2.1k, R_4 = 3.9k, R_5 = 3k, V_{CC} = 6.0V dc and V_{EE} = −6V dc. Overall bandwidth was from dc to 70 MHz. Worst case output dc level drift was ±3 mV over a temperature range of −25 to 75°C. This corresponds to −0.42 μV/°C referred to the input, as amplifier gain was 140. □

Variable dc offset using a current source

by P. Bruce Uhlenhopp
NOAA
Boulder, Colo.

In many circuit applications it is useful to offset ac signals at varying dc levels. Some designs call for offsetting signals that are to be fed into a moderately high input impedance. A dc offset can be added to an ac signal without significant interaction by using a high-impedance current source to provide the desired level shift as illustrated in the figure.

Transistor Q_4 and its resistive network form a high-impedance adjustable current source. The input signal and the offset signal are fed to the base of Q_3 which drives Q_1 and Q_2, a complementary-symmetry emitter follower.

For the values shown, the level can be shifted about ±7V dc. This is particularly useful in a video circuit whose dc level controls the intensity of a CRT. DC level can be controlled from a panel-mounted potentiometer since the high-frequency video signal does not pass through the pot.

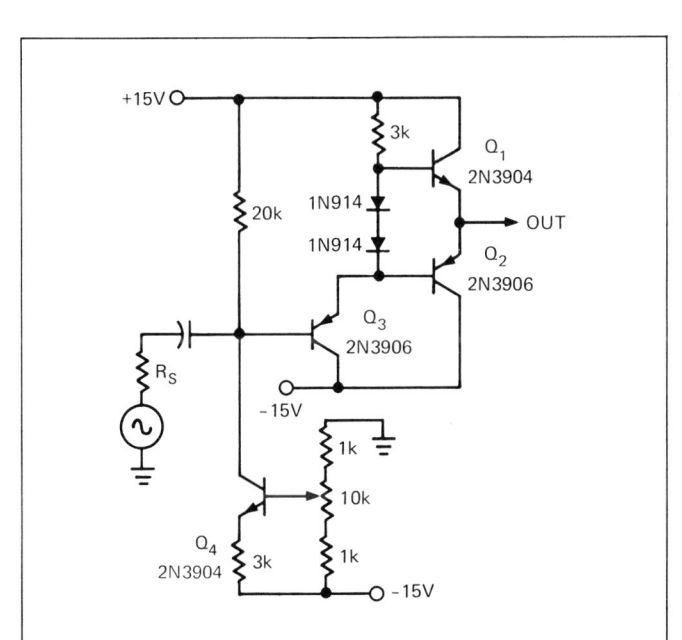

This simple circuit, using an adjustable current source, allows a video signal to be offset ±7V dc.

Track-and-hold amplifier

by Richard S. Burwen
Analog Devices, Inc.
Norwood, Mass.

The low input current and high slewing rate of the AD503J make it a useful amplifier for track-and-hold applications. The circuit described here will track a ±10V input signal at frequencies up to 4 kHz.

When the track-hold gating signal changes from +5V to zero, the series FET, Q_3, opens and the input voltage is retained on capacitor C_1. The output amplifier A_2 provides a high input impedance so that C_1 will not discharge rapidly.

Drift rate is determined primarily by the OFF leakage current of Q_3, which tends to be greater than that of amplifier A_2. With a leakage of 100 pA, the drift rate is 10mV/sec and the rate doubles every 10°C. Lower drift rate and higher accuracy (at the expense of a slower acquisition time) can be achieved by increasing the value of C_1. This capacitor should be a type (such as polystyrene or "Teflon") having low dielectric absorption.

Use of a low-pinchoff-voltage FET, as specified for Q_3, allows the circuit to handle ±10V input voltages with supply voltages of only ±15V.

In the "track" mode, the gating signal is at +5V, Q_1 and Q_2 are cut off, and the gate of Q_3 assumes the voltage at the output of A_1. Thus the FET is zero-biased over the entire input-voltage range and has an ON resistance of less than 100Ω. Resistor R_5 adds to the ON resistance so as to better isolate C_1 from A_1 to prevent ringing.

In the "hold" mode, both Q_1 and Q_2 conduct, thus pulling the gate of Q_3 to −15V. The capacitor voltage stops tracking at the time when the gate voltage of Q_3 reaches a value 3V below the source voltage, or about 100 nsec after the track-hold gating signal changes to zero.

Because of gate-drain capacitance in Q_3, the gating-signal swing causes a small charge (proportional to the change in gate voltage, i.e. 15V plus the input voltage) to be delivered to C_1. The charge causes a small offset in the "hold" mode (relative to the "track" mode) which is typically less than 10mV over the full ±10V input range.

There are also small settling transients in amplifiers A_1 and A_2, so it takes up to 2 msec for the output to settle to within 1 mV of the final value. The settling time for an input step of 10V in the "track" mode is less than 15 msec to within 1 mV of final value.

An added error caused by the capacitor's dielectric absorption may reach 3 mV if the input does not remain constant long enough in the "track" mode before changing to "hold."

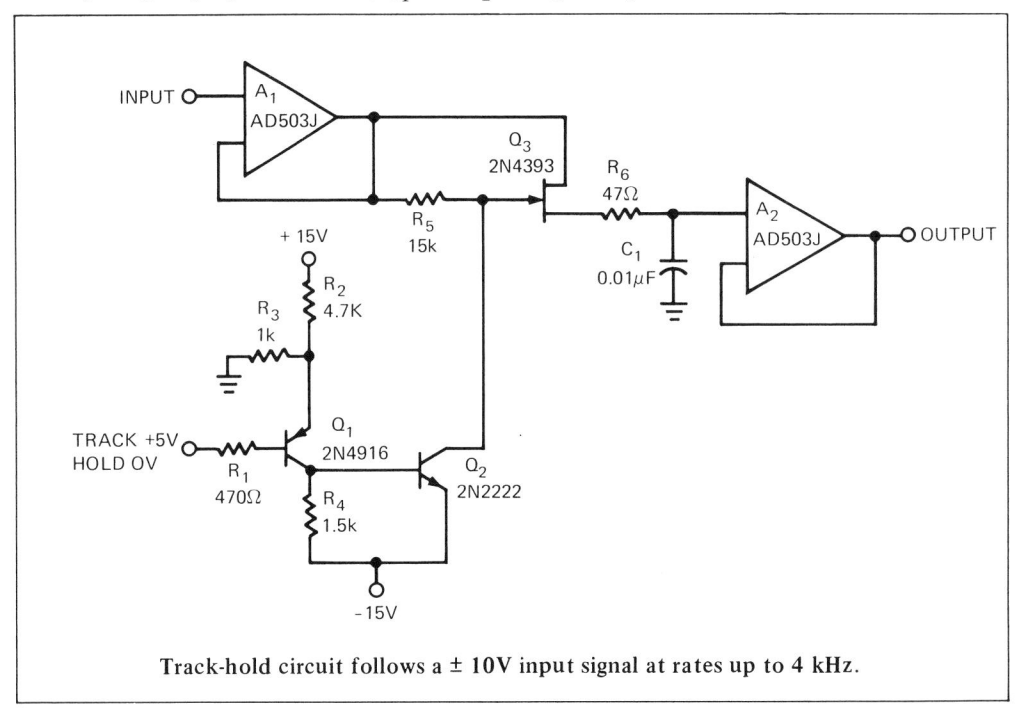

Track-hold circuit follows a ± 10V input signal at rates up to 4 kHz.

Reversible Linear Counter

by W. J. Godsey
Hayes International Corp.
Birmingham, Ala.

This circuit provides uniform discrete steps in output voltage for pulses at the input terminals. Negative pulses at A cause positive steps in the output voltage. If terminals A and B are joined, positive pulses will cause negative steps, and negative pulses will cause positive steps in the output voltage. The input pulses, however, must be short and of small enough amplitude so that overshoot of the trailing edge will not drive the counter in reverse.

Operation is as follows: initially the voltage between the base of Q_2 and the reference bus, will be the voltage across C_3 ±0.7 volts (conducting potentials of diode D_1 and the base-emitter diode of Q_2). The voltage at the collector of Q_1 will be zero, since Q_1 is not conducting.

A negative pulse into terminal A will cause Q_1 to conduct, and the collector voltage will approach V_{CG}. Capacitors C_2 and C_3 form capacitive voltage-divider, but C_3 apparent = C_3 actual (β + 1), where β is the current amplification factor of Q_2. The charge which flows into C_3 is (β + 1) times the charge which flows out of C_2. When the input pulse is removed, the collector voltage of Q_1 returns to zero in a time determined by R_2. The charge which must flow back into C_2 flows from C_3 through D_1 and is approximately equal to the charge which was forced into the base of Q_2. The net charge left in C_3 is β times the charge from C_2.

Assuming that the voltage step across C_3 is small compared to supply voltage, this voltage increase,

$$E_{step} = \frac{\beta \times C_2 \times V_{CC}}{C_3}$$

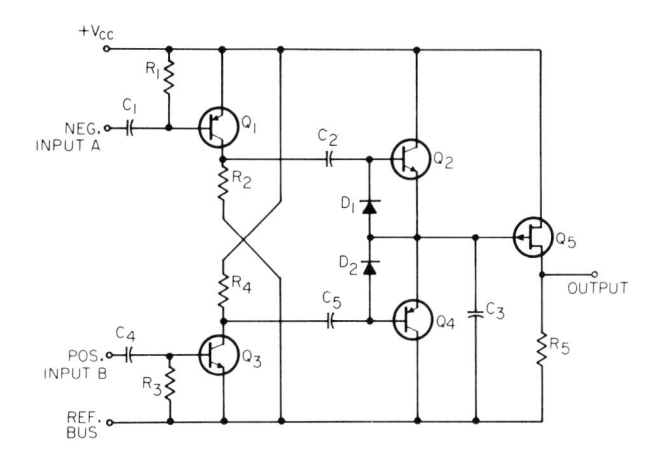

Counter circuit in which negative pulses at A give positive steps in output voltage. Counting is reversed by applying positive pulses at B.

$C_2 \times V_{CC}$. This relationship holds for all values of voltage across C_3 until the voltage across Q_2 is less than 1 V.

Positive pulses into terminal B cause exactly the same action, via Q_3, C_5 and Q_4, but in the opposite direction. Coincident pulses cancel, so this counter can add and subtract simultaneously. FET Q_5, is an emitter follower, which prevents loading of C_3.

The counter can be reset to zero by clamping the base of Q_2 to the positive supply bus, or the base of Q_4 to the reference bus. This circuit can also be used as a net frequency indicator by shunting C_3 with a resistor. The average output voltage will then be proportional to the frequency of pulses at A less the frequency at B.

Low-power, multiple-input comparator for ac/dc inputs

by Merle Converse
Southwest Research Institute
San Antonio, Texas

Three or more ac or dc inputs can be compared directly with the comparator shown in the figure. The input with the largest amplitude yields a positive output state. The circuit consumes very little power — only 2 mW for the 3-input example shown, an audio-frequency voltage comparator. Each input is rectified and applied to the base of a transistor which forms one side of a Schmitt trigger. Both sides of each section are coupled to the emitters of the corresponding transistors in the other sections. Bias voltages are chosen so that only one input transistor can be on, and two output transistors must be on. Thus, one output transistor must be off, providing one high output at +7V. This high output will always be from the section having the highest input amplitude.

Circuit operation is very similar to that of a two-transistor Schmitt trigger. Suppose that *Input 1* has the largest amplitude. Q_1 will be held on by the input signal, and Q_2 will be off. Because of the Q_1 emitter current in the 10-kΩ emitter resistor, Q_3 and Q_5 are biased off, holding Q_4 and Q_6 on. Now, suppose that the *Input 2* amplitude increases. As Q_3 begins to turn on, Q_4 is biased off by the coupling from the Q_2 collector. Q_1 must turn off because of the increased emitter voltage as Q_3 turns on, and Q_2 must turn on because of coupling from the Q_1 collector.

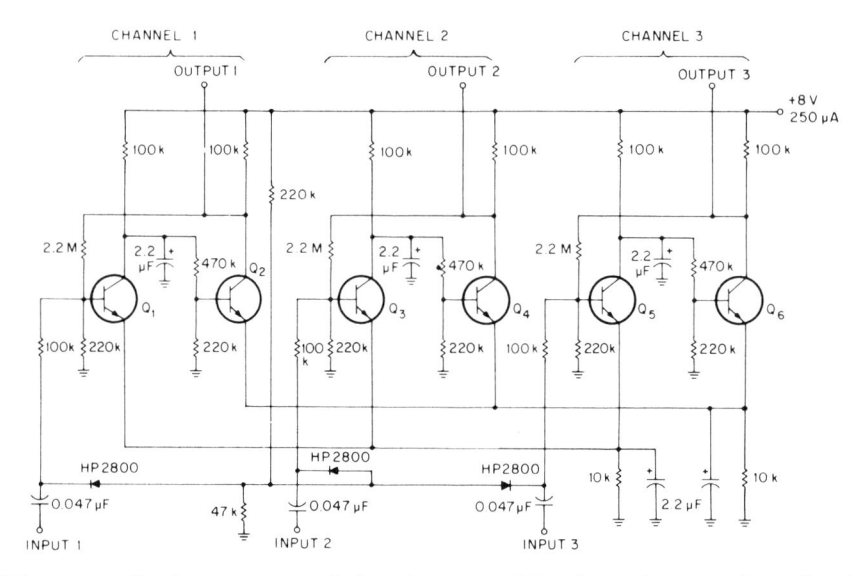

This 3-input amplitude comparator is based on repetitive 2-transistor sections that lend themselves to construction of an n-input comparator. The transistors are 2N4286.

SECTION II
Digital and Pulse Circuits

End edge ambiguities with two ICs

Dennis McLaughlin and **Carlo Fanstini**
Naval Research Lab, Washington, DC

A 74155 dual 2-line-to-4-line decoder/demultiplexer and a 4-bit binary counter generate a family of strobes which can free your designs from edge-ambiguity problems. **Fig. 1** shows the generator.

A binary counter's internal stages use trailing edge triggering. However, a counter which counts on the leading edge of the driving clock can be used to advantage. The driving clock pulses of **Fig. 2** strobe the 74155, Q_A acts as the data input, and Q_B and Q_C act as select lines. The edges of the various 2Y-pulses do not coincide with each other, with the edges of the various 1Y-pulses or with the edges of Q_B, Q_C or Q_D. The result is hazard-free strobing.

After you become familiar with the basic idea, the advantages will be apparent. These are a few: 1) a reduced IC package count, 2) a great number of output sequence variations from different 74155 input configurations and 3) a possibility of more output lines by using a 74154 4-line-to-16-line decoder/multiplexer.

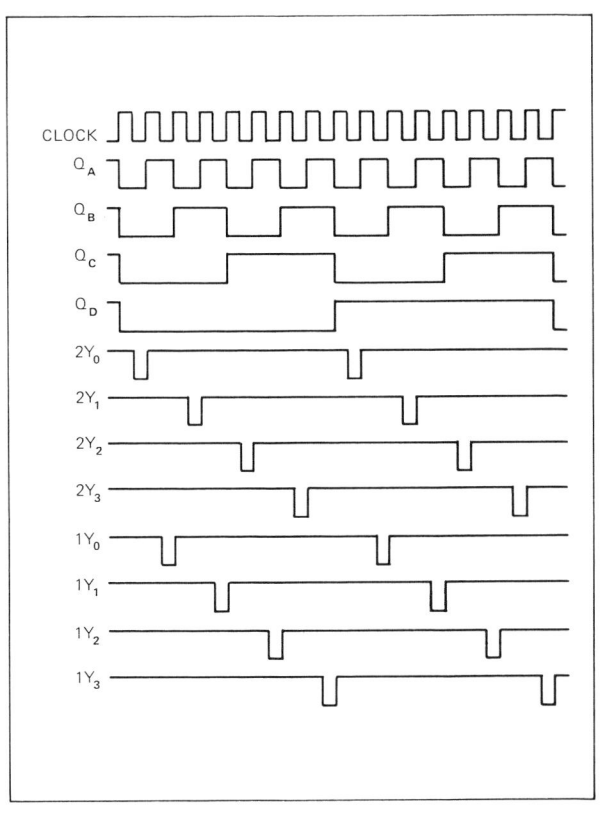

Fig. 2—**Waveforms show hazard-free strobing** and gating possible with this circuit.

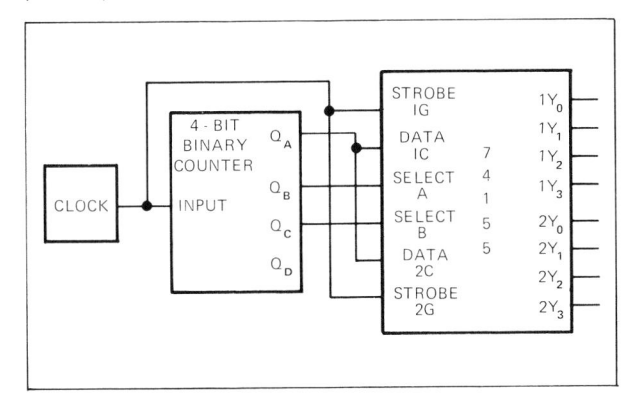

Fig. 1—**A 4-bit counter and a 74155 decoder** team up to produce a strobe generator without edge ambiguities.

One shot multivibrator has delayed output

Louis Sims
Key Tronic Corp. Spokane, Wash.

This circuit will generate an output pulse after a delay of approximately one microsecond. Delay $\cong 0.36\, R_A C_1$, where $R_A = R_1 \| 4k$. The 4k is the internal equivalent resistance of the TTL 7413 gate. R_1 may be chosen much smaller than 4k to "swamp" out the effects of the large manufacturing tolerance of the 4k resistance and to arrive at a more accurate delay time.

The pulse width is given by: Tpw $\cong 0.36\, R_B C_2$, where $R_B = R_2 \| 4k$.

This can be made much longer than the triggering pulse, if desired. Again the same considerations apply in the selection of R_2 as for R_1. Trigger pulses that are shorter than the delay time are rejected and will cause no output.

Recovery of the delay circuit is Beta limited by the 7403 output transistor. Full recovery of the pulse timing circuit is determined by the time constant for $R_3 C_2$.

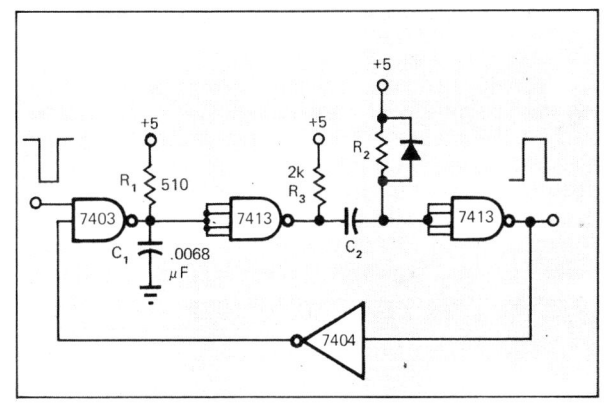

Delayed pulse generator, in this case set for approximately 1 μsec, can be set for other delay times by varying R_A and C_1. Output pulse width is set by varying R_B and C_2.

Control signal determines modulo of IC counter

by Ivars P. Breikss
Honeywell
Denver, Colo.

Many applications require digital counters whose modulo can be easily changed by applying external logic signals. One specific application occurs in electronic digital clocks.

In this application, the counter controlling the least-significant hours digit must have a modulo of ten as long as the most-significant hours digit is zero or one. When the most-significant hours digit becomes two, the least-significant hours counter must operate with a modulo of four. This, of course, is because there are 24 hours in one day.

The circuit shown in Fig. 1 gives the switchable 10/4 modulo needed for this application. In addition to counting in either mode, the circuit provides a CARRY signal which forms the input for another divider stage. Fig. 2 shows the external connection when the circuit is used for the least-significant hours stage of a digital clock.

Here is how the circuit works: When the X input is logic 0, gates G_1 and G_2 are never enabled. So flip-flops FF_3 and FF_4 do not receive clock pulses. The first two flip-flops, FF_1 and FF_2, function as a conventional modulo-4 counter. Also, the output of G_4 is the inverse of B, therefore the output of I_5 provides the correct modulo-4 CARRY signal.

When the X input is logic 1, the outputs of I_1 and I_2 are identical to B and A respectively. I_1 then supplies clock pulses to FF_2, and I_2 provides clock pulses for FF_4.

In this mode, all four flip-flops count in binary sequence. They give 1-2-4-8 weighted binary-coded outputs at A, B, C and D respectively.

Components G_3 and I_3 supply a logic 1 to the J input of FF_4 when FF_2 and FF_3 are both in the "true" state (count 8). When the next falling edge of A occurs (end of count 7), FF_4 also becomes "true." Finally, the succeeding falling edge of A (end of count 9) resets FF_4 to its "false" state, thus giving logic 0 on each of the lines, A, B, C and D.

Inverter I_5 provides an output identical to D. This is the correct CARRY signal for a modulo-10 counter.

The "1" input of FF_1 is normally connected permanently to +5 Vdc.

Fig. 1. Modulo of this counter circuit is either **10 or 4** depending on the logic level at point X. Outputs A thru D are 8-4-2-1 BCD format.

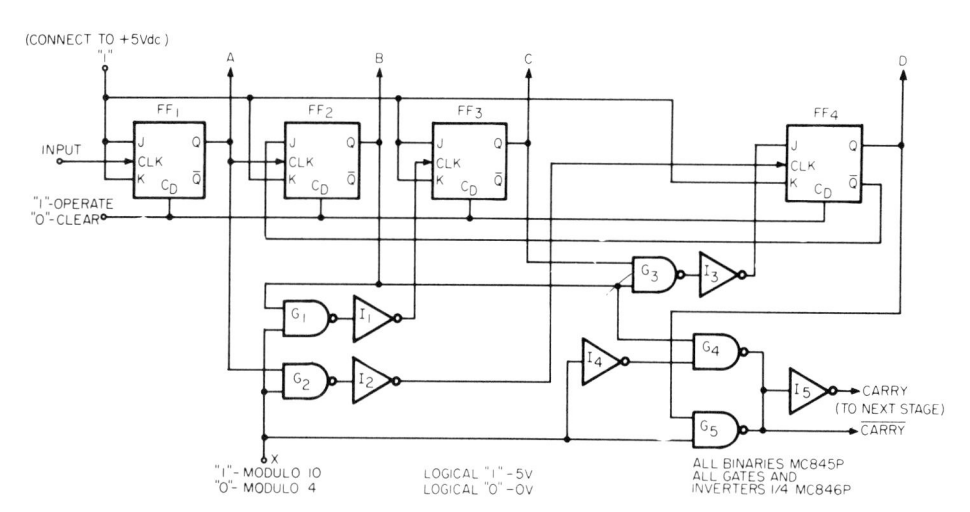

TRUTH TABLE

(Modulo 4)

A	B	Decimal Equiv
0	0	0
1	0	1
1	1	2

TRUTH TABLE

(Modulo 10)

A	B	C	D	Decimal Equiv
0	0	0	0	0
1	0	0	0	1
0	1	0	0	2
1	1	0	0	3
0	0	1	0	4
1	0	1	0	5
0	1	1	0	6
1	1	1	0	7
0	0	0	1	8
1	0	0	1	9

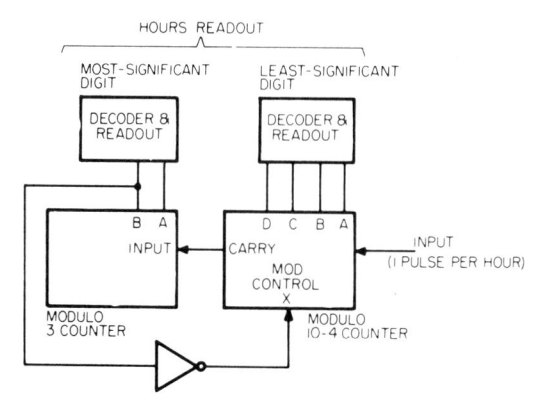

Fig. 2. One possible application for the modulo 10/4 counter is in a digital clock. The least significant hours stage receives input pulses from the most significant minutes stage. The hours counters then count to a total of 24 before resetting.

Single transistor improves CMOS one-shot

Paul Ghelfan
M.G. Electronics, Rehovot, Israel

Monostables implemented with CMOS gates usually dissipate a surprising amount of power during their active state.

They also exhibit considerable performance variations from unit to unit and over the temperature range.

These disadvantages are due to the long, unstable transition time which results when an RC-shaped waveform interacts with the smooth threshold of a CMOS device.

During these transitions, both the p and n devices in the gate are active, forming a shorting path to ground.

The illustrated circuit overcomes these shortcomings, as the transistor satisfies the need for a sharp threshold to generate fast transitions.

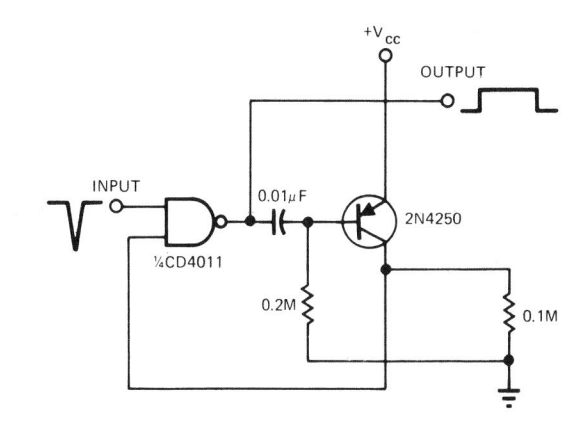

Short transitions provided by discrete transistors lower power dissipation of this CMOS monostable.

Pulse width is determined here essentially by the RC time constant.

Simple bi-quinary to BCD converter

By Charles Engelberg
American-Standard
New Brunswick, N. J.

BI-QUINARY CODES have found wide acceptance in digital systems because bi-quinary counters offer advantages over the more conventional binary-coded-decimal (BCD) counters. Bi-quinary-to-decimal decoding requires far fewer gates than BCD-to-decimal decoding. Also, the required counter fan-out is lower.

Where decimal display devices are used, it is preferable to use a bi-quinary counting system for the display and to convert the bi-quinary output to BCD for driving peripheral equipment like data printers.

The circuit shown converts bi-quinary information to BCD, using only six diodes and four resistors per decade. In the truth tables logic *1* equals zero and logic *0* equals +15 Vdc. Looking at the truth table for the bi-quinary input, we can see that a logic *1* occurs at the 2 or 3 inputs (2 + 3) only during states 2 and 3. Similarly, a logic *1* appears at the 4 or 5 inputs during states 4 and 5.

Now we must differentiate between the odd and even numbers, to get zero and one information. We do this by monitoring the first flip-flop in the decade counter. This yields the required odd-even logic signal for input to the converter.

BCD CODE					BI-QUINARY CODE					
	1	2	4	8		Odd 0	2	4	6	8
						Even 1	3	5	7	9
0	0	0	0	0	0	0	1	0	0	0	0
1	1	0	0	0	1	1	1	0	0	0	0
2	0	1	0	0	2	0	0	1	0	0	0
3	1	1	0	0	3	1	0	1	0	0	0
4	0	0	1	0	4	0	0	0	1	0	0
5	1	0	1	0	5	1	0	0	1	0	0
6	0	1	1	0	6	0	0	0	0	1	0
7	1	1	1	0	7	1	0	0	0	1	0
8	0	0	0	1	8	0	0	0	0	0	1
9	1	0	0	1	9	1	0	0	0	0	1

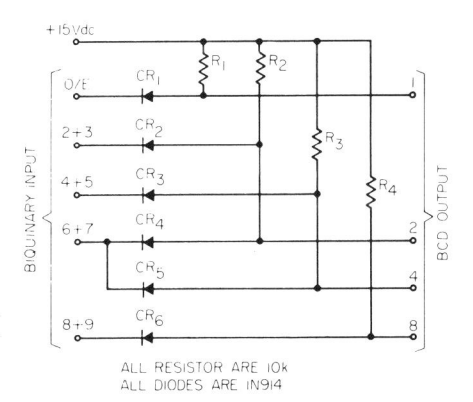

Code converter for coupling decimal displays to BCD devices like data printers.

Each time an odd number occurs, logic *1* appears at the O/E terminals of the converter. For example, when state six is present, the 6 + 7 terminal is grounded causing the converter binary outputs 2 and 4 to be clamped to ground through CR_1 and CR_5. But, when state seven is present, the O/E terminal is grounded as well as the 6 + 7 terminal. This causes converter outputs 1, 2 and 4 to be clamped to ground.

Inputs to the converter are derived from the decade counter, decoder, and cold-cathode indicating tube that constitute a typical decimal display system. Odd-even information comes from the first flip-flop of the decade counter, while information relating to pairs of decimal digits comes from the five cathodes of the indicating tube. Diodes CR_2 thru CR_6 must have reverse-breakdown voltages greater than the cathode-prebias voltage of the indicating tube. This is usually in the range 65 to 120 V.

The converter circuit supplies four-line-per-decade BCD (1-2-4-8) to output devices like data printers. Logical *0* is +15 V, with a 10-k source impedance. Logical *1* is +1.5 V, with a maximum sink current of 2 mA.

Digital and Pulse Circuits 51

One shot triggers on both edges

by **Ury Priel**
National Semiconductor Corp. Santa Clara, Calif.

A modification to the external circuitry can make a monostable-multivibrator IC trigger on both edges of an input pulse. The required connections are shown in **Fig. 1**. This circuit is much simplier than other techniques which typically require several gates in addition to the basic one-shot IC.

Circuit operation can be best understood by referring to **Fig. 2** which shows the input portion of the circuitry on the IC chip. The one-shot fires when the second gate in the input circuit turns on, or in other words when Q_1 turns on.

Note that because of the added external diode, the second gate has a threshold approximately 0.7V lower than that of the first gate. Therefore, as the input voltage at point A rises, it will first reach the threshold of the second gate (whose output is Q_1). This causes a race condition because of the different propagation delays.

The timing diagram **(Fig. 3)** shows the relationships between the waveforms in the circuit of Fig. 2.

At time t_1, all the inputs to the second gate are high and thus point C goes low. When the input signal (point A) crosses 1.3V, point B goes low and thus the second gate turns off at time t_2.

On the falling side of the input pulse (point A), we have a similar situation. Before the input signal drops to 1.3V, point B is low and point C is high. When A reaches 1.3V, B goes high and C goes low. When A reaches 0.6V, the second gate turns off and C returns to the high condition.

The pulses at point C will always trigger the one-shot portion of the IC. Thus the complete circuit triggers from both edges of the input pulse.

The external RC network (R_1 and C_1) determines the output pulse width of the one-shot. The DM7850 has a minimum holding time of 40 nsec, therefore the minimum pulse width at point C must be 40 nsec. Because of this limitation, the circuit described here is restricted to applications where the pulse rise and fall rates are greater than 60 nsec/V (40 nsec/0.7V). This is the price that must be paid for circuit simplicity.

A further limitation of the circuit is reduced noise immunity resulting from the lowered threshold of the second gate. A bypass capacitor at the input can help correct this problem.

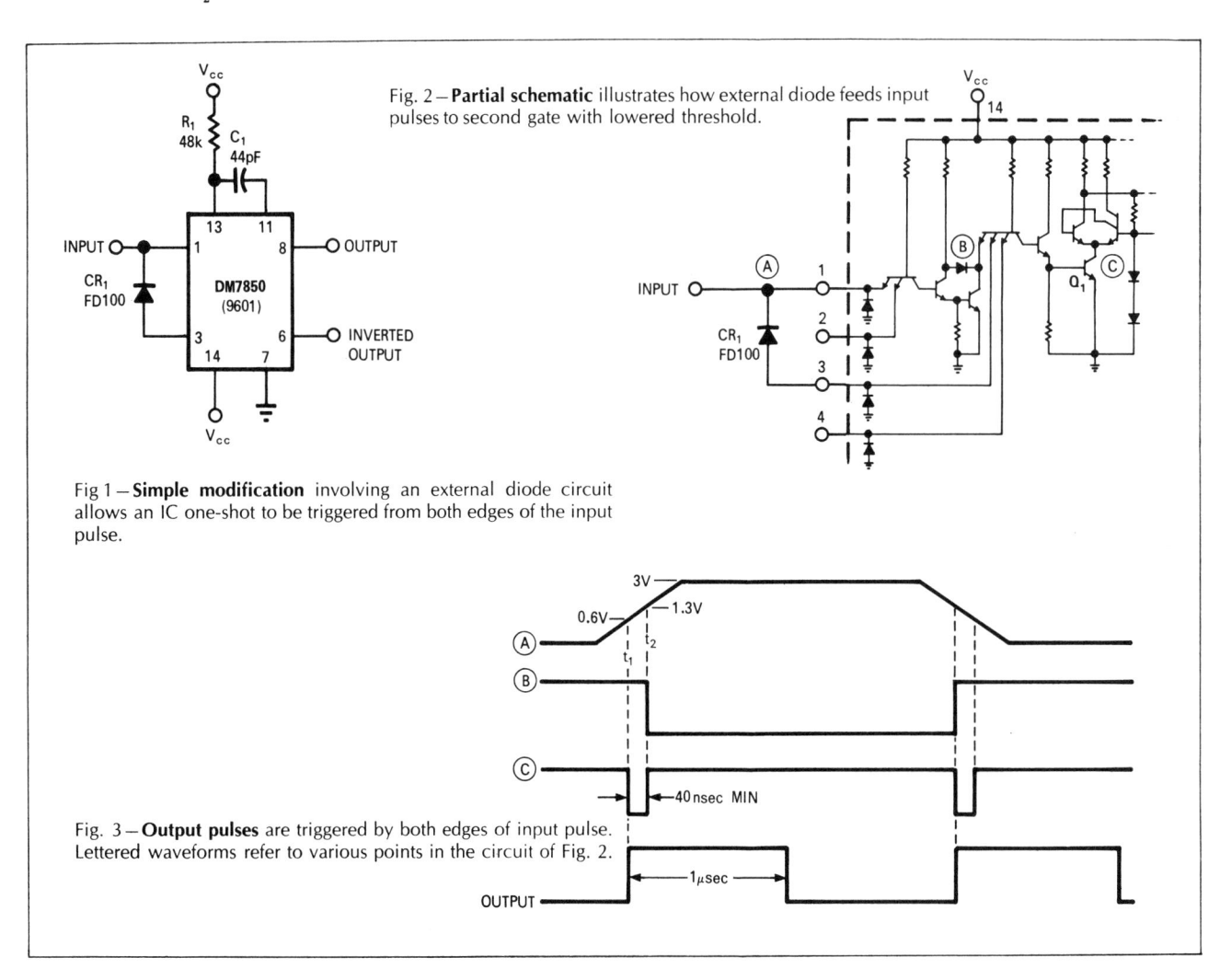

Fig. 2 – **Partial schematic** illustrates how external diode feeds input pulses to second gate with lowered threshold.

Fig 1 – **Simple modification** involving an external diode circuit allows an IC one-shot to be triggered from both edges of the input pulse.

Fig. 3 – **Output pulses** are triggered by both edges of input pulse. Lettered waveforms refer to various points in the circuit of Fig. 2.

Phase-shifter for phase locked loops

J. S. Krikorian, Jr.
Warwick, RI

Phase locked loops can be utilized as synchronous AM demodulators or tone detectors. Other phase locked loop applications require a measurement of signal strength or a means of determining signal acquisition. The quadrature detector illustrated in **Fig. 1** is employed to obtain such signals. Frequently, an RC phase-shift network is used to obtain the 90° phase-shift. An alternate method of phase-shifting can be implemented using logic circuits, as illustrated in **Fig. 2**. The output of this circuit is especially suitable for use with FET gate-chopper multipliers or hard-limiter logic multipliers, while the input is compatible with a VCO that has a square wave output within its voltage limits.

The logic phase-shift circuit uses a flip-flop to divide the frequency by two. The input, f, and the output, f/2, of this flip-flop are fed into an equivalence logic circuit to obtain a square wave, f/2-90°, that is shifted in phase by 90° with respect to the output, f/2, of the flip-flop. An exclusive-OR in place of the equivalence logic circuit would also provide a 90° phase-shift. To compensate for the frequency division by two, the VCO center frequency has to be doubled.

Because of its insensitivity to frequency changes, this circuit is useful for wideband phase locked loop applications. High-frequency accuracy is dependent on the speed of logic circuitry used in the implementation. Along with the associated circuitry for the quadrature detector, the logic phase-shifter can be incorporated into a self contained phase locked loop IC package.

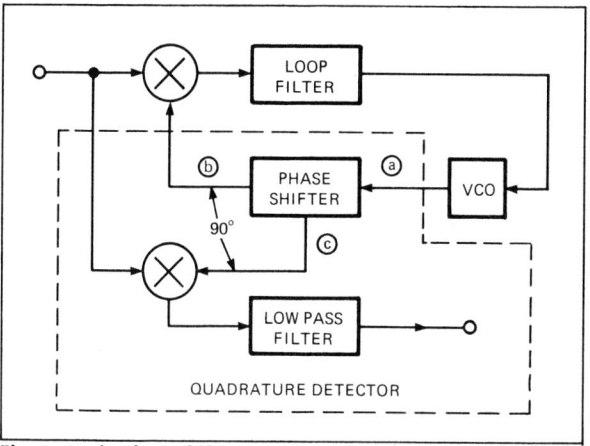

Fig. 1—Logic phase-shifter provides 90° phase-shift signal for use by the quadrature detector.

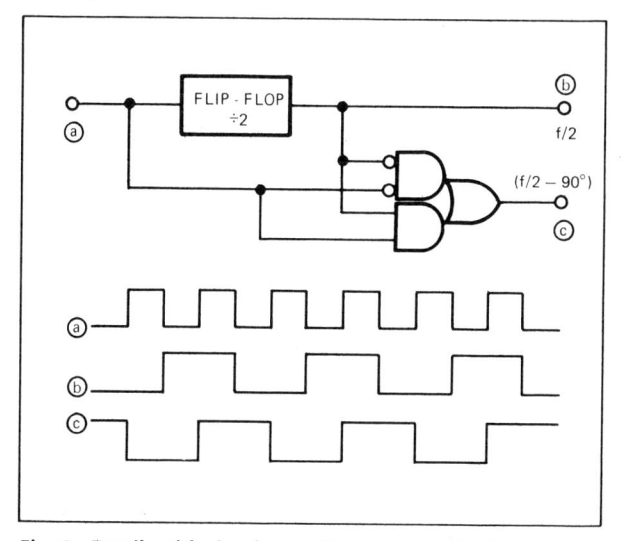

Fig. 2—Details of logic phase-shifter showing flip-flop, equivalence circuit and associated signals.

Flip Flop Operated by Input Signal NOR

By Luigi Mercurio
L.R.E. Olivetti
Milan, Italy

THE USUAL FLIP FLOP, with resistive inputs (RS type), is set (reset) when one of the input signals on the SET (RESET) side goes to its high level. The modified flip flop described here is set (reset) when all the input signals $x_{s1}, \ldots x_{sn}$ ($x_{R1}, \ldots x_{Rm}$) are low. From a logical standpoint, it is set (reset) by the NOR-function of its input signals instead of the OR function. To implement the same function with the usual flip flop, two extra transistors would be required.

When Q_s is on, the current I_s flows through D_{s2} into the base of Q_s, while I_r flows through D_{R1} into the collector of the same transistor. To drive Q_s OFF, i.e to change the state of the flip flop, the current I_s must be at least temporarily zero. The same considerations apply to the other side of the flip flop since the circuit is symmetrical.

We can add the usual SET or RESET inputs (dotted in figure) by coupling the signals to the base through resistors.

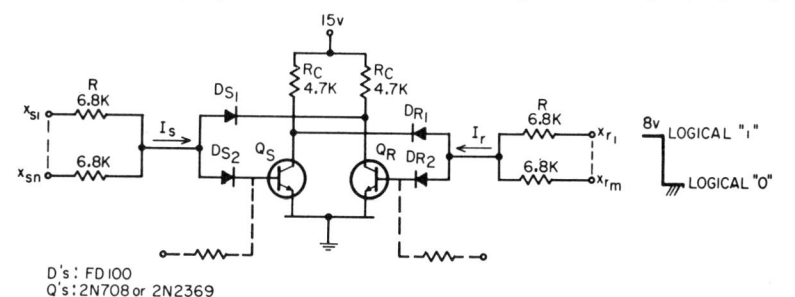

Modified flip flop is set or reset when all input signals are low.

Improved tunnel-diode threshold circuit has adjustable hysteresis

by Otakar A. Horna
Research Institute for Mathematical Machines
Prague, Czechoslovakia

It's well known that a tunnel diode TD_1 can be connected between the base and emitter of a transistor Q_1 to form a simple and stable threshold circuit, as shown in Fig. 1. A major disadvantage of this simple circuit is the large hysteresis of the switching characteristic, shown in Fig. 2.

Figure 2 shows that the tunnel diode switches to its high-voltage state when input current i_1 is greater than the diode's peak current I_p. But the diode doesn't switch back to the low-voltage state until i_1 is less than the sum of the diode's valley current I_v and the transistor's base current I_b.

Figure 3 shows an improved circuit in which the hysteresis can be varied over a wide range from positive values to zero. The circuit can also be adjusted to give negative hysteresis though, of course, this is an unstable state.

When the diode TD_1 is in the low-voltage state, Q_1 is non-conducting and bias current i_2 flows through resistors R_L, R_4 and R_3 into the diode. This continues until the input current i_1 reaches the value I_p-i_2. The diode then switches to the high-voltage state and Q_1 saturates. The saturation voltage of Q_1 is only a few hundred millivolts, therefore no bias current i_2 can flow into the diode. Before TD_1 can switch back to the low-voltage state, i_1 must be less than $I_v + I_b$.

If we increase i_2, by adjusting R_4, we can reduce the value of i_1 required for the diode to switch to its high-voltage state. The value of i_1 for back-switching remains unchanged. So we can reduce the hysteresis of the circuit. This is shown dotted in Fig. 2.

The current values for the three possible hysteresis modes are given by the following expressions:

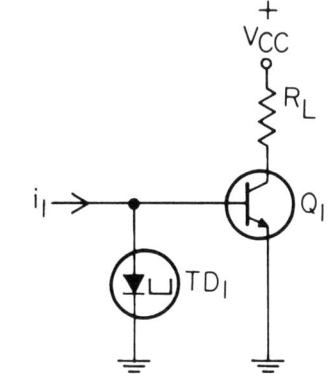

Fig. 1. Widely-used tunnel-diode threshold circuit has the disadvantage of high hysteresis.

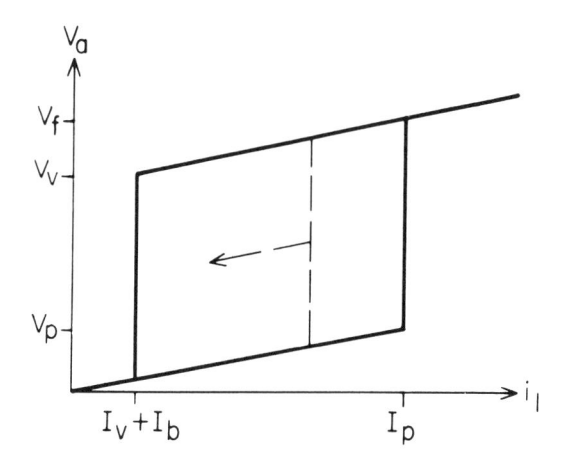

Fig. 2. Hysteresis loop of tunnel-diode threshold circuit. Dotted line shows how hysteresis can be reduced by lowering the value of i needed for switching to the high-voltage stage.

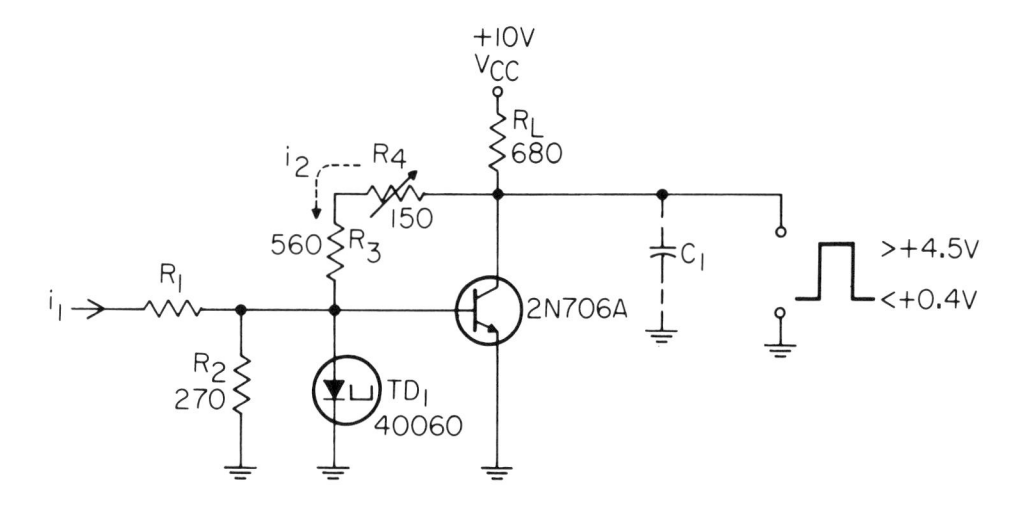

Fig. 3. Improved threshold circuit with adjustable hysteresis. The magnitude of current i_2 determines whether the circuit hysteresis is positive, zero or negative.

For positive hysteresis,
$$i_2 < I_p - I_v - I_b \quad (1)$$
For zero hysteresis,
$$i_2 = I_p - I_v - I_b \quad (2)$$
For negative hysteresis,
$$I_p - I_v - I_b < i_2 < I_p \quad (3)$$

If the circuit has negative hysteresis, it will oscillate when the input current i_1 is greater than $I_p - I_2$ but less than I_b. The period of oscillation is determined by the switching time constants of Q_1 and by external capacitance C_1 (see Fig. 3).

For normal operation with positive hysteresis, temperature stability of the lower threshold is determined by currents I_v and I_b. Tempco can be greatly improved by first choosing a tunnel diode having a high I_p/I_v ratio and then adding a parallel resistor R_2. The value of R_2 should be selected so that the resulting I_p/I_v ratio is only 3/5, for R_2 and TD_1 together. This technique reduces the temperature coefficient more than four times without appreciably reducing the switching speed of Q_1. Complete compensation can be achieved by using a negative tempco resistor in place of R_2.

The semiconductors used in the original version of the Fig. 3 circuit were of European origin. The tunnel diode was a GaAs type with an I_p of 10 mA (similar to RCA 40060). Q_1 was a high-speed switching transistor type BSY62 (similar to 2N 706A).

The circuit was developed for interfacing within a digital computer. Output levels are compatible with TTL or DTL ICs.

Timer produces time delays from μsecs to hours

Y. Kurahashi, Exar Integrated Systems, Inc.
Sunnyvale, Calif.

Commercially available one-shot IC's are usually limited to applications requiring relatively short delays of a few milliseconds or less. By adding several external components, including one transistor, a one-shot IC can be used to generate timing periods of up to one minute. A new multifunction IC, the XR-220, can further simplify the external connection and provide greater versatility.

The functional block diagram and basic external connections for timing applications of the XR-220 are shown in **Fig. 1**. The monolithic device contains a linear current source and reference circuit, voltage comparator, flip-flop, timing switch, buffer amplifier and a driver amplifier. The connection shown for a timing circuit requires only three external components: a timing resistor, R, timing capacitor, C, and load resistor, R_L.

Two outputs are available. Terminal 10 is a low-current output which is normally "LOW" and TTL compatible. An external resistor is required to provide a load for the open collector, which is internally coupled to the next stage.

Terminal 12 is normally "HIGH," and is also TTL compatible. Although somewhat slower, it offers a 100 mA output capability in either "HIGH" or "LOW" states.

With the component values shown in **Fig. 1**, the circuit has a delay period of 30 sec. The period can be extended to several hundred seconds by increasing the value of R. The practical limits of R range from 3 kΩ to 1 MΩ for a 5V power supply. The circuit is non-retriggerable. The timing cycle is initiated by grounding terminal 5 or applying a positive pulse to terminal 6.

The amount of the current delivered is dependent on the value of R. This yields a timing period of precisely 2 RC. The factor of 2 is due to the internal arrangement of the reference voltage. When the voltage across capacitor C reaches the threshold voltage (approximately 2.4V) the voltage comparator triggers the flip-flop which returns to its previous state. Thus, terminal 10 returns to the "LOW" state and terminal 12 to the "HIGH" state. To reset the timing cycle, terminal 7 is grounded.

The maximum length of the timing period is limited by three main factors: The dc leakage current of the capacitor, the input bias current of the internal comparator, and the leakage current of the analog switch. Of the three, capacitor leakage is the dominant factor.

The 100 μF aluminum tubular capacitor used in this circuit has a maximum leakage of 1 μA. This means that the value of R can be no greater than 1 MΩ. If a higher power supply voltage is available, one can increase both the timing capacitor and resistor values and generate delays that exceed one hour.

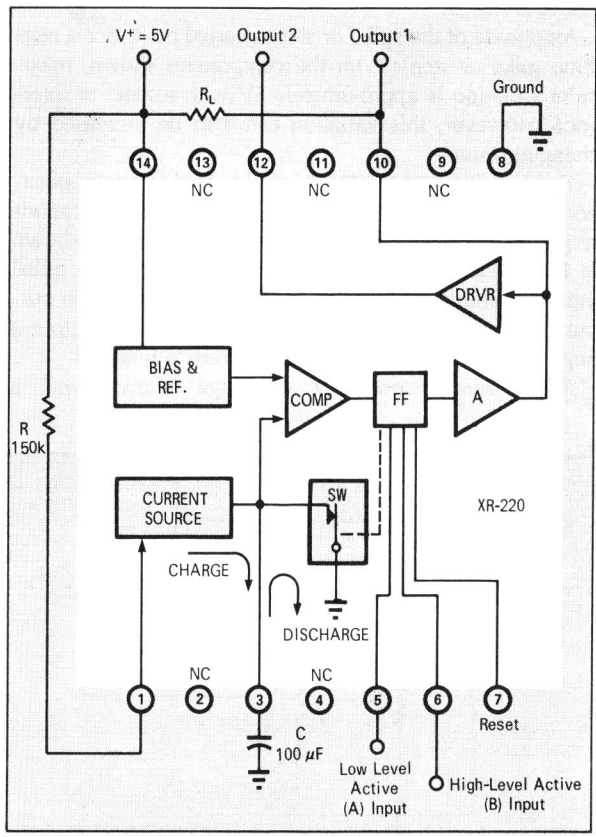

Fig. 1—Monolithic multifunction IC, the XR-220, requires only three external components to generate timing pulses. In this configuration the internal constant current generator is programmed, by R, and charges C linearly until the voltage across C reaches the threshold of the comparator, ending the cycle.

Digital and Pulse Circuits 55

2 TTL packages convert BCD up-counter for down counting

Jack Sellers
Mostek Corp., Carrollton, Tex.

This circuit illustrates a unique, inexpensive approach to providing a choice of up or down count from a normal up counter. This avoids the requirement of the more expensive up/down counters where a simple down count is desired. The trend to MOS multi-counter circuits makes this circuit a practical means to obtain low-package-count and still provides both up- and down-count capability from the standard up-counter.

The circuit does not provide full up/down counter capability; that is, one cannot count up to a certain count and reverse direction to count back down. However, many applications for up/down counters are such that they are always operated in an up-count or always in a down-count mode. Several applications, including digital voltmeters, require the up/down mode to change on a cyclic basis with this change occurring following a reset condition. It is for these applications that this circuit is intended.

Two packages are all that's required to achieve this conversion of a up-counter (e.g. Mostek's MK 5002 P/MK 5007 P) to a down-counter from the reset condition.

The circuit shown accepts true BCD in and provides the following output dependent on the state of the U/D control:

U/D = Logic 0 (LOW) allows the data at the input to appear in the same true logic state at the output also.

U/D = Logic 1 (HIGH) adds the complement of the number 9 (0110) to the BCD information at the input. Adding 0110 transposes a 9 and 0, 8 and 1, 7 and 2, 6 and 3, 5 and 4.

True BCD is provided at the output in both cases. During 9's complement addition, the following is true: No carry is generated in Bit 1; a carry is generated when Bit 2 is logic 1 and the carry is added to Bit 4; a carry from Bit 4 is generated anytime Bit 2 carries or Bit 4 is a logic 1 with this carry added to Bit 8.

A mode change at any time other than reset would not necessarily provide a useable count output.

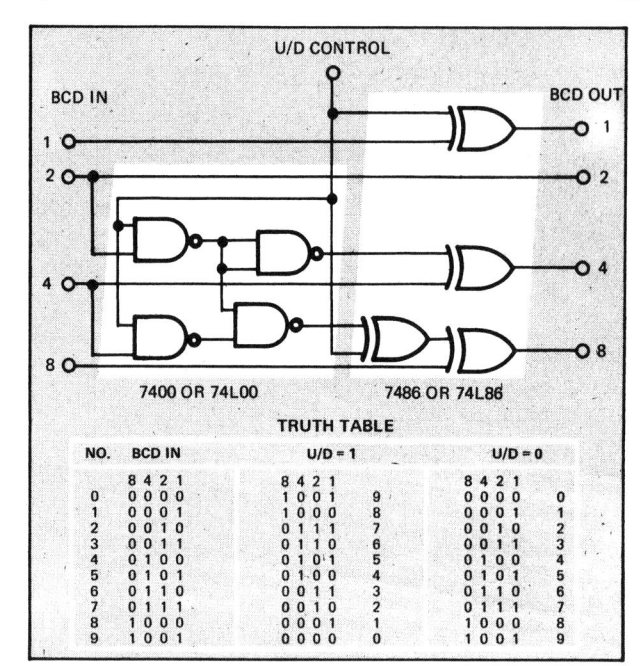

Standard or low-power TTL can be used to implement this 9's compliment BCD generator. The low power versions will be more desireable if you are driving an MOS BCD up-counter. As seen in the truth table, only one control bit is required to set the conversion circuit for up or down counting.

Simple FET Timer

By James Geekie
Siliconix
Sunnyvale, Calif.

THIS SOLID-STATE TIMER uses a FET constant-current source to eliminate time-period errors due to unregulated power supplies and line transients.

In the "reset" switch position, the timing capacitor charges to the pre-set positive voltage on the divider network. The timing cycle begins when the switch is moved to the "time" position. The positive capacitor plate then is grounded and a negative voltage equal to the original positive capacitor charge is transferred to the base of bipolar transistor Q_2. The output voltage v_o rises to $+12$ V.

The FET, connected as a self-biased constant-current supply, now begins to linearly remove this negative charge in a time $t = CV/I_D$, where C = capacitance, V = reset voltage + $V_{BE(SAT)}$, and I_D = FET constant drain current.

The timing cycle ends when Q_2 turns on, and drops the output voltage v_o to $V_{CE(SAT)}$. Sharp turn-on action is provided by a regenerative action increasing available FET constant current as Q_2 turns on.

To achieve overall zero or near zero temperature coefficient for the timer, the FET constant drain current is set by R_1 to the FET nominal zero T.C. drain current, about -10 μA. Operated thus, the timer shown has a time period approximately 0.1—50 sec controlled by R_2.

Total component cost for the circuit using a U-110 FET is estimated at $12.76 in small quantities and $9.97 in 100 and up quantities. With a lower cost, lower g_m U-146 FET, component costs are $10.76 in small quantity and $7.42 for 100 and up.

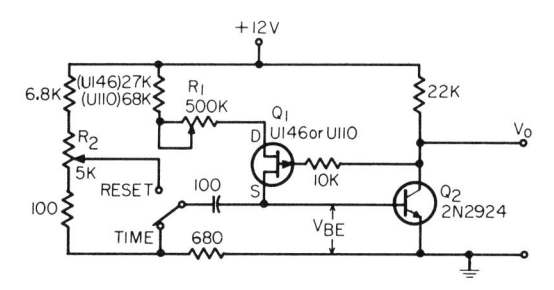

Low-cost, wide-range FET timer circuit.

Single-digit BCD adder uses 3 ICs

Dennis W. Wood
The Boeing Co. Seattle, Wash.

A single-digit BCD adder can be implemented with only three dual in-line packaged ICs. If more digits are required, basic single-digit circuits can be cascaded, with the *carry out* from one stage connected to the *carry in* of the next stage.

As can be seen in the schematic, a single-digit BCD adder uses two 4-bit binary full adders and a multiple-input AND-OR gate.

The two BCD digits at the input are first added together in IC_1. If the sum of the two digits and the *carry in* is ≤9, there is no need for a carry to the next stage, and the binary output of IC_1 is simply the BCD sum.

The B inputs to IC_2 are wired such that IC_2 adds a count of zero to IC_1's output if there is no carry, and adds a count of six if there is a carry.

If the sum of the two inputs and the *carry in* is ≥10, the circuit must generate a *carry out* to the next BCD-digit stage. The multiple AND-OR gate, IC_3, detects the specified input conditions and generates the required *carry out* signal.

The necessary input conditions for producing a *carry out* are satisfied whenever one of the following states occurs for IC_1: S_8 and S_2 are true, or S_8 and S_4 are true, or the carry C_4 is true.

Because the output of IC_1 is a binary sum instead of a BCD sum, a further correction is needed. This involves adding an extra count of six to the output of IC_1, thus forcing the binary counts of 10 through 19 to a count of 0 through 9. The procedure is summarized in the table.

Using similar techniques, it is possible to build a BCD-digit subtractor instead of an adder. Also, the radix of the adder can be changed to allow the addition of BCD minutes and seconds instead of straight digits. In this case, the circuit should force a carry at a sum of six or greater and the output of IC_1 should be corrected by adding a count of ten to it.

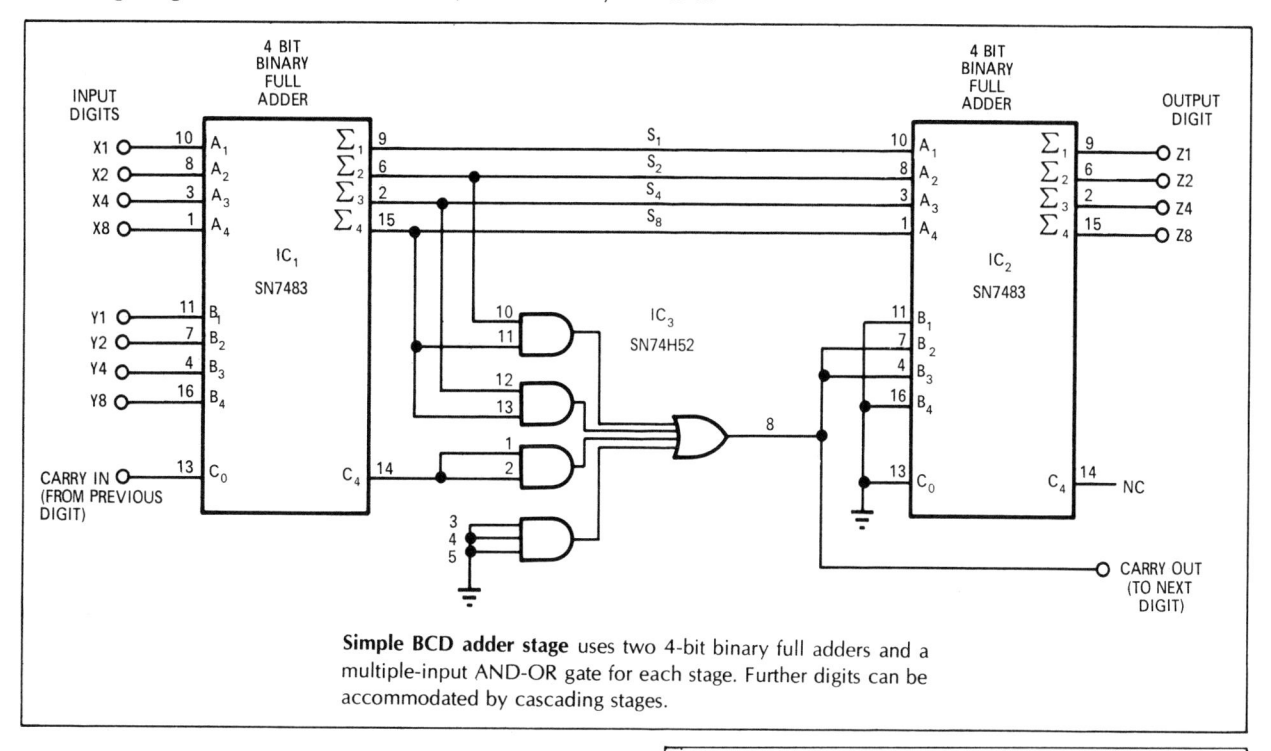

Simple BCD adder stage uses two 4-bit binary full adders and a multiple-input AND-OR gate for each stage. Further digits can be accommodated by cascading stages.

Generating a carry

SUM OF INPUTS AND "CARRY IN"	ADDER 1 OUTPUTS $C_4 S_8 S_4 S_2 S_1$	ADDER 2 OUTPUTS $Z_8 Z_4 Z_2 Z_1$	"CARRY OUT" TO NEXT DIGIT
0	0 0 0 0 0	0 0 0 0	0
1	0 0 0 0 1	0 0 0 1	0
2	0 0 0 1 0	0 0 1 0	0
3	0 0 0 1 1	0 0 1 1	0
4	0 0 1 0 0	0 1 0 0	0
5	0 0 1 0 1	0 1 0 1	0
6	0 0 1 1 0	0 1 1 0	0
7	0 0 1 1 1	0 1 1 1	0
8	0 1 0 0 0	1 0 0 0	0
9	0 1 0 0 1	1 0 0 1	0
10	0 1 0 1 0	0 0 0 0	1
11	0 1 0 1 1	0 0 0 1	1
12	0 1 1 0 0	0 0 1 0	1
13	0 1 1 0 1	0 0 1 1	1
14	0 1 1 1 0	0 1 0 0	1
15	0 1 1 1 1	0 1 0 1	1
16	1 0 0 0 0	0 1 1 0	1
17	1 0 0 0 1	0 1 1 1	1
18	1 0 0 1 0	1 0 0 0	1
19	1 0 0 1 1	1 0 0 1	1

Digital phase-locked loop with loss-of-lock monitor

by Charles A. Herbst
Comfax Communications
Garden City, N.Y.

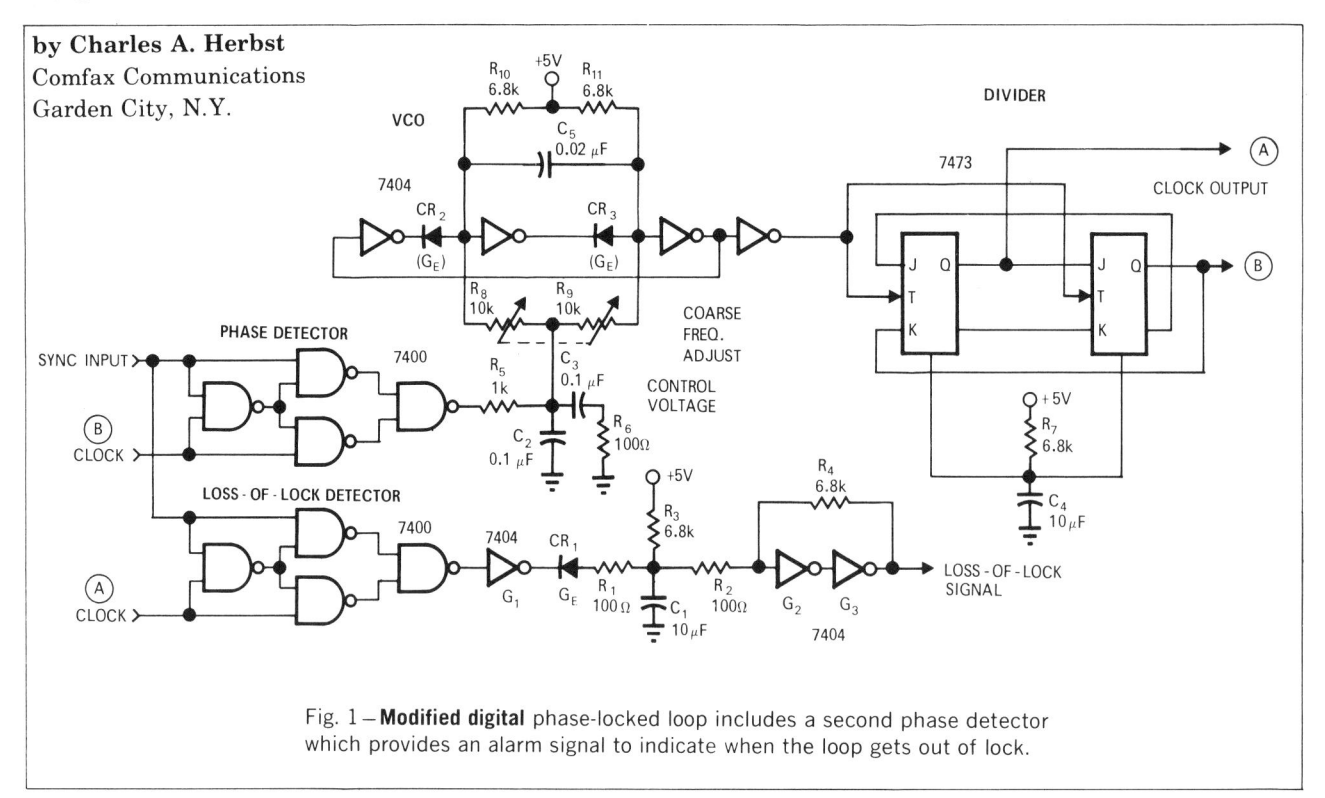

Fig. 1—**Modified digital** phase-locked loop includes a second phase detector which provides an alarm signal to indicate when the loop gets out of lock.

A phase-locked loop (PLL) can provide a convenient method of synchronizing the clock frequencies of two pieces of equipment. But this type of circuit has the disadvantage that it becomes a free-running oscillator if it goes out of lock.

The digital PLL, shown in **Fig. 1**, includes a monitor circuit. This provides an alarm, or disables other circuitry, when phase lock is lost or when the phase error between the internal clock and the external sync signal exceeds a predetermined limit.

The voltage-controlled multivibrator (VCO) in the circuit operates at four times the desired clock frequency. This VCO drives a two-stage, switch-tailed, ring counter that provides a two-phase internal clock signal.

The "B" phase of the clock is compared with the incoming sync signal in an exclusive-OR phase detector. When the two signals are locked 90° out of phase with each other, the phase detector produces a "zero" error voltage of about 2V (see **Fig. 2**) which forms the control voltage for the VCO. As the phase difference be-

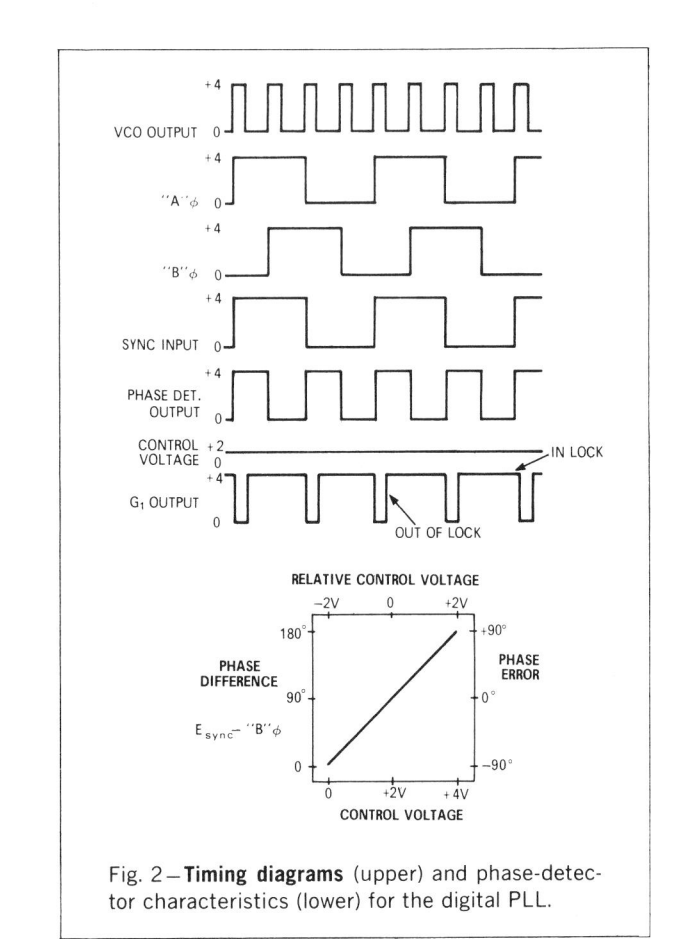

Fig. 2—**Timing diagrams** (upper) and phase-detector characteristics (lower) for the digital PLL.

tween the two signals varies ±90° from the normal lock condition, the control voltage swings approximately ±2V from the "zero" value and in a direction which tends to reduce the phase error and bring the loop into lock.

A second exclusive-OR gate compares the "A" phase of the clock with the incoming signal. This "loss-of-lock" phase detector determines when the two signals are in step with each other. When they are in phase, the output of G_1 stays at a high logic level, indicating an "in-lock" condition. As the two signals begin to get out of phase with each other, small negative-going pulses (whose widths indicate the magnitude of the phase error) begin to appear at the output of G_1, causing C_1 to commence discharging through R_1 and CR_1. If the pulses get wide enough to discharge C_1 below the trigger point of the Schmitt trigger (G_2 and G_3), then an "out-of-lock" condition will be indicated. Failure of either the VCO or sync signal will also give an "out-of-lock" indication.

The network R_7 and C_4 provides a power-on clear circuit which ensures that the ring counter starts in the correct state. The network R_5, R_6, C_2 and C_3 is a low-pass filter which determines the dynamic response of the loop and removes carrier ripple from the control voltage. Capacitor C_5 sets the VCO center frequency, while C_1 sets the trigger point for the phase-error alarm.

Two TTL gates drive very long coax lines

Robert W. Stewart
Control Data Corp., Santa Ana, Calif.

Need a great TTL coax driver and receiver? This circuit has all the good features you want and haven't been able to find. It will transmit information via coax or twisted pair over long lines at bit rates exceeding 10 Mb per sec. And the best part is its cost, that of standard logic gates, plus a few resistors.

The distance over which this driver/receiver will work is a function of the channel used (coax or twisted pair) and the data bit rate, due to both rolloff and phase distortion in the channel.

Safe operating maximums for two types of coax can be derived from **Fig. 1**. Twisted pair typically has a characteristic impedance in the range of 50 to 120Ω and has approximately the same loss characteristics as RG 174 below 10 MHz. For a convenient rule of thumb, the channel attenuation should not exceed 10 db for the data clock frequency. That is, if the clock frequency is 5 MHz, the overall channel attenuation should not be greater than 10 db at 5 MHz to support a data rate of 5 Mb per sec.

In **Fig. 2**, the device types have been chosen to provide two drivers or two receivers per IC package. The unused input in the receiver could also be connected as a strobe.

Success of the circuits depends upon operating about the threshold point of the receiving gate. This threshold is very accurately determined by connecting output to input of an adjoining gate on the same IC chip. The diffusion process by which the ICs are fabricated insures that the bias gate and the receiving gate have identically the same threshold and will track each other with temperature and voltage. Being a linear feedback amplifier, the bias gate, G_3, exhibits a low output impedance and can easily terminate the channel load resistor. Operation of the receiver is enhanced by forcing the driver to operate symmetrically about the receive threshold voltage. When the input to the driver is high, the driver output becomes saturated at near ground potential and when the input is low, feedback around the driver limits its output to twice the driver threshold voltage. Although different from the receiver threshold, close matching is not necessary and only becomes important for channel losses greater than 20 db. The logic input to the driver circuit should be a standard TTL gate with a positive swing of 3.0V or better and a saturated zero level of about 0.25V.

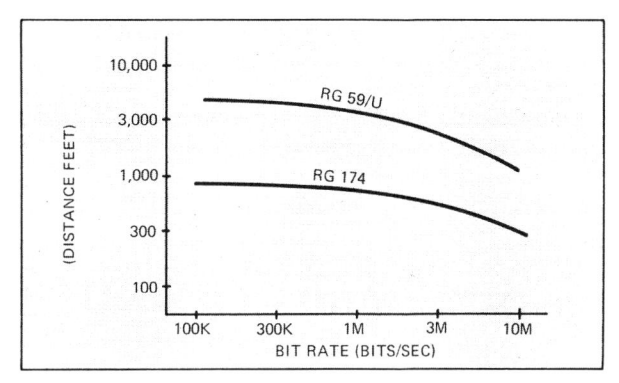

Fig. 1–Performance curves for driver circuit when used with RG59 or RG174 coax. Twisted pair lines approximate the performance of RG174 at data rates below 10 MHz.

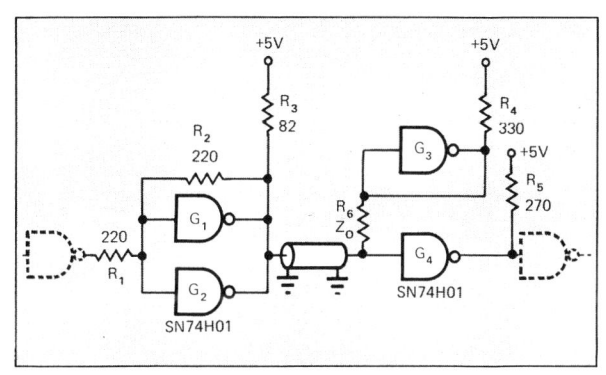

Fig. 2–Driver/receiver pairs require only 4 TTL gates, and a few resistors. R_6 is the coax termination resistor.

If you're at all concerned about operating these gates in their linear mode, a quick check of the open collector gate circuits will show that power dissipation is being kept within reasonable bounds and that nothing catastrophic can happen.

Double-edge pulser uses few parts

David Giboney,
Hendrix Electronics, Londonderry, NH

Occasionally the need arises to generate a pulse from each edge (every 180°) of a waveform. This can be done with the circuit of **Fig. 1** which will react to each transition of an input clock regardless of polarity. An exclusive-OR gate is used as a "programmable inverter" to return point C to a quiescent, LOW state following each transfer of data through the 5474.

One particular application of this circuit is for frequency doubling. It is desirable in these applications for the input waveform to be symmetrical, since the output of the circuit is equal to the propagation delay of the flip-flop plus the delay of the exclusive-OR gate. **Fig. 2** shows a method of frequency multiplication times 4 which is not sensitive to input symmetry; however, the input frequency must be known and unchanging since one-shot multivibrators are used to install symmetry in the waveform.

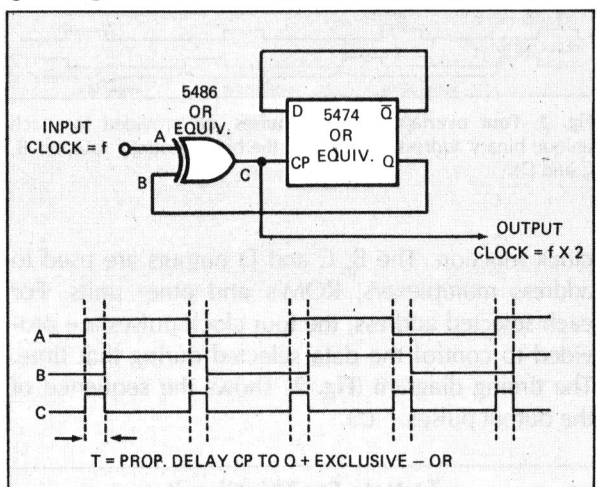

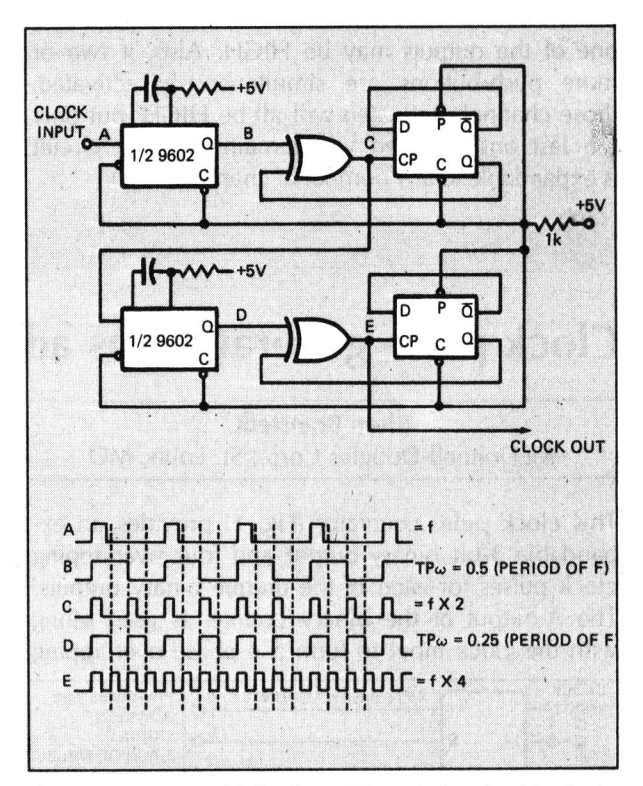

Fig. 2—**Frequency multiplication x 4** is carried out by this circuit, which is sensitive to symmetry of the input waveform.

Fig. 1—**Output pulse is delivered at C** for each negative or positive transition of the input clock pulse.

Latching circuit provides noise immunity

by **Steven R. Martin**
Chicago, Ill.

This circuit is especially useful for noise rejection, since any noise spike or other undesired pulse whose duration is less than a predetermined minimum length will not erroneously set the latch. The minimum length of a pulse that will activate the circuit is determined by the RC combination. If R is selected at 10 kΩ, then C is determined by C = T/3.424 (for C > 10^3 pF and nonpolarized (where C is in pF, and T (in nsec) is the desired minimum length of a pulse that will activate the circuit. (For C < 10^3 pF or C being polarized, you should check the specification sheet on the Fairchild 9601 monostable multivibrator.)

A high to low transition at pin 1 of the 9601 causes the output to produce a high to low pulse of length T. If the input pulse is longer than T, two highs will be seen at pins 4 and 5 of the 846 causing a low at its output, thus setting the latch. If the input pulse is shorter than T, pin 5 of the 846 will go low before pin 4 goes high and the circuit will not activate.

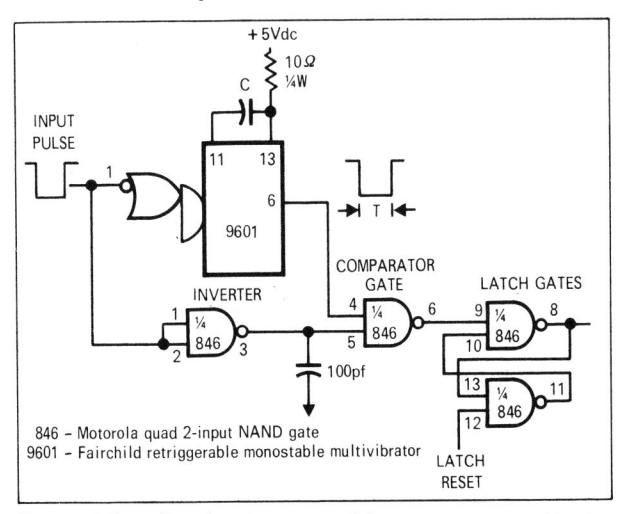

Fig. 1—**False triggering** is prevented by generating a pulse, T, equal to the minimum input pulse. Pins 4 and 5 of the 846 must be high simultaneously to set the latch, but this cannot occur unless the input pulse equals or exceeds time "T".

High speed circuit converts binary to BCD

John T. Hannon
Computer Sciences Corp., Huntsville, Alabama.

The complex logic used in today's digital systems presents a great demand for high speed binary-to-BCD converters. There are several methods presently in use but most of them are not well suited to the high speed requirements of present logic.

This circuit combines high speed, low cost and readily-available components to implement the add-three-and-shift method of binary-to-BCD conversion. **Fig. 1** presents the basic four-bit binary-to-BCD converter. The three NOR gates determine whether the number is five or greater. If it is, the gate output goes high and causes a ONE to be applied to the two least significant bits of the binary adder. This adds a binary three to the number in the adder. The outputs of the adder are the BCD equivalent of the binary number.

This method can be easily expanded to as many bits as

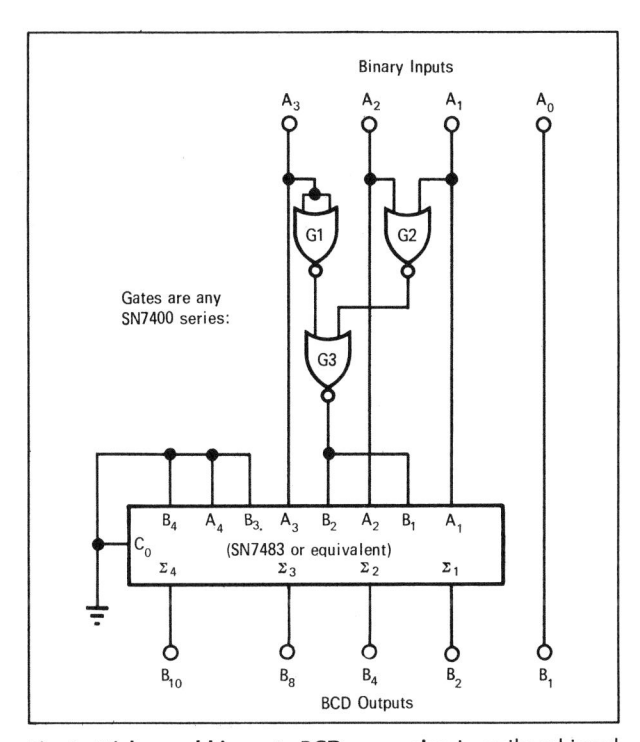

Fig. 1—High speed binary to BCD conversion is easily achieved in this add-three-and-shift method. The basic 4-bit converter uses only 3 NOR gates and a 4-bit binary adder.

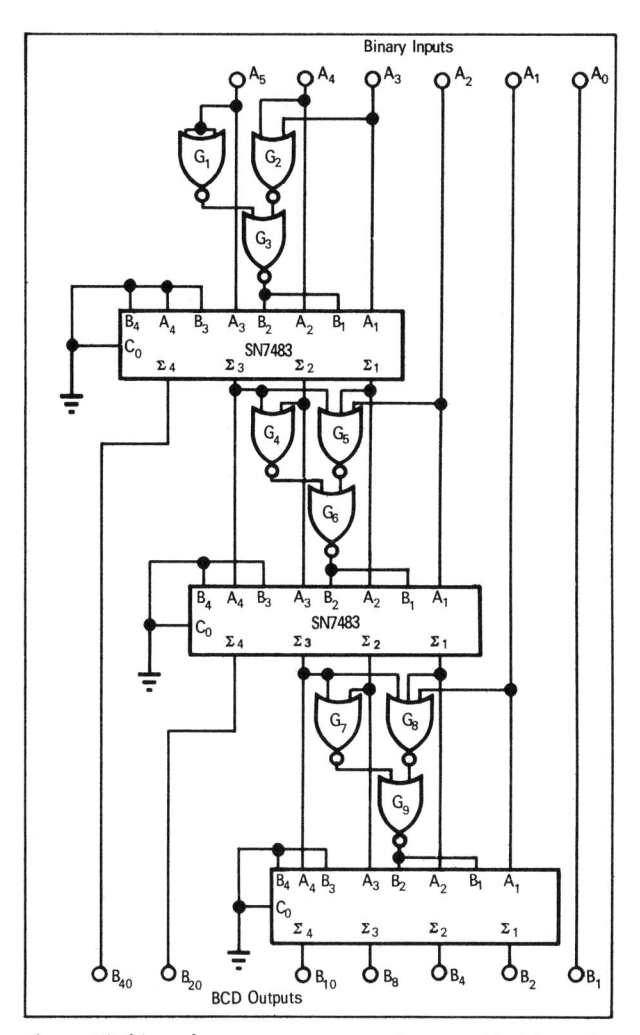

Fig. 2—Six-bit or larger converters can be assembled from the basic 4-bit circuit by hard-wire shifting the outputs from each detector/adder stage to the inputs of the next stage.

necessary. The outputs from each detector-adder stage are hard-wire shifted to the next stage to implement the next step in the detect, add and shift cycle. **Fig. 2** illustrates the connections for a six-bit converter.

Note that NAND Gates could be used for the five-or-greater detector, but by using NOR gates, one less gate is required for each detector.

The advantages of this type of circuit are low cost and easy availability of the circuits. Since there is no clocking or shifting, as in usual add-and-shift method, there are no timing problems. The speed of the converter is limited only by the propogation delay of the circuits, consequently speeds in the megahertz range can be easily obtained.

Flip-Flop Has Improved Rise Time And Stability

By Douglas Brooks
Hughes Aircraft Co.
Culver City, Calif.

THE BASIC FLIP-FLOP of Fig. 1, has an inherently slow rise time due to the charging of C_2 through R_2. This is particularly true if the flip-flop is to have a relatively long period and yet draw little power. This requires large values for C_1 and C_2. A second drawback is the dependence of the half-period on the value of $E_{o\ max}$, which in turn depends on the load. If the load is variable or unknown, stability is lost.

By the addition of a few, inexpensive components, (Fig. 2), the performance can be greatly improved. Output V_2 has a very fast rise time, since it is isolated from the RC time constant by D_1. The fall time of V_2 is slow until D_1 becomes forward biased, at which time the regenerative switching action takes place. (The larger the value of R_7, the faster will be the initial rate of fall; otherwise the value of R_7 is not critical.) The zener diode is chosen such that $E_{o\ max}$ is less than $V_{1\ max}$ so that the switching of E_o occurs only along the fast slopes of V_2. Using a C of 22 μf and a frequency of less than 15 cps, rise and fall times on the order of tenths of microseconds are easily obtained. Since the timing components of the flip-flop are well isolated from the output, the frequency will remain constant for any load.

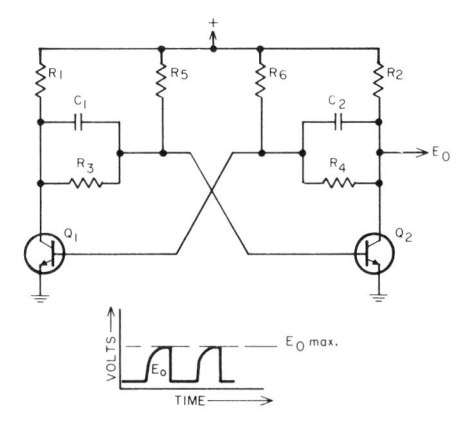

Fig. 1. Basic flip flop rise time is limited by time constants of $R_2 C_2$.

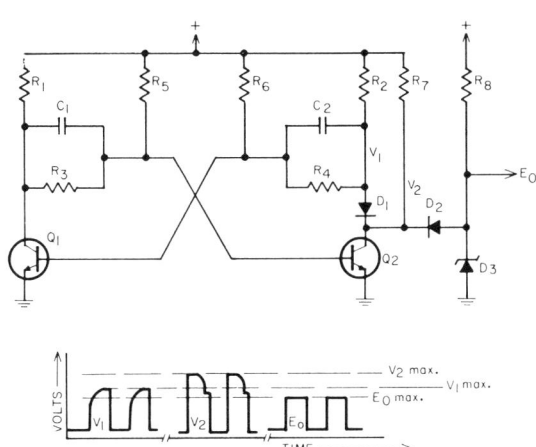

Fig. 2. Modified output circuit isolates output pulse from RC time constant.

DC to DC One-Shot Starting Circuit

By Marvin Shapiro
General Electric Co., RSD
King of Prussia, Pa.

IN THIS CIRCUIT, components R_1, C_1, D_1, R_2 and Q_1 function as a one-shot turn-on circuit connected to a typical Royer circuit. Turn-on is independent of the rate of application of the supply voltage, V_s.

At time t_o, D_1 is cut off, and Q_1 supplies turn on current to the base of Q_2. This current is limited by R_1. Note that excessive starting current will inhibit oscillations since it divides between the transistor base and the feedback winding, and as such, represents bucking amp-turns.

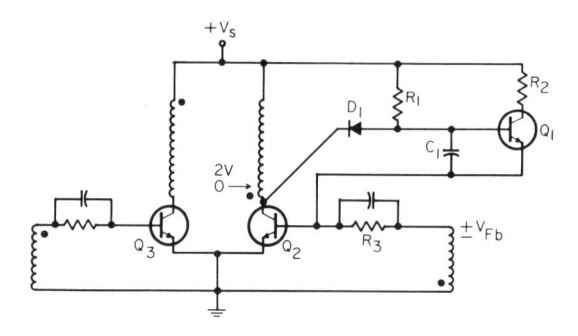

Fig. 1. DC to dc one-shot starting circuit.

TIME CONSTANT ≅ $R_1 C_1$
$R_1 \gg R_3$ $R_1 C_1 \gg t_1$
$V_q < 0.7\ V$

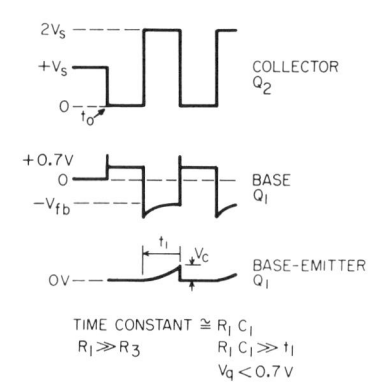

Fig. 2. Waveforms.

When Q_2 turns on, D_1 clamps the base of Q_1 to 0.8 v. while the base of Q_2 clamps the emitter of Q_1 to 0.7 v. Thus Q_1 is cut off. During the alternate half cycle, C_1 charges toward $+V_s$ via R_1. If the $C_1 R_1$ time constant is long compared with a half-cycle, C_1 cannot charge to $+0.7$ v and Q_1 does not conduct. The function of C_1 is to force the base of Q_1 to follow the emitter of Q_1, which goes to $-V_{fb}$ during the alternate half cycle. (See Fig. 2.) This same scheme is applicable to initializing digital circuits where the frequency of operation is high compared with $R_1 C_1$.

J-FET generates clear or preset commands for low-power TTL

J. E. Buchanan, Westinghouse Electric
Corp. Baltimore, Maryland

A set or reset signal is often required to insure that a counter assumes a required state following power turn-on.

For the more common types of low-power TTL counters, a P-channel J-FET, two resistors and a capacitor as shown in **Fig. 1**, provide a set or reset circuit that will insure that the counter starts up in the required state.

When the logic supply (+5 volts) is first applied, the J-FET (Q_1) gate-source and gate-drain voltages are near zero, allowing Q_1 to turn fully ON to provide a sink path to ground for the clear or preset line. (The low-power 54L TTL counter requires a low-logic-level signal that is capable of sinking up to 0.18 mA of current per set or reset signal).

After power is applied, Q_1 remains ON for a given time, depending on the RC time constant of the gate RC network. As the gate voltage approaches the pinch-off level, the circuit begins functioning as a voltage follower; and as the gate voltage further increases so does the output, until a final value is reached that is near a pinch-off level and

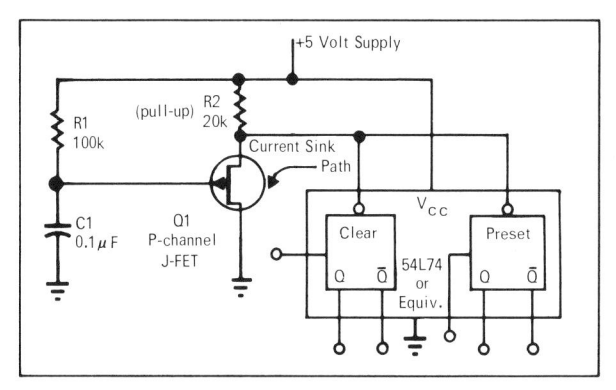

Fig. 1 — A clear or preset circuit for low power TTL counters requires only one active component and assures that all counters will be properly set upon application of power to the circuit.

below the logic supply level. A low pinch-off, low on-resistance P-channel J-FET, such as the 2N2844 (V_p max. = 1.7 volts, R_{DS} typical = 560 ohms), is required. Individual selection of the FET may be required to insure both the low ON-resistance and low pinch-off level.

Improved One-Shot Output Circuit

By Thomas G. Ellestad
General Electric Co.
Sunnyvale, Calif.

IT IS SOMETIMES NECESSARY to take the output of a one-shot from the timing side to obtain the correct polarity output pulse. This is normally accomplished with a resistive divider (R_1, R_2) and capacitive speed-up as shown in Fig. 1. A serious limitation of this circuit is the large variation in output pulse width resulting from the use of Q_1's collector voltage to drive the output amplifier. Because of the limited current available to recharge C_1, the trailing edge of Q_1's collector voltage pulse is a ramp. The point on the ramp where Q_3 turns on, and hence the output pulse width, is a function of the divider accuracy, the supply voltages, the beta of Q_3, and the load.

In the improved circuit, Fig. 2, the timing current serves to discharge C_1 during the active timing period and supplies base drive to the output amplifier during the recovery period. This is accomplished by replacing the clamp diode on the right side (CR_2 in Fig. 1) with the base-emitter junction of the output amplifier.

Transistors Q_1 and Q_2 form an astable multivibrator with Q_2 normally biased on. Transistor Q_3 is also normally turned on by the current through R_4. The positive-going leading edge of the input pulse, differentiated by C_1, turns on Q_1. The resulting negative-going pulse at the collector of Q_1, coupled through the timing capacitor, C_2, turns off Q_3 and holds off Q_2. Resistor R_4 discharges C_2, turning off Q_2 at the end of the active timing period. As Q_2 turns on, Q_1 turns off. The positive-going pulse at Q_1's collector, coupled by C_1, turns Q_2 off harder and supplies speed-up drive to turn on Q_3. Transistor Q_3 is now held on by the current through R_4, and C_2 is recharged through the base emitter junction of Q_3 by R_2.

For the power supply voltages and component values shown, the active timing period is:

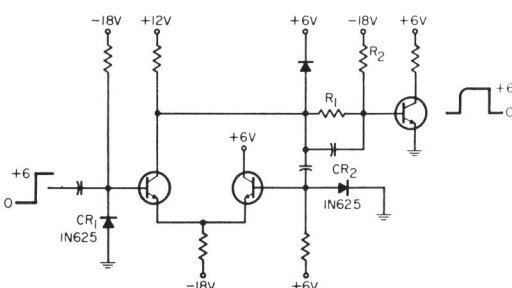

Fig. 1. Basic one-shot output circuit.

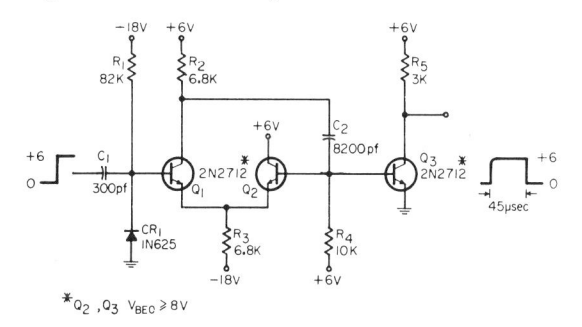

Fig. 2. Improved one-shot output circuit.

$$T_a = R_4 C_2 \ln \frac{6.6}{12.6} = 0.646 R_4 C_2$$

The recovery period is:

$$T_r = 5 R_2 C_2$$

If Q_3 is a low-leakage silicon transistor, the timing performance will not be affected.

The improved circuit, besides using less power, gives improved timing accuracy with the use of fewer components.

A synchronized phase locked loop

Robert Bohlken
Honeywell Inc., Hopkins, MN

A problem that exists in measuring the frequency of short-signal bursts is the long acquisition time of phase locked loops.

A short time from signal acquisition to phase lock was obtained by synchronization of the VCO to the input phase. This allows correction pulses to be developed in the correct polarity only. Lockup times of less than 10 cycles of input are achieved with a 1 to 10 kHz using an idling frequency of 12 kHz for the VCO (PRESET). A high idle frequency is used because it is easier to slow the VCO from a high to low frequency.

Input signals are compared to the synchronized VCO at exclusive-OR A, gating of the error pulses by gate F and flip-flop G-H allows I or J to drive current pulses of the correct polarity into storage capacitor C_1. The voltage correction on C_1 is proportional to the width of the error pulses and can be controlled by the value of R_2 and R_3. The voltage on C_1 is buffered with a FET input op amp, AD501, which drives the VCO with a constant current inversely proportional to C_1 voltage. When the control gate is turned OFF, Q_3 and Q_4 are turned OFF allowing the VCO to run at lockup frequency which is equal to input frequency.

The VCO in provided with a fast discharge time. When voltage first appears across R_4, one-shot M turns on Q_1 which dumps C_2 much faster than Q_5 could. The sawtooth of the VCO is synchronized by use of pulses derived from the positive edge of the input, into the OR gate of one-shot M. The output of M is also fed into J-K flip-flop P which is synchronized into the proper state. The output of P is at VCO frequency which can be fed into a counter, or the analog voltage on C_1 can be ultilized. Range of frequency can be changed by varying R_1. D_1, D_2 and D_3 provide limiting VCO voltage.

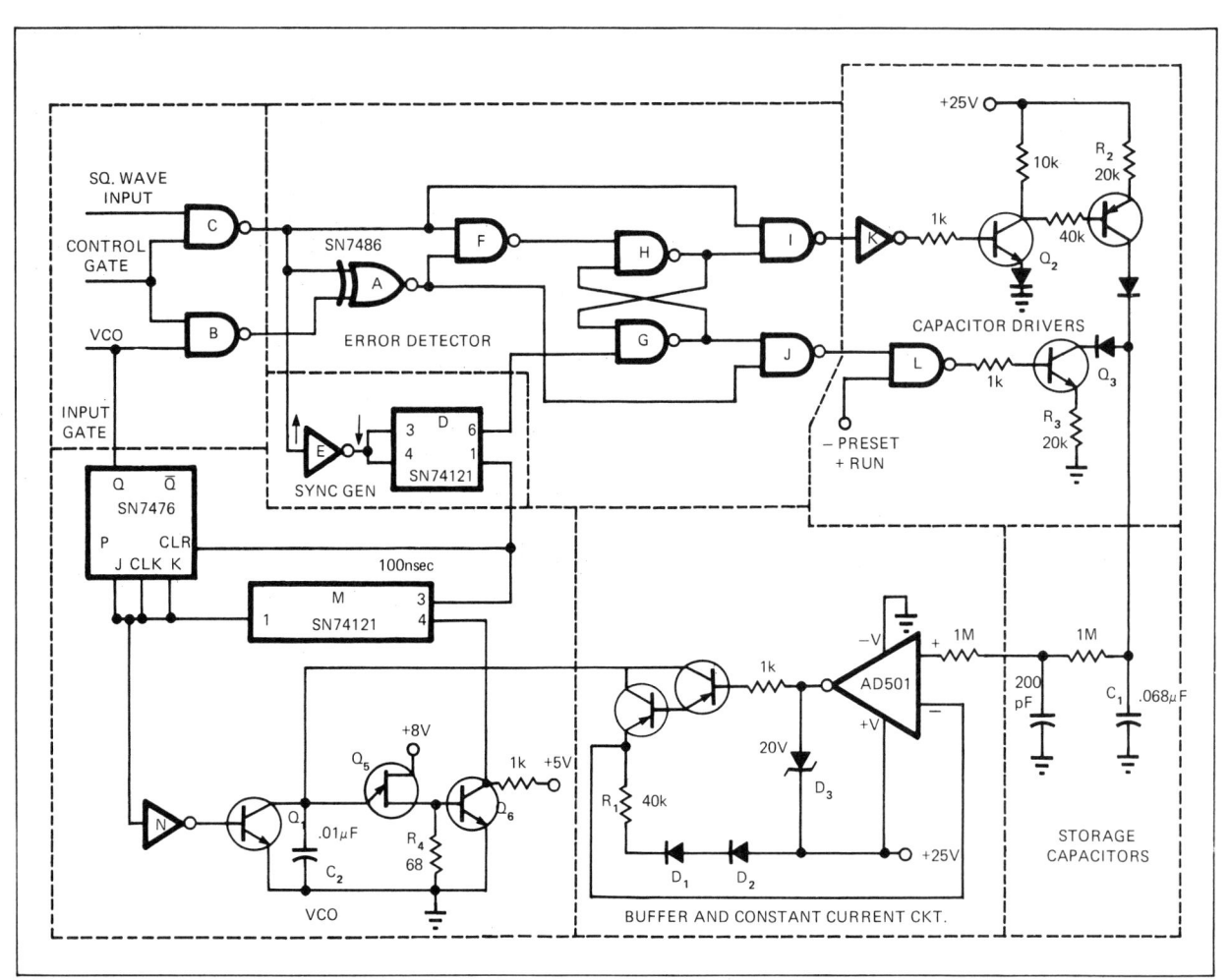

Phase locking in less than 10 cycles of the input is achieved by idling a high frequency and "pulling down" the VCO, rather than the more common approach of idling at mid range.

Fast-Recovery One-Shot Multi Gives 10:1 Width Control

By Jack Rogers
Eldorado Electronics
Concord, Calif.

MOST ONE-SHOT multivibrator circuits that have short recovery time compared with the output pulse width use additional transistors and diodes to provide a discharge current that is much greater than the charging (timing) current. This disadvantage of extra components is not present in the circuit shown here, which is currently in use in a commercial radar range unit and a pulse analyzer.

The pulse width can be readily varied from 0.1 μsec to 10 msec in decade ranges by using appropriate timing capacitors. Risetime of the output pulse is essentially constant at about 10 nsec over the entire width control range. Further, very high duty factors (90 percent for shorter widths and 97 percent for maximum widths) can be realized because the recovery time is nearly a constant fraction of the pulse width.

With an npn and a pnp transistor, both transistors in the multivibrator are simultaneously on or off, as contrasted with the usual case in which one transistor is on while the other is off. In the circuit shown, Q_1 and Q_2 are off during the pulse-width period, the longest time interval in the circuit. In addition to achieving high duty factors, use of the npn and pnp configuration provides a large discharge current path for the timing capacitor, C, without the requirement for additional components.

A negative-going trigger is applied to the base of Q_2 via CR_1 and CR_2 to turn Q_2 off. The resulting positive change at the collector of Q_2 appears at the base of Q_1 to quickly cut off Q_1, and the emitter of Q_1 is restrained from following the base of Q_1 by timing capacitor C. Thus, Q_1 remains in the cutoff state and regeneratively completes turning off the base of Q_2. Both Q_1 and Q_2 remain off until C charges to the voltage level present at the base of Q_1. This level is set by R_5, the width control, to determine the time required for the emitter and base voltages of Q_1 to equalize.

When the emitter and base voltages of Q_1 are equal, Q_1 and Q_2 conduct as heavily as they can (limited primarily by their internal resistances) to discharge C to its quiescent voltage level. Timing resistor R_2 functions to maintain sufficient current in the emitter of Q_1 so that the circuit remains on over the entire range of R_5 after the heavy discharge current dies out. CR_3, a disconnect diode, permits the output at the collector of Q_2 to swing a full 12 V regardless of the setting of R_5.

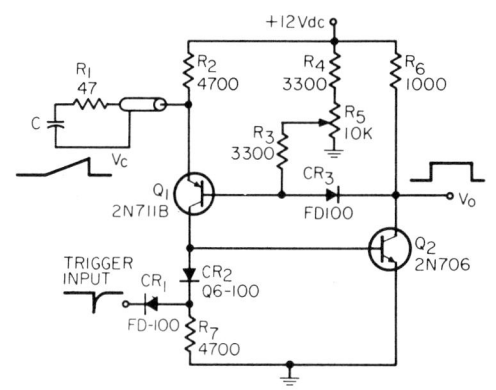

Opposite polarity transistors allow fast recovery with broad width control.

With the exception of timing capacitor C, component values are as indicated on the schematic. The value of C is determined from the following relationship: C (pf) = Maximum Pulse Width (μsec) × 10^2. The value of C for a decade range of 1 to 10 μsec, for example, is 1000 pf.

In addition to having a fast recovery time, this circuit exhibits several other features: essentially constant trigger sensitivity over the complete width range (a pulse width of many milliseconds can be generated with a trigger width of 50 nsec); only one power supply is required; no large recharge current flows through the power supply to cause transients; a width control range of 10:1 is readily obtained with a remote (up to 10 ft) control; and timing capacitors that determine the decades over which the pulse width is variable can be remotely (up to 10 ft) located on a range switch by using coaxial cable with the shield grounded at the circuit but not at the switch.

Optical tape-marker detector

by Charles A. Herbst
Dumont, N.J.

DETECTING beginning and end markers on digital magnetic recording tape is often difficult because of varying ambient light conditions and blank tape reflections. Manually adjusting the sensitivity of the light detector is not a practical solution. This self-compensating sensor does the job automatically by comparing short-term light variations (the signal pulse) against long-term variations (ambient light) in order to detect the presence of the desired signal.

In the diagram B_1' is a light source which illuminates the moving tape. Q_1 is a phototransistor which detects the light reflected from the tape marker. R_3C_1 is a low-pass filter having a time constant of about 5 times the expected incoming pulse width (10 ms). The filter stores the long-term light level without reacting to the short signal pulse. R_4C_2 is a low-pass filter with 1/20 the time constant of the incoming pulse width. This filter reduces spurious noise without deteriorating the incoming pulse. R_5 provides a slight positive bias to hold the output of A_1 in negative saturation. R_7 and D_1 provide a level conversion for DTL-gate G_1.

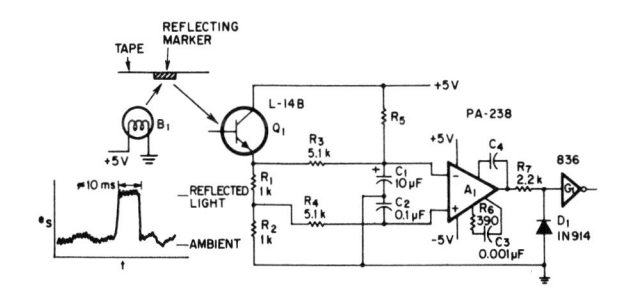

A phototransistor, op amp and DTL gate are used to construct an optical end-of-tape sensor.

Digital and Pulse Circuits 65

Low-cost manual pulser

By Thomas Carmody
AIL, Div. of Cutler Hammer
Deer Park, New York

THIS SIMPLE low-cost circuit provides manually-initiated set/reset voltages and clock pulses. It is useful for testing many types of digital circuits such as flip-flops, counters and adders. The circuit eliminates the effects of switch bounce that could cause false triggering of circuits under test. Total cost (excluding the optional monostable) can be less than $3.00.

The basic circuit gives complementary set/reset voltages at the collectors of Q_1 and Q_2. If a monostable circuit is connected to one of the collectors as shown, the complete circuit also provides manual clock pulses. The monostable is triggered by changes of state of the collector voltage. In the circuit shown, the monostable gives a single positive-going pulse each time the Q_2 collector voltage changes from "low" to "high."

The circuit works as follows: With the switch in the position shown, Q_1 base is clamped to ground. So Q_1 is cut off. Base current from the V_{cc} line flows through R_3 to turn on Q_2. The collector of Q_2 then approaches ground potential thus latching off Q_1.

When the switch is moved to the other position, Q_2 turns off and Q_1 turns on. With 2N706 transistors, total latching time is less than 200 ns. This is much shorter than the duration of contact bounce for a typical toggle switch. Spurious input pulses, caused by contact bounce after initial switch closure, have no further effect on the circuit. This is because all input pulses are of the same polarity (positive-going for the circuit shown).

Another possible circuit modification is to use a momentary pushbutton switch, instead of the toggle switch shown. Then the circuit generates a short pulse each time the switch is pushed. Rise time of Q_1 determines rise time of the collector voltage and hence the delay of the output pulse from the monostable. If the delay is excessive, a faster transistor such as 2N709 can be substituted for the 2N706.

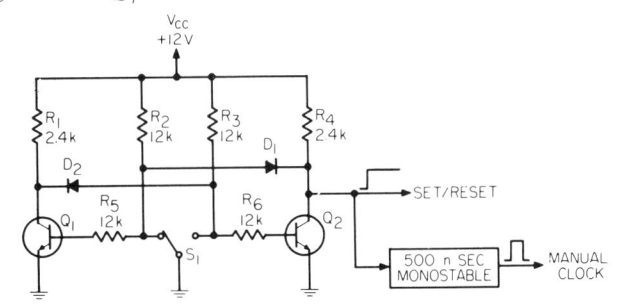

Manual no-bounce pulse generator for testing digital circuits.

Combination Schmitt Trigger-Monostable Multivibrator

By Gilbert Marosi
Friden, Inc.
San Leandro, Calif.

THREE TRANSISTORS in a complementary configuration combine the function of a Schmitt trigger and a monostable multivibrator.

The advantages of the circuit are two-fold: the triggering level is accurately controlled and the output pulse width is independent of input drop-out since the circuit is regenerative.

In the circuit, as soon as the input level crosses a threshold established by P_1, then Q_1 and Q_2 turn on. R_5 increases the regeneration of the circuit by providing additional turn-on current to Q_1. Q_3 is normally on and Q_2 normally off; therefore, C_2 is charged as shown in the figure. As soon as Q_2 turns on, the positive step through C_2 turns off Q_3. Positive feedback from Q_3 through R_7 back to the base of Q_2 insures that if the level at the input drops below the triggering level of Q_1, then Q_2 will still remain turned on. C_2 discharges through R_8 until Q_3 turns on, thus terminating the cycle.

The figure shows the Schmitt trigger to be made up of Q_1 and Q_2, and the monostable of Q_2 and Q_3.

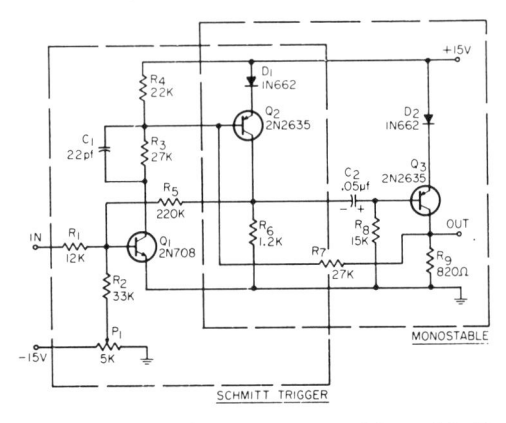

Schmitt trigger, (Q_1, Q_2) and monostable multi (Q_2, Q_3).

Wide Range Monostable Multivibrator

By Gilbert Marosi
Friden Co.
San Leandro, Calif.

The addition of one transistor to a linear one-shot increases the range through a ratio of 150 to 1. In the circuit shown, Q_6 is the extra transistor. Its associated circuitry is shown within the dashed lines.

Q_5 is normally on, Q_1 and Q_3 are off. Q_6 is forward-biased and therefore provides a steady base current to Q_5. A positive pulse coming in through C_5 cuts off Q_5 and turns on Q_1 and Q_3. This positive step is transmitted through C_1, which was charged to approximately -15 v and maintains Q_5 off during the one-shot timing cycle. C_1 charges through constant-current source Q_3 at a rate determined by the setting of P_1 and the voltage across it, approximately 5.7 v. When the voltage at the collector of Q_3 has reached -1 v, Q_5 starts turning on, Q_1 begins to turn off, Q_6 starts conducting, turning Q_5 on harder. This cycle of events is regenerative so that Q_5 rapidly saturates and cuts off Q_1 which in turn cuts off Q_3.

The circuit effectively separates the charging current through C_1, which varies the timing of the one shot, from the biasing current turning on Q_5, which should be constant. The current through Q_3 has to be only sufficient to set Q_5 into the active region and thus start regenerative action.

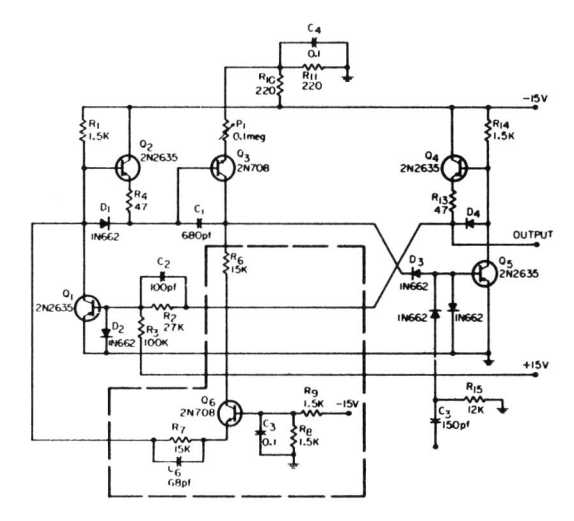

Added transistor Q_6 increases range of monostable.

Variable Schmitt, Amplitude Comparator

By Miles A. Smither and William E. Zrubek
NASA Manned Spacecraft Center
Houston, Tex.

The circuit shown can be used either as a Schmitt trigger with a variable trigger voltage (where V_B determines the trigger voltage), or as an amplitude comparator (where the amplitude of V_A is compared with the amplitude of V_B). These functions are obtained by using a minimum-hysteresis Schmitt trigger[1] and applying the voltage V_B to the base of Q_2 through R_B.

Features of the circuit are:

☐ A variable trigger voltage is readily obtained without changing the load on the input signal.

☐ A regenerative comparator is obtained having negligible hysteresis.

☐ The circuit is simple and inexpensive, and its operation is readily predictable.

☐ The circuit readily provides high speed operation.

Formulas for the resistor values can be found in the reference. The parameters used were:

$V_T = 7.0$ V, $V_c = 7.0$ V, $V_{s1} = 24$ V. $V_{s2} = 0$ V, $R_E = 1$ K and $R_B = 6$ K.

The resistor values shown are within 2 percent of the calculated values; 1-percent resistor tolerance was used throughout.

The voltage comparator formula giving the trigger voltage at point A (V_{TA}) as a function of V_B is linear:

$$V_{TA} = V_T + \frac{R_E}{R_E + R_X} \cdot \frac{R_T}{R_T + R_B} V_B$$

where V_T is the trigger voltage with $V_B = 0$ and R_T is the parallel combination of $(R_L + R_2)$ and $[(\beta_2 + 1)(R_E + R_X)]$. This reduces to:

$$V_{TA} = 7.000 + 0.116\, V_B$$

for the values shown. The curve shows the experimental plot of V_{TA} vs. V_B to be in excellent agreement with that predicted by the formula. The hysteresis for the circuit shown was 30 mV.

Reference:
1. William E. Zrubek, "Minimum-Predictable-Hysteresis Schmitt Trigger," EEE, Dec. 1963, pp. 40-43.

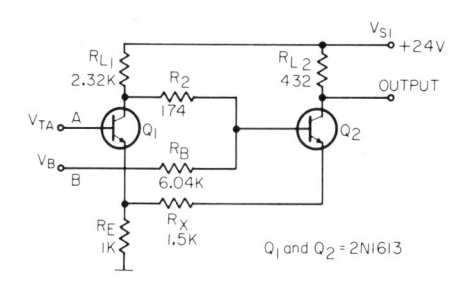

Variable Schmitt trigger, amplitude comparator in which V_B controls trigger level or is compared with V_{TA}. Comparison of calculated and experimental triggering points (below).

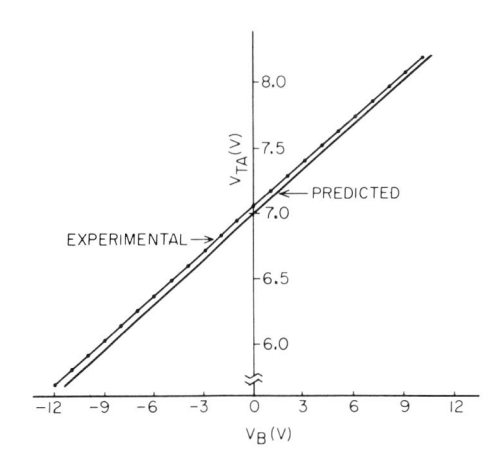

Digital and Pulse Circuits 67

Improved one-shot multivibrator using ICs only

by Clinton H. Kopper
Jet Propulsion Labs.
Pasadena, Calif.

A PREVIOUSLY published circuit[1] showed how flip-flop ICs could be connected with external gates to form a one-shot multivibrator. But the circuit had a disadvantage. If the time delay through the gates were insufficient, triggering of the flip-flop could be unreliable. An improved circuit shown in Fig. 1 eliminates this problem.

This circuit uses one flip-flop and one or more gates. A falling clock pulse sets the flip-flop after a short internal propagation delay. When the Q output of the flip-flop reaches about 1.5 volts, the gate threshold is reached. After a propagation delay, the gate's output falls, thus resetting the flip-flop. The gate propagation delay allows adequate time for the flip-flop to complete its "set" transition which is already well under way when the Q output reaches 1.5 V.

As shown in Figures 2 and 3, the output pulse of $\overline{Q}$ is narrower than Q. This is a characteristic of the flip-flop circuitry. The $\overline{Q}$ is set directly by the falling voltage on R_D. This setting of the $\overline{Q}$ output is internally coupled to the Q output of the flip-flop to cause it to reset. The time lag required for the Q output to reset after $\overline{Q}$ has set is equivalent to one gate delay. Hence the Q output pulse is the widest.

This circuit will work with flip-flops and gates of most logic families. If, for some reason, the resulting output pulse is not wide enough for the application, width can be increased by adding gates in series with the single gate shown. If an even number of gates is required, then the $\overline{Q}$ output should be used to drive the gates.

Reference
1. Theodore Shepertycki, "IC one-shot needs no external resistors or capacitors," *EEE*, December 1968, p. 102.

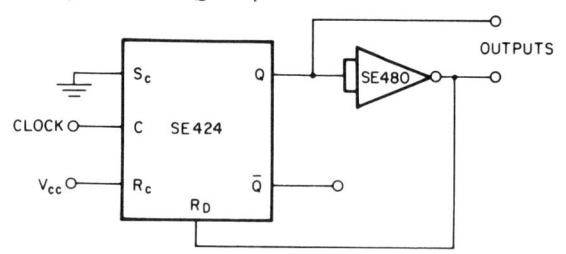

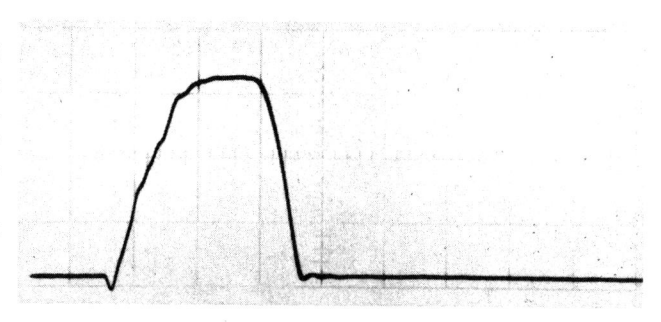

Fig. 2. The Q output of the flip-flop. (Vert. = 1.0 V/cm, horiz. = 50 ns/cm, Vcc = 4.0 V.)

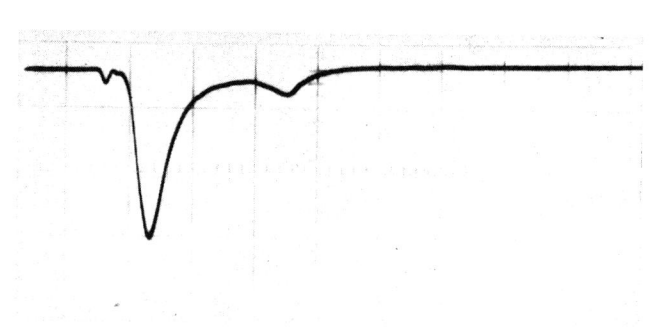

Fig. 3. The $\overline{Q}$ output of the flip-flop. (same scale as Fig. 2.)

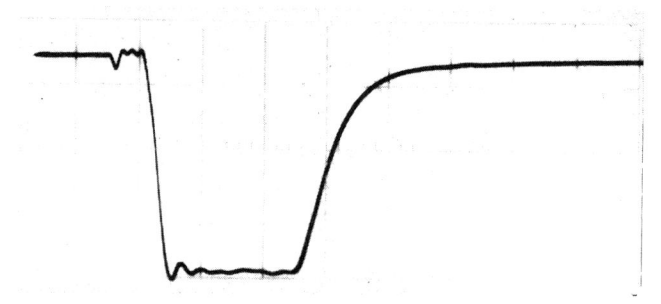

Fig. 4. The output of the SE480 gate. (same scale as Fig. 2)

Fig. 1. One-shot circuit uses just one flip-flop and one gate. Circuit triggers reliably for a wide range of gate delays.

Low-Hysteresis Trigger Circuit

By Duncan B. Campbell
GM Defense Research Laboratories
Santa Barbara, Calif.

A CONVENTIONAL DIFFERENTIAL AMPLIFIER pair, with a constant-current source replacing the emitter resistor, can be used as a level detector with small hysteresis. Capacitor C forms the positive feedback path to provide discrete switching action.

As the base of Q_1 becomes more positive than the base of Q_2, Q_1 starts to turn on, thus taking a larger portion of the current from the constant-current source, Q_3. This has the effect of turning Q_2 off. The positively going voltage at the collector of Q_2 is coupled through C to the base of Q_1, thus turning Q_1 on harder. A reverse action occurs as the base of Q_1 becomes more negative than that of Q_2.

The hysteresis is low because of the balanced nature of the differential amplifier and can be made even lower by increasing the amplification of the amplifier. The circuit works satisfactorily up to at least 60 kc.

R can be adjusted if it is desired to switch when Q_1's base is either positive or negative of Q_2's base. This has the effect of adjusting the symmetry of the output when triggering on a sine wave with the base of Q_2 grounded.

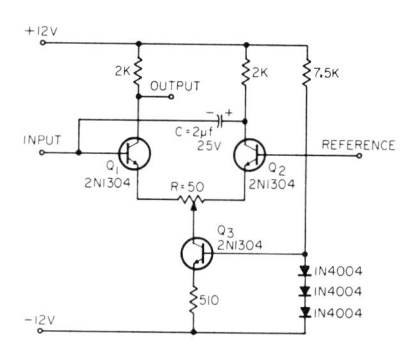

Low-hysteresis trigger circuit.

A programmable combinational logic circuit

Richard Merrell
Zenith Radio Corp., Chicago, IL

The June 5, 1973, issue of EDN featured an article entitled "Multiplexers Synthesize Boolean Expressions". My circuit shows multiplexers used in much the same way. However, rather than requiring the design of a new circuit for each new output function, this circuit requires only the programming of the function by proper setting of a series of switches. This circuit is usable as a design device which can be as valuable to the logic designer as resistor and capacitor substitution boxes are to others in the industry.

An 8-input multiplexer and 16 switches can be connected to make a programmable unit which can generate any output function for a combinational logic circuit having 4 inputs. As shown in the accompanying figure, each of the 16 switches corresponds to one of the 16 combinations possible with 4 inputs. Each switch has ZERO and ONE positions. To use the unit, examine the Truth Table and, for a given combination of D_0-D_3, set the corresponding switch to ZERO or ONE according to whether the output is to be ZERO or ONE.

The switches shown in the figure are programmed for the output function given in the following table:

Truth Table				
D_3	D_2	D_1	D_0	OUTPUT (Z)
0	0	0	0	0
0	0	0	1	0
0	0	1	0	0
0	0	1	1	1
0	1	0	0	1
0	1	0	1	0
0	1	1	0	1
0	1	1	1	1
1	0	0	0	0
1	0	0	1	0
1	0	1	0	0
1	0	1	1	1
1	1	0	0	1
1	1	0	1	0
1	1	1	0	1
1	1	1	1	0

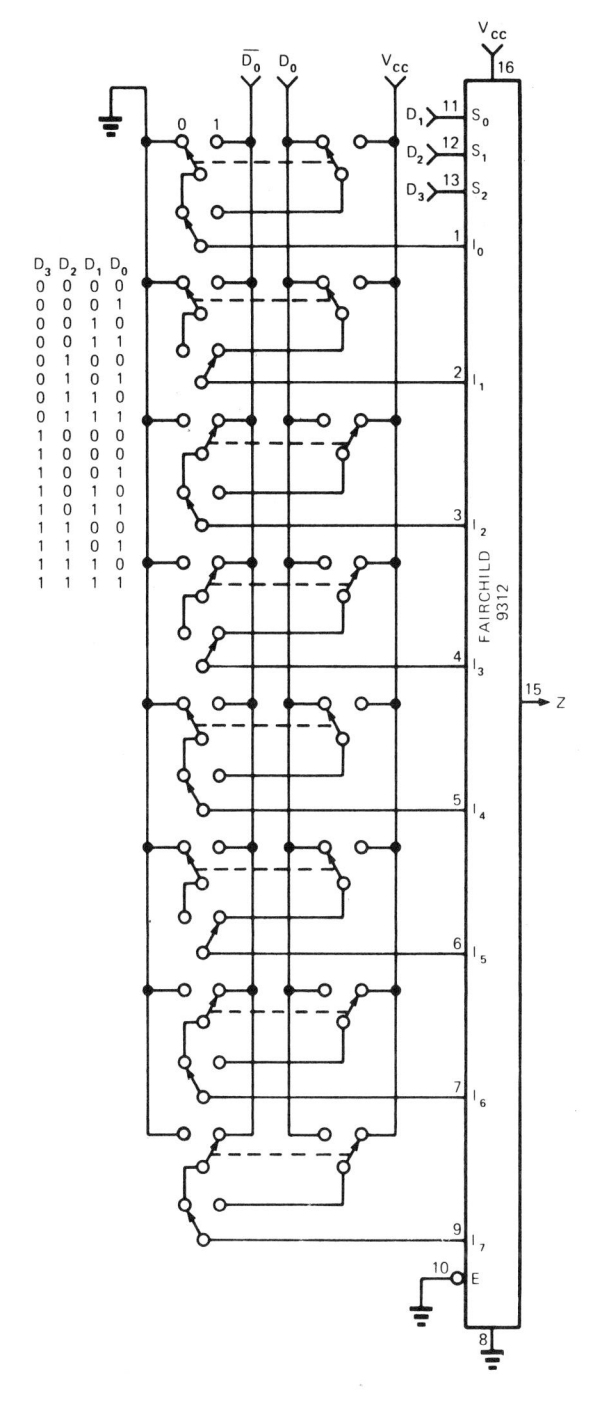

Programmability of this combinational logic circuit (which uses a single multiplexer IC and can handle up to 4-variable function logic) makes it a valuable breadboarding tool.

Note the partitioning of the Table. For $D_3=D_2=D_1=$ZERO, the output is ZERO regardless of D_0. Also, for $D_3=D_2=D_1=$ZERO, the output of the multiplexer is the same as the I_0 input. With the first and second switches set to ZERO, the I_0 multiplexer input is connected to ZERO (ground), which accomplishes the desired result. For $D_3=D_2=$ZERO and $D_1=$ONE, the desired output is seen to duplicate D_0; with the third and fourth switches set at ZERO and ONE, respectively, the I_1 multiplexer input is switched to D_0. For $D_3=D_1=$ZERO and $D_2=$ONE the output is the same as $\overline{D}_0$, and the switches feed the $\overline{D}_0$ input to I_2. For $D_3=$ZERO and $D_2=D_1=$ONE, the output is ONE regardless of D_0; in this case, the I_3 input is connected to ONE (V_{CC}). This covers the four output combinations obtainable for one pair of switches. The remaining 4 pairs of switches follow the same procedure.

Extraneous pulse detector

by Tony Randazzo
Atlantic Research
Costa Mesa, Calif.

THE FIGURE shows a combination of a cross-coupled latch and a JK flip-flop that together form an extraneous pulse detector. Given a reset command, Q goes low and the clock input is inhibited by the "0" present at the K input coming from the latch. When the input pulse goes low, the latch permanently changes state allowing any transients of the input pulse to trigger the JK flip-flop. In this situation, a "1" will be retained at the Q output and the failure indicated by lamp L_2.

Pulses and transients down to 25 ns can be detected by the circuit if series MC800P ICs are used.

This circuit indicates an extraneous pulse by turning L_1 and L_2 on. A "clean" pulse turns on L_1 only.

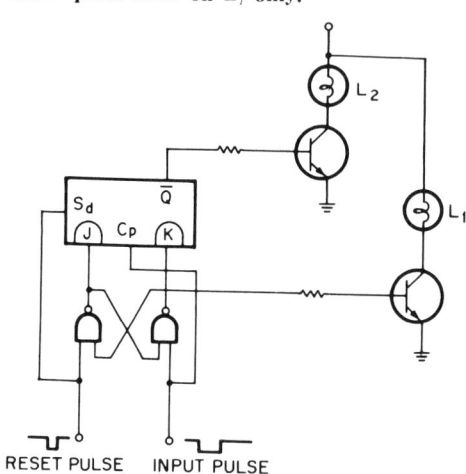

Variable delay blanking-pulse generator

by Donald E. Norris
SCI Electronics
Huntsville, Ala.

This circuit produces a variable width blanking pulse at the end of a selectable delay time after being initially triggered by an input pulse.

With the component values shown in the schematic, the circuit can produce output pulses having widths ranging from 8 µsec to 12 µsec. The delay period is adjustable from 26 to 36 µsec.

The circuit can easily be modified to provide different ranges of output pulse width and delay. Components C_2 and R_4 control delay time while C_4 and R_8 control the width of the blanking pulse.

When a logic 0 is applied at E_1, output E_2 from the input gate goes to logic 1. This turns on Q_1 and triggers the first "one-shot," IC_2, which generates a delayed pulse at point E_4. The trailing edge of the pulse from IC_2 then triggers Q_2 which, in turn, triggers the output "one-shot," IC_3.

The first "one-shot" controls delay time, and R_5 allows fine adjustment of delay. The second "one-shot" controls the width of the output pulse, and R_9 allows continuous adjustment of pulse width.

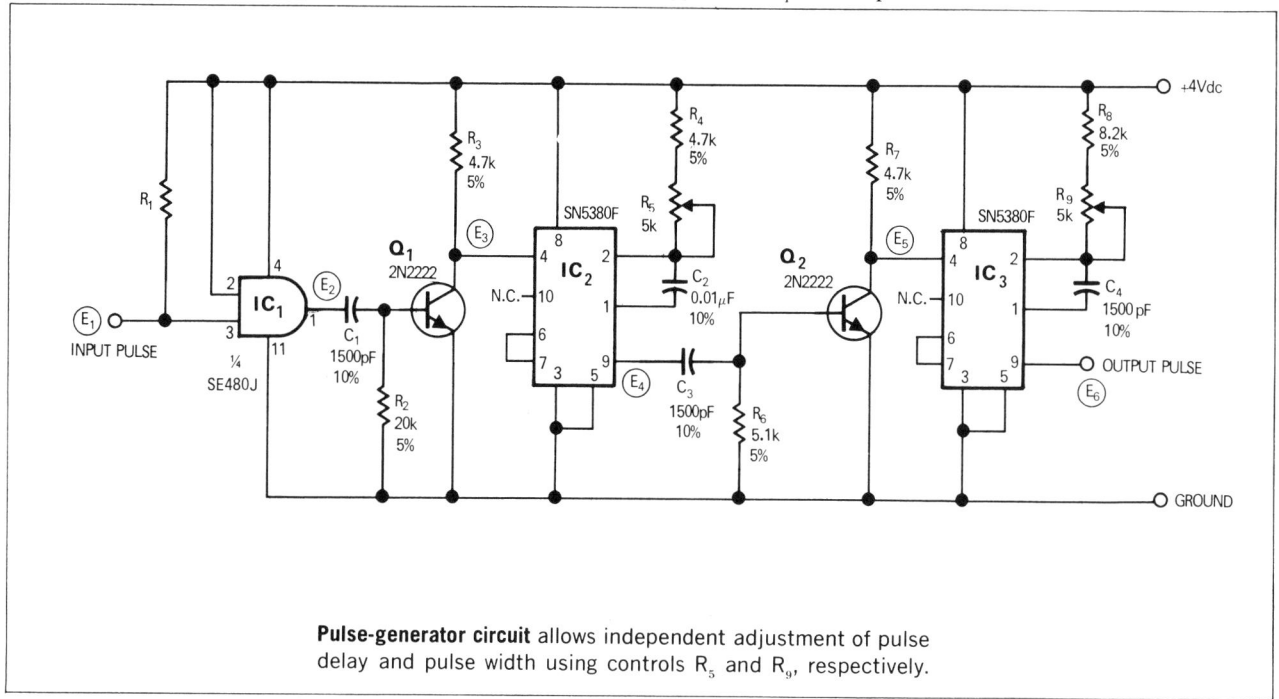

Pulse-generator circuit allows independent adjustment of pulse delay and pulse width using controls R_5 and R_9, respectively.

Anti-coincidence circuit prevents loss of data

Entry by J. H. Burkhardt, Jr.
Union Carbide Corp., Oak Ridge, Tenn.

One hazard in the use of bidirectional counter circuits, in which up and down data occur randomly and independently, is that an up pulse may occur simultaneously with a down pulse. If the maximum frequency of the separate data pulses is known, this "Anti-coincidence Circuit" can be used to separate these random up/down pulses and assure that no information is lost.

Each falling edge of the clock toggles flip-flop FF1. The outputs of FF1 are square waves with a frequency of one-half that of the clock. These complementary square waves are used as clocks for FF2 and FF3. The reset inputs of RS flip-flops FF4 and FF5 are held high by the normally disabled FF2 and FF3. Data pulse D1 will set FF4, and D2 will set FF5, enabling FF2 and FF3. Recall that the clocks of FF2 and FF3 are complements, so that these flip-flops will toggle 180 degrees apart.

If FF2 toggles first, FF4 will be reset and FF2 will be disabled until another D1 pulse sets FF4. Then FF3 will toggle and reset FF5, which disables FF3 until another D2 pulse is received. The high-to-low transition of outputs 1 and 2 will always be separated by at least one clock period regardless of the relationship of D1 and D2, provided their maximum frequencies are always less than half that of the clock.

In general, any type of 2-input positive NAND gate may be used to form FF4 and FF5, and any type J-K master-slave flip-flop may be used for FF1, FF2 and FF3. The selection of logic type is made on the basis of circuit requirements, cost and availability.

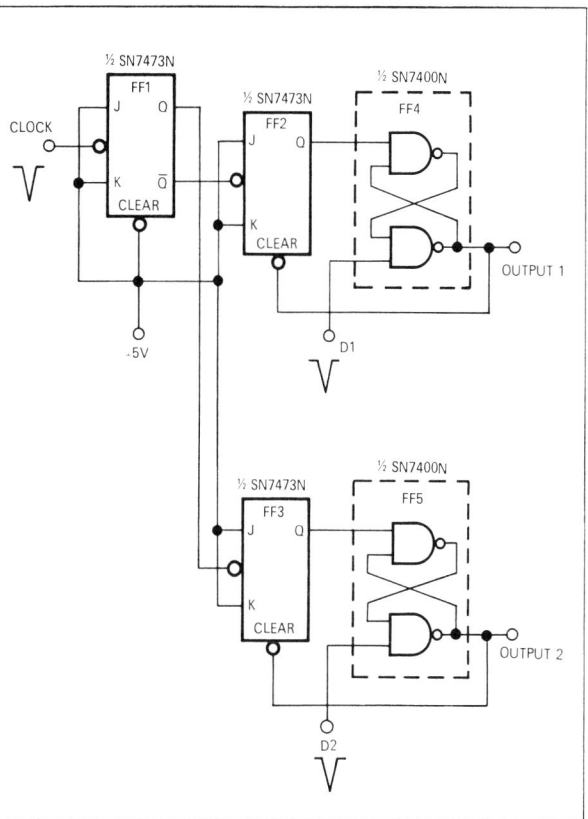

Outputs 1 and 2 will always be separated by at least one clock period, even if inputs D1 and D2 occur simultaneously. This assures no loss of information in bidirectional counters.

Stable Threshold Circuit With Low Hysteresis

By Robert M. Muth
General Electric Company
Syracuse, N.Y.

IF A REQUIREMENT arises for a circuit that determines if a millivolt-range signal exceeds an adjustable threshold the circuit shown can be easily designed to meet the need. The circuit is similar to a standard Schmitt trigger circuit.

The forward gain of the amplifier has been increased by adding a second differential-amplifier stage. This is done by adding only two low-cost transistors and three resistors. The positive feedback is taken from the output of the second differential amplifier to the base of the input differential amplifier.

The additional gain allows:
- ☐ A stable hysteresis as low as 2 mv compared to 100 mv with a standard Schmitt trigger circuit.
- ☐ Either polarity output due to large resistance of the feedback resistors.
- ☐ Superior threshold stability with temperature due to ability to match R_7 with the input resistance.
- ☐ Increased gain and superior temperature stability of the input differential amplifier due to ability to place the threshold adjustment in the collector circuit of the amplifier.

With the resistor values shown in the circuit, the hysteresis is 5 mv, the temperature drift is less than 100 $\mu v/°C$, the upper threshold voltage can be adjusted from +20 to −10 mv, and rise and falltimes are less than 0.2 μsec.

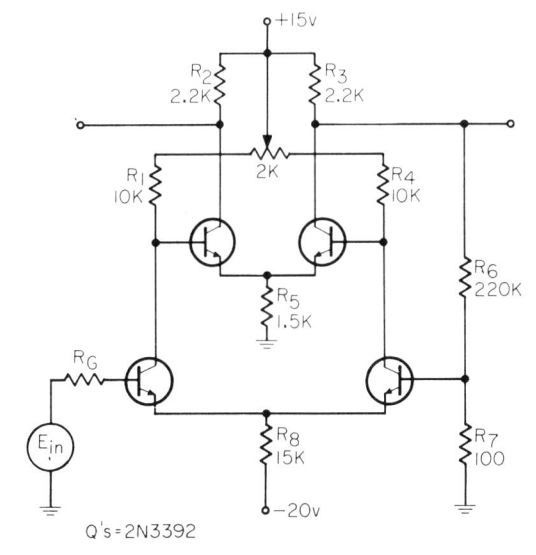

Low-level threshold detector.

Pulse generator-to-CCSL interface

by David Y. F. Lai
Lawrence Radiation Lab.
Livermore, California

A SIMPLE CIRCUIT matches the negative output of a pulse generator to the positive-input requirement of an IC in the compatible current-sinking-logic family. It maintains a constant current of at least −1.6 mA for the "0" input to the CCSL and its offers wide latitude in choice of component values.

In the basic circuit of Fig. 1, the base of npn transistor Q_1 is grounded through R and the collector (pin A) is connected (at pin 1) to the CCSL component through a constant-current regulator diode. The generator delivers an output that varies between ground ("0" state) and an adjustable negative voltage ("1" state).

Ideally, the generator output and the required CCSL input are as shown in Fig. 2. To see the various voltage-current relationships, one can apply a negative-going voltage ramp in place of the generator output. In place of the CCSL component, one can use a milliammeter to +5 V (typical for CCSL) as shown in Fig. 1 if A is connected to 2.

When the ramp starts at ground, Q_1 is biased off and no current flows through D_1. As the ramp voltage increases and reaches about −0.7 V, the diode and transistor begin to conduct. Initially the diode has a low resistance so the current through it is governed by the transistor. When the regulating-current level of the diode is reached, the diode resistance rises rapidly, keeping diode current fairly constant. As the ramp continues to rise, the transistor saturates. Fig. 3 shows the voltage-current relationships of this test circuit.

It's convenient to use 1 kΩ for R so the base current in milliamperes will have the same numerical value as V_2 in volts. But this can lead to heating because, though the collector current saturates at 2 mA, the base current continues to rise as the ramp voltage increases. If we reduce the base current in the test circuit by using 10 kΩ for R, we modify the collector current as shown by the dashed line in Fig. 3.

Better yet, we can limit the base current to a maximum of 0.2 mA by using a regulator diode (pin B to pin 5) in place of the base resistor in Fig. 1.

The maximum voltage we apply to this circuit depends on the diode used for D_1. The V-I relationship for the 1N5305 is shown in Fig. 4. Constant-current operation occurs at the plateau between 3 V and the peak operating voltage which, in this series, is 100 V. The lower limit of the plateau depends on the diode, being lower for lower-current diodes. Above POV, the current increases sharply. Since the voltage across D_1 is always larger than that across D_2, the voltage applied to the circuit must not produce a drop across D_1 greater than POV.

The 1N5303 regulates at 1.6 mA, which is the current required for a single CCSL input. Additional inputs must be connected (as in Fig. 1 with pin A to pin 3) by using a separate diode for each input. Since all input current must pass through the pulse generator, the generator must be able to sink at least the combined currents.

If it's desirable to set the constant current at different levels, instead of using one regulator diode for each input we can use a potentiometer and FET (pin A to pin 4 in Fig. 1) to provide a constant current that's variable.

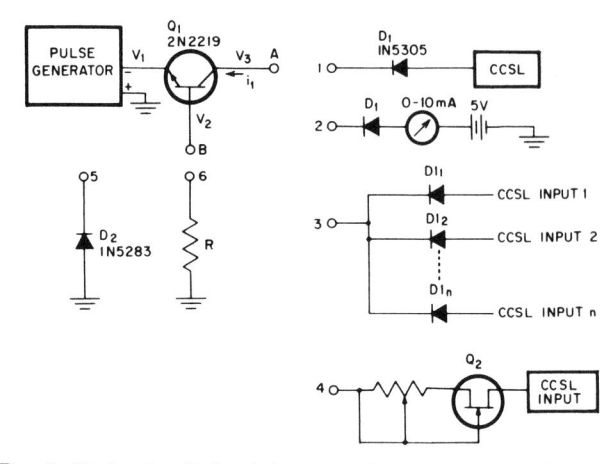

Fig. 1. Basic circuit for interconnecting a negative-going pulse generator to CCSL ICs. Many variations provide flexibility.

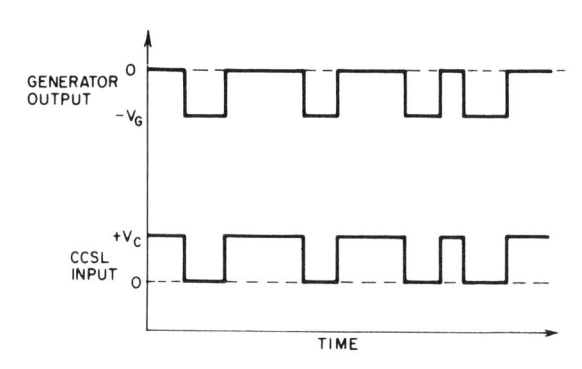

Fig. 2. The generator output must be shifted above the zero axis to provide a suitable CCSL input.

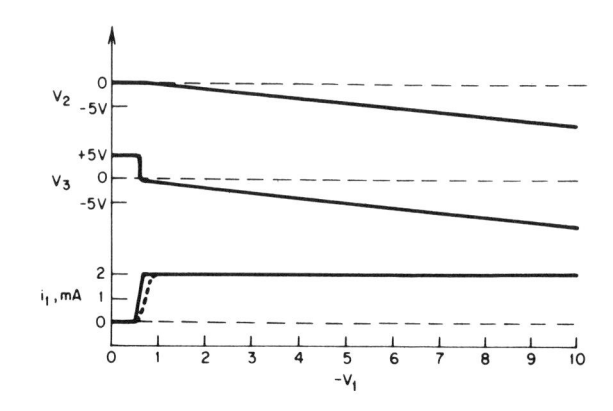

Fig. 3. Voltage-current relationships for a test circuit in which a negative-going ramp replaces the pulse generator and a meter and power supply replace the IC.

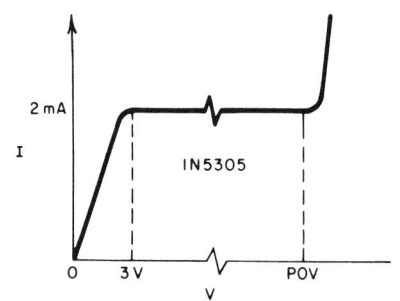

Fig. 4. V-I relationship for the 1N5305. The diode must never be operated beyond its peak operating voltage.

Eliminating False Triggering in Monostable Multis

By Herbert Cohen
Frequency Engineering Laboratories
Farmingdale, N. J.

THE ADDITION OF three diodes and one resistor to a conventional monostable multivibrator permits an increase in the timing resistor, R_t, without making the circuit susceptible to false triggering.

The monostable should be able to accommodate large timing resistors for two reasons: a wider variation of pulse width can be achieved without changing C, and larger R/C ratios can be used, thereby increasing the duty cycle.

In the conventional monostable, large values of R_t cannot supply enough base drive to Q_1 to provide adequate noise immunity when the monostable is in the quiescent state, making the circuit susceptible to false triggering.

In the modified monostable shown, substantial base drive is supplied by R_b when in the quiescent state. A negative-going trigger step at the input turns Q_1 off, driving Q_2 into saturation and diverts Q_1 base current through D_1. At this time the junction between C and D_3 goes negative. Both D_2 and D_3 are now cut off. C must now charge through R_t to V_{be} of Q_1 before Q_1 can turn on.

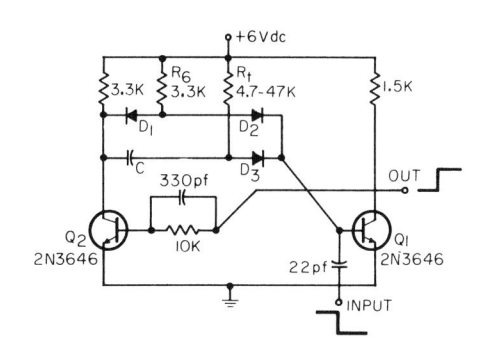

Extra components allow better triggering operation in mono stable multi.

When Q_1 turns on, Q_2 cuts off, allowing R_b to supply hard base drive to Q_1. The monostable will stay in this state until intentionally triggered.

The maximum value of R_t depends on the β of Q_1. The value of R_t may be extended by using a high-β Q_1 or by driving Q_1 with an emitter follower.

Pulse-Peak Indicator

By John C. Rich
General Electric Co.
St. Petersburg, Fla.

THIS CIRCUIT indicates the peak of a fast voltage pulse to within one of several predetermined voltage ranges. These ranges are established by tunnel-diode level-sensing circuits and indicated by a series of "exclusive-or" dual-coil reed relays.

The circuit shown indicates the peak of an incoming negative voltage pulse within one of the following ranges: 10-15 volts, 15-20 volts, 20-25, and 25 volts and above. Minimum switching voltage is 2 v. The circuit is divided into four similar channels.

Assume that an 18-v voltage pulse is applied. The signal is large enough to momentarily switch tunnel diodes D_1 and D_2 to a high-voltage state, but not large enough to switch D_3 and D_4. Hence, Q_5 and Q_6 will be latched "ON."

The contacts of K_1 will be open because both coil A and coil B of K_1 will be carrying current and the coils are arranged so that their magnetic fields are in opposition. The contacts of K_2 will be the only ones closed because only coil A of K_2 is carrying current, but not coil B; hence, no flux cancellation. The contacts of K_3 and K_4 will be open because neither coil A or B in K_3 or K_4 will be carrying current. The reset switch must be operated to turn all silicon controlled switches "OFF" before the next pulse can be measured. The circuit operation is similar for other voltage levels.

The number of ranges may be increased by adding additional channels. The voltage range of one channel can be narrowed by appropriate adjustment of the tunnel-diode level sensing potentiometers. Polarity reversal can be obtained by reversing the tunnel diode, changing to an npn transistor, and coupling to the anode gate of the silicon controlled switch. Multiple contacts could be incorporated in the reed relay assembly or the single reed relay contact could be used to energize a multiple contact relay.

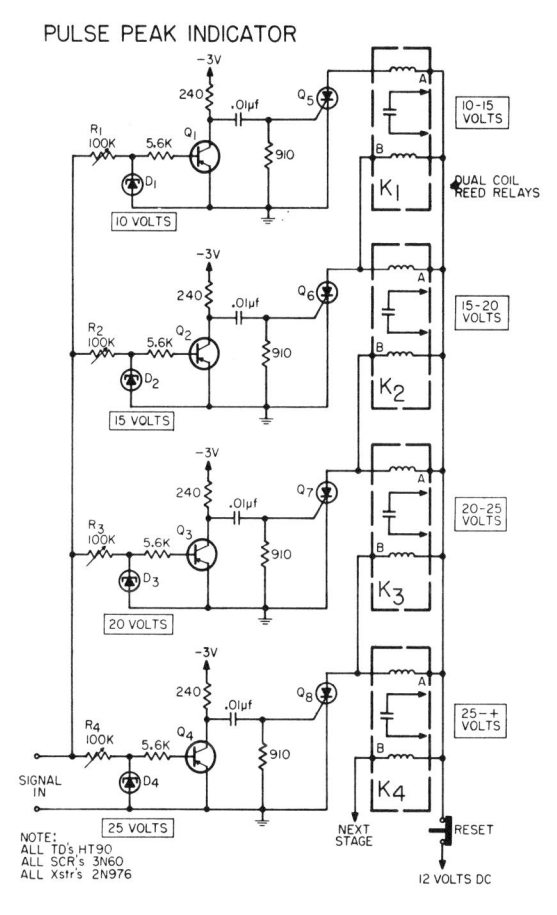

Pulse-peak indicator circuit.

The tunnel-diode level sensing circuits have functioned satisfactorily to detect the peak of pulses resembling half sine-wave shapes having a 1-μsec width at 50-percent amplitude points. The level adjustment has remained constant to within ± 2 parts per thousand at room temperature for several weeks.

Digital and Pulse Circuits 73

Flip-flop with pulse and level outputs

by Charles McBrearty
Control Data
Norristown, Pa.

This novel circuit has greater versatility than the conventional RS flip-flop. In addition to Q and $\bar{Q}$ outputs, the circuit of Fig. 1 provides an output trigger pulse (T) each time the flip-flop changes state, and buffered Q and $\bar{Q}$ outputs insensitive to noise induced onto the output lines.

Logically equivalent output signals are available at Q, Q_1 and Q_2. On the Q line, negative-going noise can cause the flip-flop to reset. If the output signals are taken from either Q_1 or Q_2, this problem is avoided. Output Q_2 is isolated from the flip-flop by one 937 stage while $\bar{Q}_1$ is isolated by two 937 stages.

$\bar{Q}$, $\bar{Q}_1$ and $\bar{Q}_2$ are also logically equivalent. Q_1 and Q_2, however, provide output isolation preventing a reset flip-flop from being set extraneously.

During either a "set" or "reset" the T pulse duration is dependent on the propagation delay of one 949 in the cross-coupled latch plus the delays of two 937s. At 25°C, pulse duration is about 60 ns.

Figure 2 illustrates waveforms present during both the set and reset transitions. During a set transition, T goes low one propagation time after Q goes high. T remains low until one propagation time after $\bar{Q}_2$ goes low. Once the flip-flop is set, all additional S-input transitions are ignored at T which remains high until the flip-flop changes to the reset state. When the flip-flop is reset, the T-output ignores further R-input transitions.

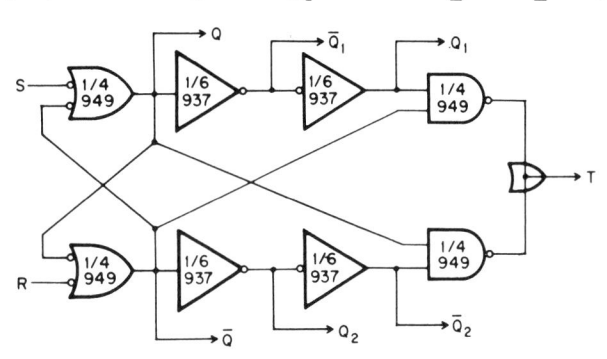

Fig. 1. This versatile RS flip-flop provides pulse and buffered outputs. It requires no discrete components.

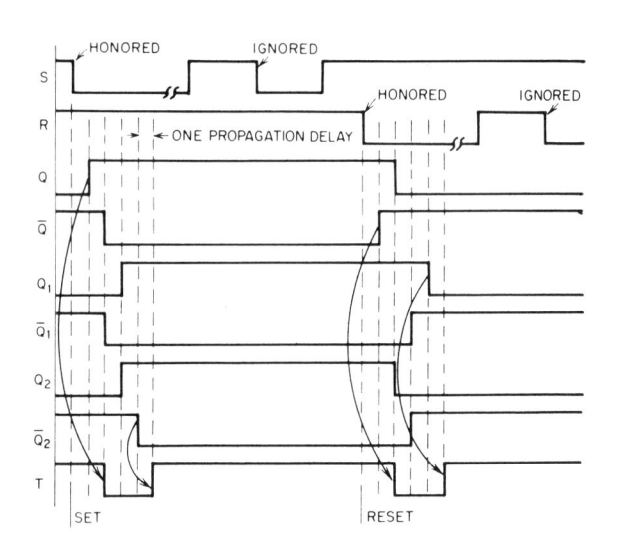

Fig. 2. Waveforms during the "set" and "reset" transitions. All amplitudes are zero or plus five volts.

Quad NAND full adder uses two ICs

William Scott Dawson
Computer Science Corp., Huntsville, AL

If in these times of Medium Scale Integration (MSI) a discrete gate circuit can be considered compact, then that is the appeal of this NAND full adder.

Tucked neatly in two DTL packages, such as the MC846P, which is a quad 2-input NAND gate package, the circuit is easily cascaded on a printed circuit board. Two three-gate exclusive NOR's (EZ-1 and EZ-2) generate the sum, which is of course the same as odd parity for the three inputs (AB and C_{in}).

AB is available at EZ1 pin 3 to OR in EZ2-6, and $A\bar{B} + \bar{A}B$ appears at EZ1-8 to NAND in EZ2-11 with the input carry C_{in}. Thus, all inputs and outputs are ones, and no inverters are required for use with MSI counters having only true outputs. Of course, DTL must be used for the wired-AND function.

Gates of EZ-1 are designated 1 on drawing and gates of EZ-2 are labeled 2.

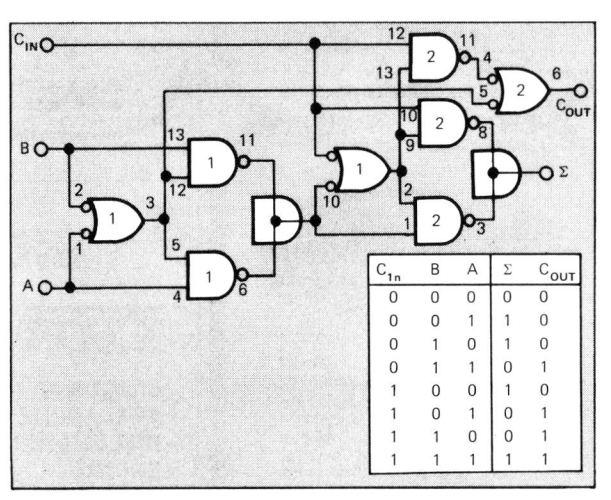

Two DTL NAND-gate packages are all that is required for this full adder.

Unbalanced to Balanced Level-Shifter

By W. J. Travis
Sprague Electric Co.
North Adams, Mass.

THIS LEVEL-SHIFTING circuit converts unbalanced (0 to +4 V) pulses to balanced (−6 to +6 V) pulses. In the "high" state, a 0.12 mA, 3.5 V (minimum) source at the input is needed to insure a +6 V output. In the "low" state, a 1 mA, 0.6 V (maximum) sink at the input gives a −6 V output. With the component values shown, the output impedance of the circuit is 90 ohms.

If the "inhibit" terminal is grounded, the output is held at −6 V regardless of the input voltage. In the original application for this circuit, fast rise time pulses were converted to slow rise time pulses, for transmission over long lines. Capacitor C was adjusted to give the required pulse shape.

Assuming the circuit is not inhibited, a "high" input will turn on Q_2 and hold Q_3 off. R_1 and R_2 control the current through Q_2. This current holds Q_4 in saturation and the +12 V supply appears across R_3 and R_4. Because Q_3 is not conducting, Q_5 is held off.

A "low" input gives the opposite effect; Q_2 and Q_4 are turned off, while Q_3 and Q_5 are turned on. Thus −12 V appears across R_3 and R_4.

The values of R_3, R_4 and R_5 may of course be changed to give different output voltages and impedances.

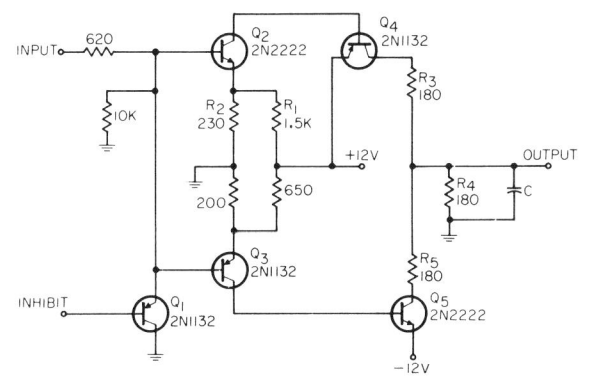

Positive pulses at input give ±6 V balanced pulses at output. If inhibit terminal is grounded, output is held at −6 V.

Long-Duration One-Shot

By Russell W. Walton
Fairchild Instrumentation
Palo Alto, Calif.

ADDING A TRANSISTOR is a solution to the problem of designing a long-duration one-shot multivibrator with integrated circuits. The low impedances of ICs usually dictate the use of large and expensive capacitors to obtain long time-constants. The circuit shown, here, however, uses a small tantalum capacitor and a commercial-grade transistor. For this circuit the total component cost is less than $3.50 in small quantities.

With the component values shown, the circuit provides pulses of up to 75-sec duration, and will operate with supply voltages as low as 2.6 Vdc. Input and output levels are compatible with standard micrologic.

The time constant is determined by R_1 and C_1. Resistor R_2 should be low enough to avoid current-starving the IC. For the μL 914 dual-gate circuit, resistor values in the range 1 k to 10 K are acceptable. R_4 provides a return path for reverse leakage current in the IC. This resistor can be bypassed with a suitable capacitor if necessary, to prevent false triggering due to ripple on the B+ line. A series resistor may be added in the feedback path between pins 5 and 7. The value of this resistor will depend on the output loading.

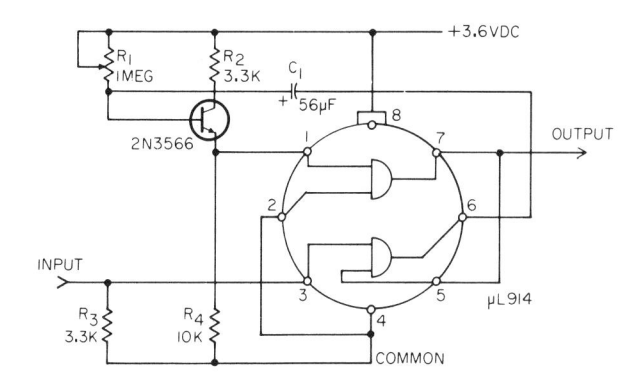

IC/Transistor circuit gives long time-constant without using large capacitors.

Synchronized one shot

by F. E. Nesbitt
Conductron-Missouri
St. Charles, Mo.

IT IS OFTEN necessary either to synchronize an input pulse to a clock or to shorten the pulse width of an input pulse. The circuit shown generates a pulse that is two clock pulses wide from an input pulse whose width is greater than 5 times the width of a clock pulse.

The circuit contains two flip-flops, designated A and B, connected as a shift register. Flip-flop A is set when the clock pulse falls and the input goes to a "1." Flip-flop A remains set until the input and clock pulse go to "0." B follows the state of A (one clock delay). Decoding $A\overline{B}$ gives the desired result (output pulse) since it occurs but once during any particular input strobe.

If it is necessary only to synchronize the input pulse to the clock and not shorten the pulse, the output of flip-flop A is sufficient.

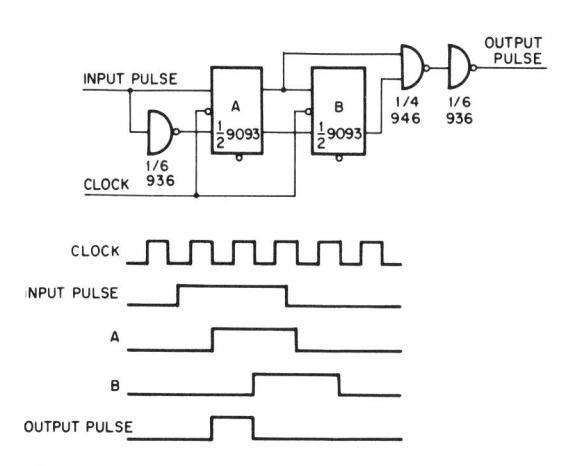

This circuit synchronizes an input pulse to the system clock. The synched output pulse is two clock pulses wide.

Digital and Pulse Circuits 75

Divide-by-N circuit has 50/50 duty cycle

David A. Scott
Naval Weapons Center, China Lake, CA

This method of creating a divide-by-N circuit that has a 50/50 duty cycle for any integer N has the advantages of ease of design and connection, and the elimination of all wired-ORs. The following procedures are used with this method:

First, write N, or products of numbers equaling N. Next, change N to a binary number written from left to right in increasing powers of two. Then, going from left to right, complement all digits up to and including the first one. With this new binary number, if the right-most digit is a ZERO, cross it off.

The trick now is to let all ONES equal circuit A and all ZEROS equal circuit B (**Fig. 1**). Connect the Q output of the last flip-flop to the extra input of each exclusive-OR gate. Connect the Q output of each circuit to the input of the next. Circuits and waveforms for symmetrical square-wave division by five and seven are illustrated in **Fig. 2** and **3**.

The proof of the circuit takes considerably more space than that allocated here. However, it is not very complicated and the circuit does work. It has been used successfully in circuit design at the Naval Weapons Center.

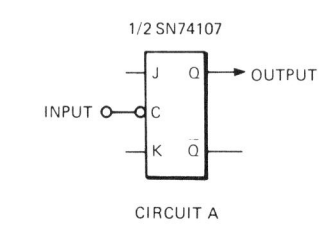

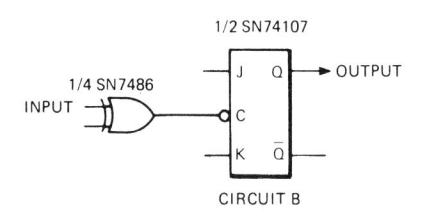

Fig. 2—**Divide-by-five circuit** and circuit waveforms showing the generation of a 50/50 duty-cycle output.

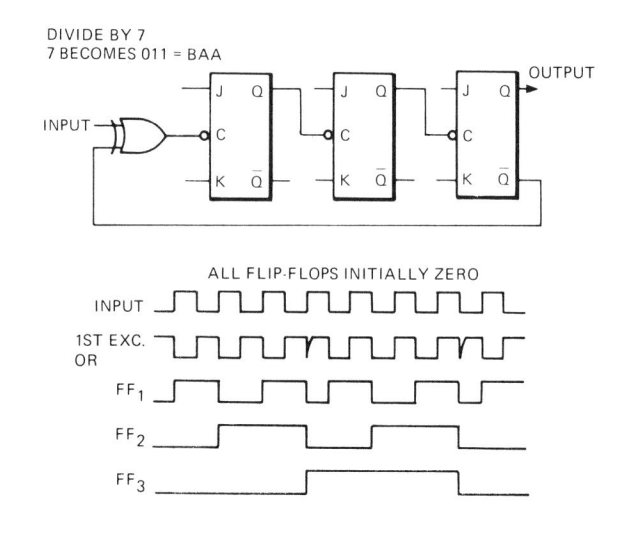

Fig. 3—**Dividing by seven or any larger number** is just a simple extension of the author's technique.

Fig. 1—**Only two types of flip-flop circuits are required for any divide by-N** circuit to provide a square-wave output.

EXAMPLES						
3 = 11	4 = 001	5 = 101	6 = 011	7 = 111	11 = 1101	BINARY FORM
01	110	001	101	011	0101	COMPLEMENT UP TO AND INCLUDING THE FIRST 1
01	11	001	101	011	0101	TRAILING ZERO DROPPED

Complementary series multivibrator

By Charles R. Bond
Electromec
Santa Clara, Calif.

The complementary series type of multivibrator has several unique properties. It provides simultaneous positive- and negative-going pulses with extremely fast rise times. Also the circuit is very simple, with a minimum number of components.

Because the two load resistors have equal value and share the same series current path, the output waveforms are coincident and of equal amplitude. With the biasing arrangement shown, there is no dc state in which the loop gain is less than unity. So the circuit is inherently self-starting, immune to lock-up, and usable over a wide range of frequencies.

The multivibrator works as follows:

Assume that Q_1 and Q_2 are initially biased off. Then C_1 charges toward +6 V through R_{b1} and R_{L1}. At the same time, C_2 charges toward ground level via R_{b2} and R_{L2}.

When the voltage across each of the capacitors, C_1 and C_2, is sufficient to forward-bias Q_1 and Q_2 respectively, the transistors begin to conduct. But the collector voltage of each transistor is cross-coupled to the base of the other transistor. So the circuit is regenerative, and both transistors change rapidly from cutoff to saturation.

The transistors conduct for sufficient time to discharge their respective capacitors. While the capacitors discharge, the transistors start to cut off. But again the action is regenerative and the transistors turn off rapidly. Thus, the circuit returns to its initial state.

Interval between pulses is given approximately by the relationship,

$$T \approx 0.7 R_b C \quad (1)$$

(where $R_b \gg R_L$).

Pulse width is determined by the time required to discharge the coupling capacitors with transistor base current. With the component values shown, pulse width is about 2 μs. Pulse amplitude is approximately half the supply voltage.

$$V_o = (V_{cc}/2) - V_{ce(sat)} \quad (2)$$

If V_{cc} is increased to obtain larger output amplitude, a diode should be placed in series with the base of each transistor to prevent emitter-base breakdown.

Another possible circuit modification is to add a capacitor from the emitter point to ground. This balances the circuit. Thus there is no need to accurately match component values or transistor parameters to achieve symmetry.

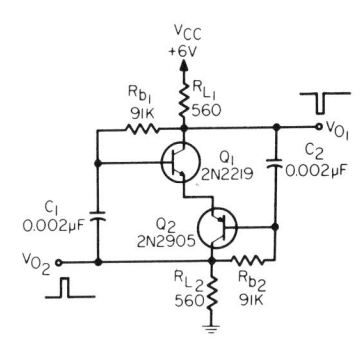

Unusual multivibrator gives fast pulses. Coincident outputs have equal amplitude and opposite polarity.

Current source improves immunity of one-shot

By Gregory L. Schaffer
Ames Research Center, NASA
Moffett Field, Cal.

One-shot multivibrators with long time constants tend to be susceptible to false triggering from power supply line noise. This problem can be overcome by using a current source to isolate line noise from the timing circuitry.

Transistor Q_1 is a current source that supplies current to transistor Q_2 and resistor R. The voltage across R is held at 1.2 volts by diode D_1 and the V_{BE} drop of Q_2. Transistor Q_3 is a low-pinch-off FET (-0.3 volt for pinch off) and is normally conducting ($V_{GS} = +0.4$ volt) to keep the current source on. When a positive trigger pulse arrives at the input of Q_4, V_{GS} is driven negative to approximately -1.0 volt and the FET is nonconducting. Q_1 is thus turned off, and this turns off Q_2. Resistor R holds the timing line at low potential while Q_1 is nonconducting. When V_{GS} of the FET reaches about -0.3 volt due to the discharge of C_T, Q_3 conducts and turns on the current source. The circuit is now back in its steady-state condition, ready for another trigger pulse.

Capacitor C is needed to prevent misfiring from high-frequency noise spikes on the supply line. A value of 0.01 μF is sufficient for most purposes. Resistor R_T should be much larger than R, for proper operation. A minimum value for R_T is around 10 k. Because of this limitation, the one-shot is not capable of short-duration pulses (less than 50 μs). However, long-duration pulses can be obtained with relatively small capacitors. A time duration of 1 hour is obtained with $R_T = 10$ M, $C_T = 330$ μF.

This one-shot will operate over a wide power-supply voltage range. With the transistors shown, V_s may be as high as 40 volts or as low as 2 volts. If V_s changes from 40 to 4 volts, the timing period varies only 10 percent. Operating at a supply voltage of 20 volts, the circuit is unaffected by 15-volt noise spikes on the power supply line.

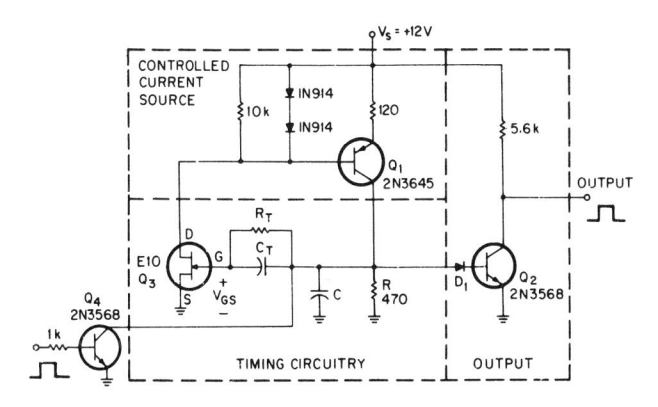

In this unusual one-shot, transistor Q_1 and capacitor C isolate the timing circuitry from transients on the dc line.

Simplified Schmitt yields fast rise time

by John H. Cone
Electronics Enterprises
Pasadena, Calif.

The basic Schmitt trigger seems such a straightforward circuit that one might ask how it could possibly be made any simpler. But for some applications, the component count can be further reduced to give improved results.

All capacitors can be eliminated, leaving a circuit with excellent performance at minimal cost. The circuit shown in Fig. 1 covers a frequency range of 10 Hz to over 100 kHz without adjustment of component values. Waveform is excellent at all frequencies, and both rise and fall times are less than 12 ns.

One practical application for the circuit is sine-to-square wave conversion. Used in conjunction with a low-cost audio generator, the circuit provides a wide-range square-wave signal source.

The speed-up capacitors, that are commonly used in Schmitt triggers, offer both advantages and disadvantages. They are useful in logic systems because they reduce the storage time of

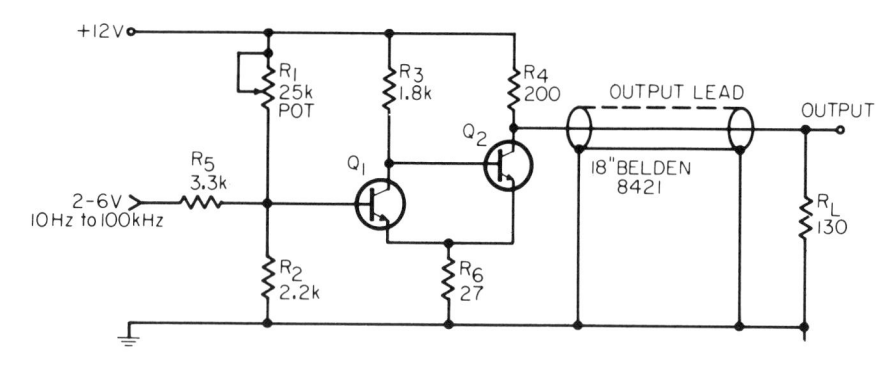

Fig. 1. Speed-up capacitors aren't necessary in this simplified Schmitt trigger that converts sine waves to clean square waves.

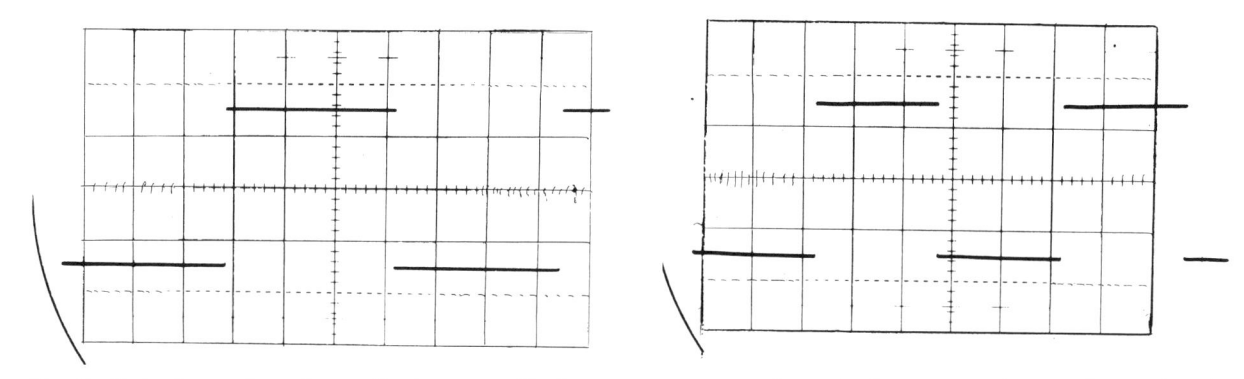

Fig. 2. Output waveform is excellent over a wide frequency range, as shown in these scope pictures. Vertical scale for both waveforms is 1 V/cm. Frequencies are 30 Hz (left) and 100 kH$_z$ (right). Horizontal scales are 5 ms/cm and 2 μs/cm respectively.

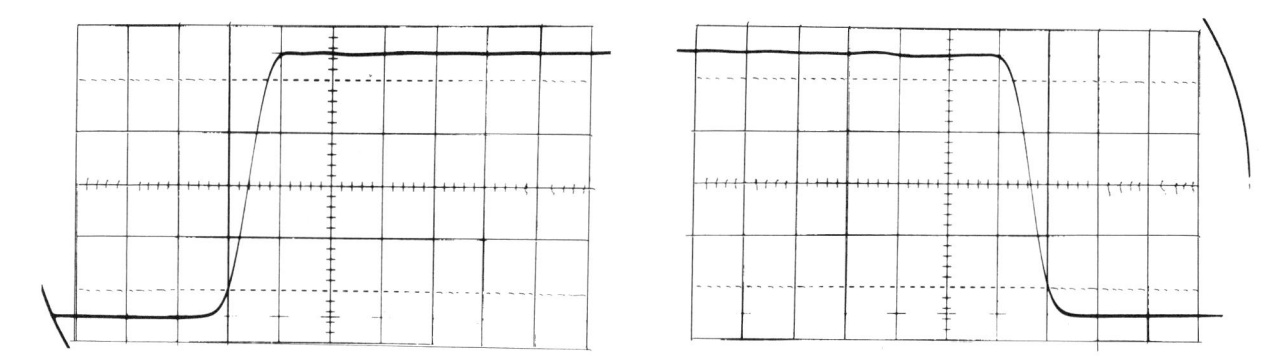

Fig. 3. At 100 kHz, rise and fall times are both less than 12 ns. Vertical scale for these pictures is approximately 0.6 V/cm and horizontal scale is 20 ns/cm. Rise and fall times are each 15 ns gross or 11 ns net (adjusting for scope response).

78 Circuit Design Idea Handbook

(continued)
the semiconductor components. So they can provide increased switching speed in applications where speed would otherwise be limited by the transistors.

But speed-up capacitors do not alter the inherent rise and fall times of the circuit and its external impedances. In fact, they make it increasingly difficult to obtain good waveforms over a wide frequency range.

Elimination of capacitors in the circuit gives near-perfect waveforms over a wide range. Figure 2 shows the output square wave at 30 Hz and at 100 kHz. Fig. 3 shows expanded views of the leading and trailing edge of a 100-kHz output signal.

One other unusual feature of the circuit shown in Fig. 1 is that the load resistor is connected at the end of the output cable. This connection gives better results than the conventional approach. The Belden cable is specified because this has lower capacitance than most widely-available flexible cables. Length should not exceed eighteen inches, but this is probably adequate for most test-bench applications.

Lead dress and layout aren't very critical but output-cable length is critical. After the cable has been cut, the exact values of R_L and R_3 should be determined for optimum fall time. The value of R_4 is then chosen for optimum rise time. This resistor should be a 1-watt noninductive (composition) type.

Finally R_1 is adjusted to give symmetrical square waves. Usually, no further adjustment is needed unless the input voltages are changed. Lower input voltages give faster rise times, and can be used when lower outputs are acceptable.

The transistors aren't too critical and low-cost types are usually adequate. Recommended types include TI 2N4418, TIS 48 or TIS 49; Fairchild 2N4275 or EN2369A; Motorola MPS2369. These are all epoxy equivalents of the 2N2369.

Clock driver for MOS shift registers

by Robert D. Hoose
and Gary L. Anderson
Trans-A-File Systems Co.
Cupertino, Calif.

Clock-driving circuitry can form one of the major expenses for serial MOS memories. The circuit shown here, however, has a large-quantity component cost of under $5. As the circuit drives 24 MOS registers, the cost per bit is less than $0.0002. The circuit operates at clock rates up to 1.5 MHz, and has been used with Signetics 1024-bit dynamic accumulators and Intel, Intersil and MIL 1024-bit serial shift registers.

Circuit operation is as follows: The input-phase clock is preshaped to a width of approximately 150 nsec. It is then steered by the Q output of the flip-flop into one DM8830 driver and by the Q̄ output into a second driver. The outputs of the drivers are capacitively coupled to an NH0025 dual clock driver which generates the 17V, 1.5A clock signal needed to drive the MOS circuits.

The portion of the circuit already described would be adequate for driving just a few MOS circuits. To drive as many as 24 circuits, additional descrete-component compensation circuitry is needed. Transistors Q_1 and Q_2 are turned on at two intervals during each cycle. When the ϕ_1 clock is ON (−12V), the ϕ_2 clock line is clamped to 5V through Q_2. Without the clamp, the cumulative parasitic coupling within the MOS devices can inject enough ϕ_1 noise into the ϕ_2 clock line to cause errors and loss of information in the dynamic accumulator.

As the ϕ_1 clock turns off (transition from −12 to 5V), Q_1 is turned on via C_1 to speed up the trailing edge of the output from the clock driver. A similar sequence applies in reverse when the ϕ_2 clock turns on. Then Q_1 clamps ϕ_1 to 5V, and Q_2 speeds up the trailing edge of ϕ_2.

The remaining components bias and drive transistors Q_1 and Q_2 (R_1 thru R_6, C_5, C_6), damp the MOS clock lines (R_7, R_8, R_9, R_{10}) and clamp the MOS clock lines (CR_1 thru CR_6, C_7, R_{11}).

Inexpensive clock driver can supply up to 24 MOS serial shift registers, thus adding under $0.0002/bit to the cost of a memory.

Level Detecting Flip Flop With Adjustable Hysteresis

By Hart Anway
General Electric Co. LMED
Utica, N.Y.

AT TIMES it may be desirable to provide a switching function at two preset levels. With the circuit shown in Fig. 1, R_2 and Q_1 determine the highest level (A in Fig. 2) while R_3 and Q_2 determine the lowest level (B in Fig. 2). With the values shown, it is possible to adjust either level anywhere between about $+8$ v and -10 v.

The input level adjustments operate as follows. Assume R_2, R_3 and R_{10} are set at $+5$ v, -5 v, and 0 v respectively as shown in Fig. 1. Also assume Q_4 conducting, Q_3 not conducting and a rising V_{in} waveform. Note, for example, that as V_{in} increases between -5 v and $+5$ v the voltage at a is maintained at the input voltage through the forward bias of CR_1. During this interval CR_3 and the Q_1 emitter-base junction are back-biased, since the voltage at a is more negative than the voltage at b.

Note that CR_3 is also back-biased during the voltage excursion between -5 v and $+5$ v. When the input voltage reaches $+5$ v (actually $5 + V_{be}$ of Q_1) current begins to flow from the $+12$ v supply through R_1, CR_3, and Q_1 emitter. This causes Q_3 base to be pulled positive, thus starting regenerative action, which changes the flip-flop state. The input voltage may go more positive than the trip point, being limited only by the PIV of CR_1 and CR_3.

After the waveform reverses and starts in the negative direction, it will pass through $+5$ v. When this happens the current through Q_1 will be cut off, but R_7 current will hold Q_3 on. Nothing further will happen until the input reaches -5 v, at which time CR_3 and Q_2 will conduct. Since R_7 will now act as the collector load for Q_2, R_7 current will be transferred from Q_3 to Q_2, thus turning Q_3 off. Regeneration then turns Q_4 on, reversing the flip-flop state. The input may go more negative than the negative trip point, being limited by the current rating through Q_2 emitter base junction, a function of V_{in}, R_3 and R_{12}. When V_{in} starts going positive, the cycle is repeated.

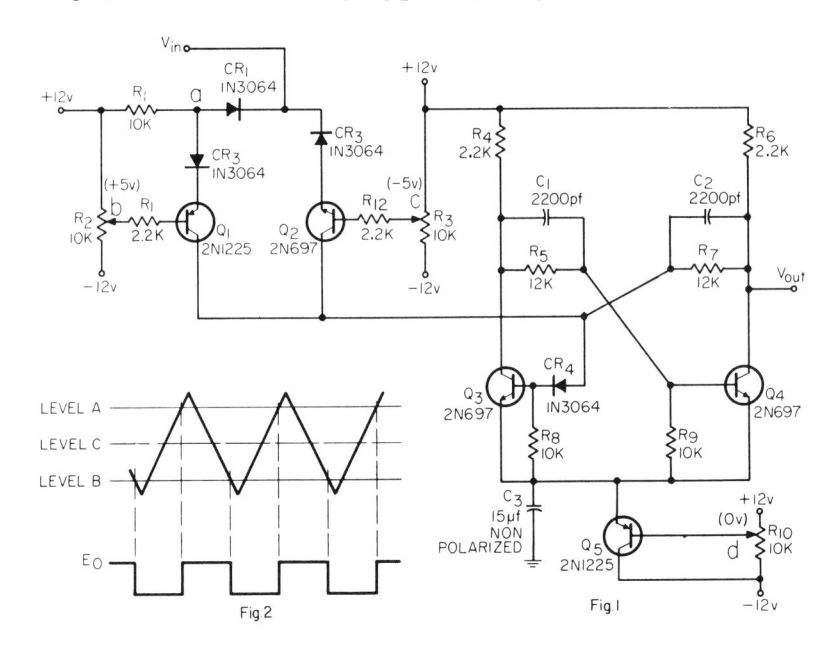

Fig. 1. Adjustable-hysteresis flip flop.

Fig. 2. Input and output waveforms.

The voltage at Q_5 emitter (C in Fig. 2) is not at all critical, but should nominally set to a value about midway between level A and level B. The criterion is that it must be set to allow Q_3 to be turned on and off with the drives from Q_1 and Q_2. It must be lower than Q_1 saturated collector by the junction drop of CR_4 and V_{be} of Q_3. This is the upper limit. The lower limit will be slightly higher than the lower trip point voltage on Q_2 collector, less than the voltage drop across CR_4 and V_{be} of Q_3.

To operate the circuit, adjust R_2 for the desired upper trip level, and R_3 for the desired lower trip level. Set Q_5 emitter to level C, about midway between the upper and lower levels.

Silicon unilateral switches detect initial event

Neal E. Pritchard,
ACDC Electronics, Oceanside, Calif.

This initial event indicator circuit is capable of monitoring several separate events simultaneously, detecting the first and storing this information. The number of separate events that can be monitored is determined by the number of stages used, which can be virtually unlimited. The circuit is especially useful in applications where the events are momentary, but greater than one microsecond.

The operating characteristics of the initial event indicator circuit is a result of the series—parallel configuration of the bistable SUS devices (D_1-D_4) in the two stages as shown below. More stages may be added. The necessary conditions for proper operation of the circuit are:

1. The applied voltage (V_{IN}) must be less than twice the switching voltage (V_S) of the bistable device (SUS) but still greater than V_S;
2. The switching voltage (V_S) must be greater than twice the forward voltage (V_F) of the SUS plus the forward voltage of the LED;
3. The IR product, namely holding current (I_H) times either resistor R_1 or R_2, must be greater than the switching voltage (V_S).

With the application of V_{IN}, assuming inputs 1 and 2 are zero volts, resistors R_3 through R_7 form a voltage divider network that determines the initial stable state of the circuit. In this state the voltage across the bistable devices is approximately $V_{IN}/2$, which is less than the required switching voltage; therefore, no current will flow in either LED.

In order to change the state of the circuit, a positive voltage (the initial event to be detected) must be applied to an input. Assuming event #1 occurs first, transistor Q_1 will saturate, increasing the voltage across D_1 until it switches to the ON state. The current through R_1 switches D_2 to the value V_F turning D_2 ON. The voltage across the input to the stages is now $2V_F + V_{LED}$, and the current through the LED is $[V_{IN} - (2V_F + V_{LED})] \div RI$. When input #1 returns to zero, the circuit still remains in this second stable state. With stage #1 ON, stage #2 cannot be switched ON. The circuit can be reset by momentarily removing V_{IN} or shorting C_1.

This circuit is especially useful in systems where a fault initiates a system shutdown, which in turn removes the fault condition. C_1 is used for noise filtering.

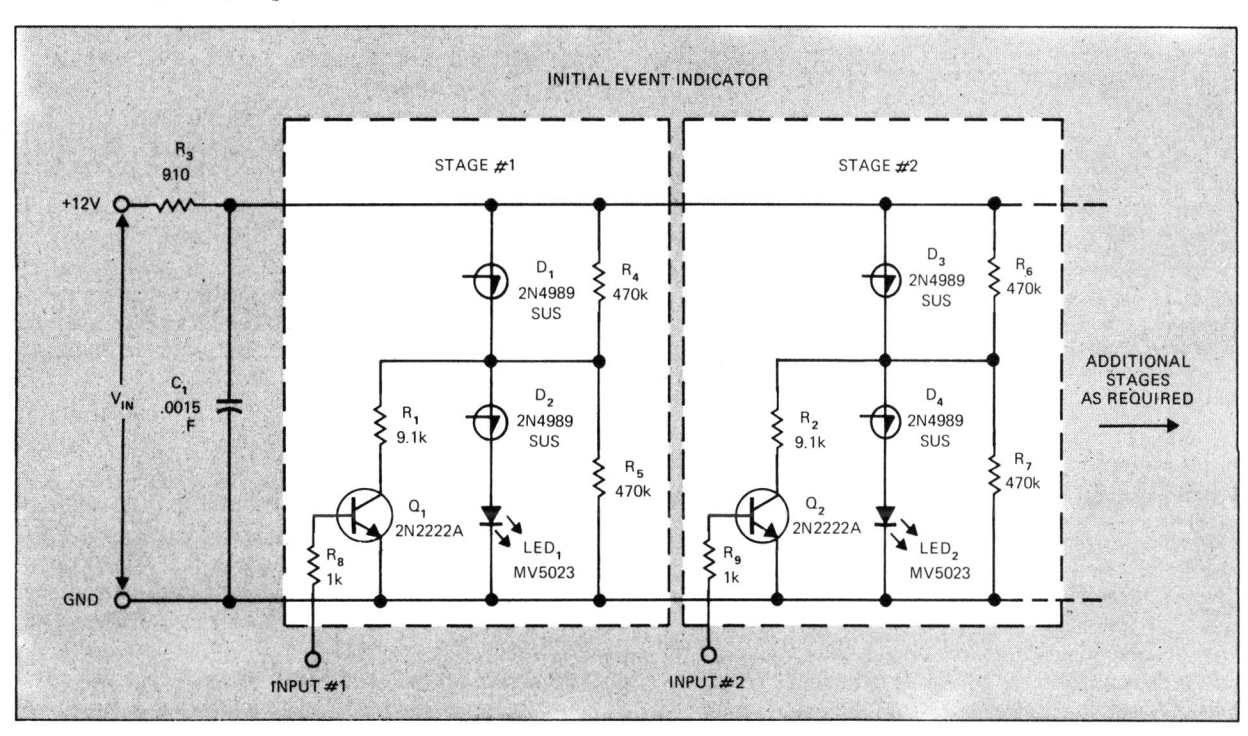

Silicon unilateral switches arranged in a series-parallel circuit will react to only the first input pulse.

Simple Pulse Phase-Splitter

By Gerald Wolff
Sylvania Electronic Systems, ARL
Waltham, Mass.

THIS PHASE-SPLITTER circuit provides bi-polar pulses which have the following characteristics: 180 deg out of phase; perfect coincidence of leading or trailing edges; approximately the same reference level; and a drive capability of saturated inverters.

Two possible connections of the circuit are shown in Figs. 1 and 2. Typical input and output pulse relationships are shown below the applicable circuit, with coincident edges of output pulse-pairs indicated by heavier lines.

Complementary transistors Q_1 and Q_2 are turned on and off simultaneously by pulses at A. Since transistor turn-on is more rapid and definite than turn-off, with its storage dependency, the circuit of Fig. 1 will produce pulses at B_1 and B_2 which are precisely timed by the positive-going edge of an input pulse at A. If precise timing by the negative-going edge is desired, the circuit of Fig. 2 would be used.

Design considerations are similar to those for a pulse inverter. It is desirable that Q_1 and Q_2 have similar switching characteristics and that their loads be balanced. For silicon transistors, saturated voltages (turn-on) at B_1 and B_2 are respectively about $+0.9$ v and $+0.5$ v in Fig. 1, -0.5 v and -0.9 v in Fig. 2.

A useful application of this circuit has been the switching of "sample-hold" gates requiring opposite-going pulses with trailing edge coincidence.

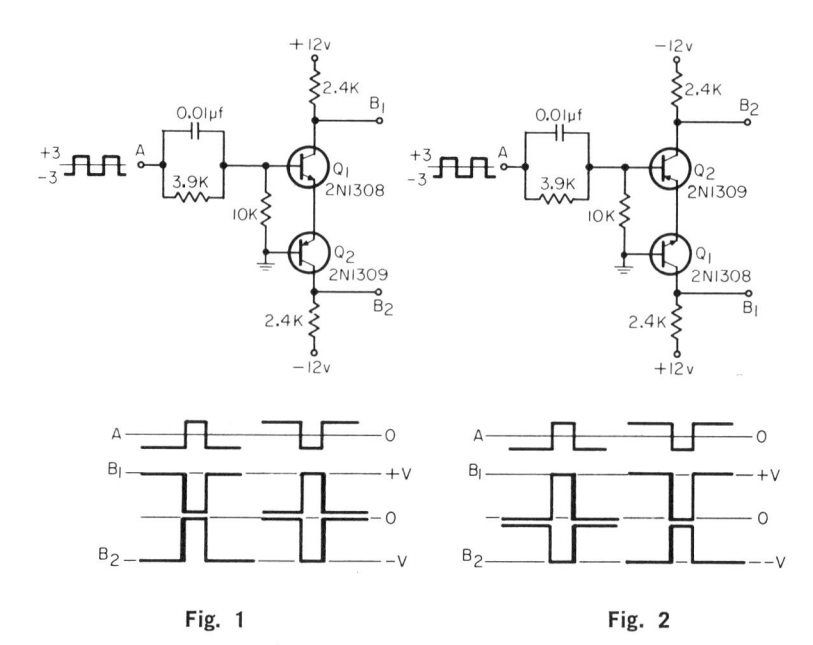

Fig. 1. Circuit for pulses coincident on positive-going edge.

Fig. 2. Circuit for pulses coincident on negative-going edge.

Retriggerable monostable

by Dragan Pantic
Denelcor
Denver, Colo.

THIS MONOSTABLE multivibrator, unlike standard designs, can be retriggered (extended) during a timing cycle. The circuit is also capable of long pulse widths (up to 60 seconds) and has a short reset time.

In the quiescent state, Q_1 is held on by R_1. A small negative voltage from the voltage divider formed by R_5, R_6, R_7, R_8 and R_9, is fed into the non-inverting input of A_1. This negative voltage drives the output of A_1 negative.

A negative trigger cuts Q_1 off and couples a positive pulse into the non-inverting input of A_1. This voltage drives the output of A_1 to a high positive voltage. This in turn is coupled back thru C2 to the non-inverting input. This positive feedback keeps A_1's output high even when the trigger pulse ends. If no other pulse is fed in, capacitor C_2 will charge from $+26$ V to -1 V. Charging time of C_2 is $C_2 (R_3 + R_5 + R_6)$. As soon as the non-inverting input of A_1 becomes slightly negative, the output of A_1 goes negative and the mono returns to its quiescent state. Capacitor C_2 discharges rapidly through diode CR_1.

If triggering pulses arrive while A_1's output is high, C_2 discharges to its original triggered state. This action initiates another timing cycle. Thus the output of the mono stays high as long as trigger pulses with a shorter interval than the mono's time constant are available. When triggering pulses stop, the mono goes through its normal timing cycle.

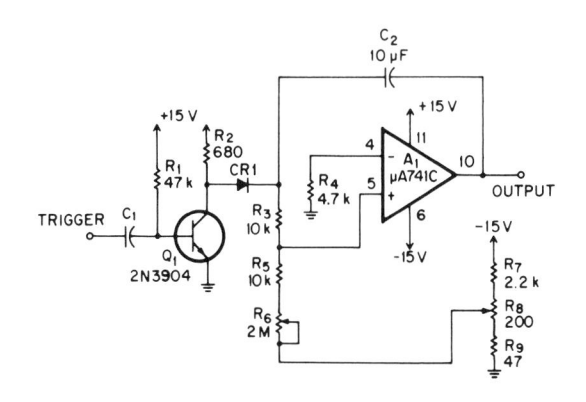

This monostable, built around an IC op amp, can be retriggered during a timing cycle. Short reset time and long timing periods are additional advantages.

High-speed, one-IC one-shot

by Alfred C. Falone
Hughes Aircraft Co.
Fullerton, Calif.

USING a single, DTL or TTL quad 2-input gate, the circuit in Fig. 1 generates a short-duration pulse that can be extremely useful as a strobe or reset signal. It's a modified set-reset latch that generates a pulse that lasts slightly longer than three propagation times, typically 40 to 70 ns.

A single pulse is generated each time a command is given, regardless of its duration. The width of the output pulse can be extended up to 500 ns by adding a capacitor, C_x, at terminal Q.

Fig. 1. (Top) An input command of any duration generates a single output pulse.
Fig. 2. Timing diagram shows pulse widths, which depend on propagation delays of the individual gates. Widths can be extended by adding a capacitor. Timing is expressed in propagation times.

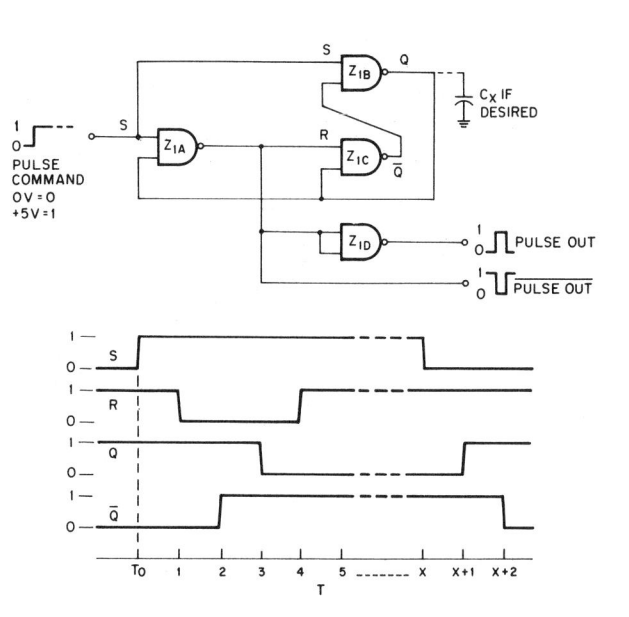

Circuit shifts asynchronous high-frequency clock sources

Scott C. Miller
Varian Data Machines, Irvine, CA

Switching between two asynchronous clock sources can be a hazardous event. The circuit shown reduces the hazard by allowing an asynchronous control signal to select between two asynchronous clock sources without producing narrow clocks or glitches and by minimizing the amount of clock time lost during a transfer.

In this circuit, there are three possible states and two possible stable states for flip-flops QA and QB (see Table). Assume that the circuit is in state 1, with the select control HIGH and clock source A enabled

State	QA	QB	Condition
0	0	0	State during moment of transfer (unstable)
1	1	0	Clock source A selected
2	0	1	Clock source B selected
3	1	1	Illegal state (unstable)

to the output. When the select control transitions to LOW, flip-flop QA will synchronously (with clock A) inhibit clock source A to the output and will release, or enable, the dc preset terminal on flip-flop QB. After flip-flop QA has released flip-flop QB, flip-flop QB will synchronously (with clock B) enable clock source B to the output. Flip-flop QB will now hold flip-flop QA in the reset state. The circuit is now in state 2 and will remain in state 2 until the select control goes HIGH again.

Flip-flops QA and QB are interlocked in such a

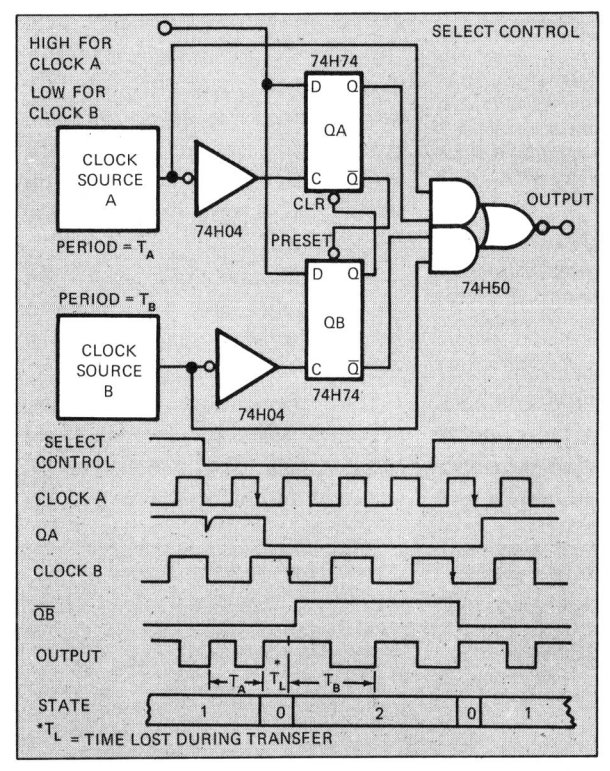

The "select control" signal determines which clock source is delivered to the output. This is done without either glitches or shortening of the output clock pulses.

manner that one flip-flop must complete a synchronous disable of its clock before the other can perform the synchronous enable of its clock. Returning to state 1 occurs in the same manner as above, establishing a clean, glitch-free transfer between two asynchronous clock sources. The average time lost during a transfer is about one-half clock period of the clock source being selected.

Digital and Pulse Circuits 83

Pulse-Train Detector and Counter

By Kirit Sheth
Honeywell
Waltham, Mass.

THE CIRCUIT shown in Fig. 1 detects the envelope of bursts of high-frequency pulses or, with a small modification, will give an output after a specified number of pulses have appeared at the input. A built-in threshold excludes noise and pulses of low amplitude.

This circuit can be used over a range of input frequencies from 1 kHz to 1 MHz with suitable component values. With the values shown, the upper limit on the input prf is 1 kHz. The input can of course be R-C coupled if desired.

Transistor Q_1 amplifies the signal from a magnetic pickup. This stage has built-in noise threshold provided by the clamp diode D_1. The collector of Q_1 goes below +5 V only after the input exceeds V_t (in this case 50 mV), enabling the current source Q_2 to charge the capacitor C. The discharge time constant of C is much larger than the input repetition rate, so V_c changes from −5 V to +2 V, thus turning on Q_3. With the given values, the circuit will operate for duty cycles greater than 0.2%. Turn-on delay for a 10% duty-cycle pulse train is given approximately by

$$T_d = nT = 0.06/(10,000 \; w/T - 17)$$
$$\simeq 0.06 \text{ msec}$$

where T_d is the turn-on delay, n the number of high frequency pulses, T the pulse-train period, and w the pulse width.

With a minor modification, the circuit can be used as a pulse counter. If the cathode of diode D_1 is returned to point "A" instead of +5 V supply as shown, the circuit "locks up" after a specified number of pulses have arrived. Capacitor C must be given time to discharge, or must be discharged manually, before the input is reapplied.

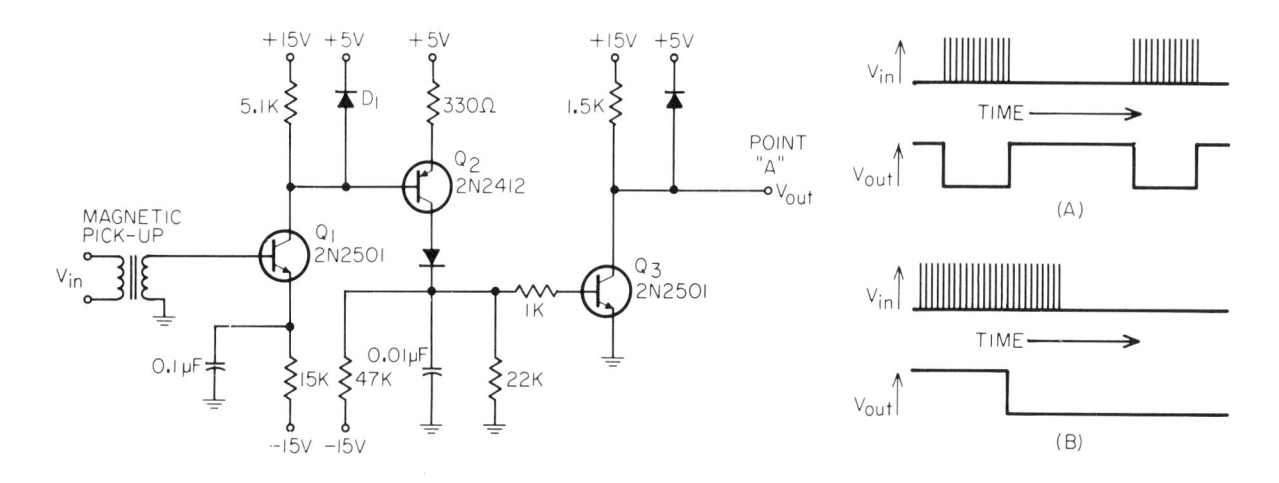

Fig. 1. Basic circuit is a pulse-train detector. If D_1 is returned to point "A", circuit will give an output after counting specified number of pulses.

Fig. 2. Input and output waveforms for detector (2) and counter (B).

Resettable one-shot with high noise immunity

by W. L. Lucas
Robertshaw Controls
Richmond, Virginia

IT'S OFTEN necessary to vary the firing time of a one-shot from a remote point. In conventional one-shot circuits, if an external timing resistor is more than a few inches from the card rack, random noise can fire the circuit. In the circuit shown here, noise won't fire the one-shot, even with the timing resistor several feet from the logic chassis. Noise immunity is equal to that of a flip-flop.

If one wants to reset this one-shot during its firing cycle, it is merely necessary to apply a logic 0 to the reset input.

If we assume the flip-flop, made of two NAND gates, is in its reset state, Q is at logic 0 and $\overline{Q}$ is at 1. The 1 from $\overline{Q}$ holds Q_1 on, which holds R_1, R_2 and C at ground, thus preventing Q_2 from firing. When a 0 pulse is applied to the trigger input, the flip-flop changes state and $\overline{Q}$ goes to 0 and Q changes to a 1. With $\overline{Q}$ at a 0, Q_1 is turned off.

C charges through R_1 and R_2. When C charges to a voltage set by the intrinsic standoff ratio of the unijunction transistor, Q_2, it fires. A voltage is developed across R_3, turning Q_3 on for approximately 500 μs. This brings the Q output of NAND 1 to ground, thus resetting the flip-flop. The Q returns to a 0 and $\overline{Q}$ returns to a 1. The length of firing time may be computed from $T = C(R_1 + R_2)$.

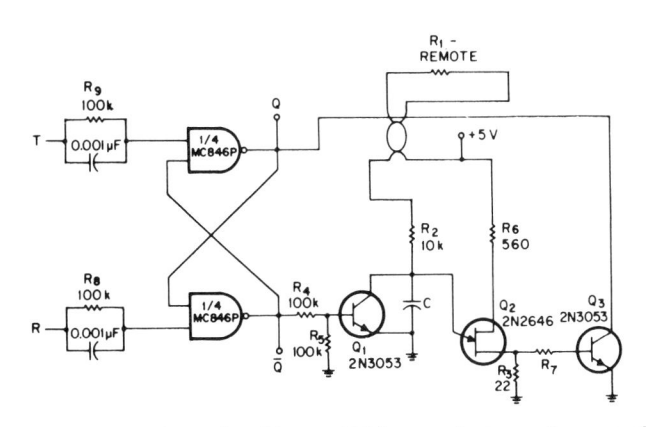

The timing resistor for this resettable one-shot can be several feet away from the circuit without allowing noise to fire the circuit.

84 Circuit Design Idea Handbook

Three-State Indicator

By Paul E. Dingwell
Hughes Aircraft Co.
Culver City, Calif.

SOMETIMES it's desirable to indicate more than two circuit conditions using a conventional lamp. This indicator circuit gives three modes; on, off and blinking.

If switches S_1 and S_2 are both open, transistors Q_1 and Q_2 form part of an astable multivibrator. With the component values shown, the lamp blinks once every second.

If S_1 is open and S_2 is closed, Q_2 turns off allowing Q_3 to conduct. Thus the lamp stays on.

With S_1 closed and S_2 open, Q_2 conducts cutting off Q_3. Thus the lamp is extinguished.

The switches can be replaced by transistors or relays, depending on the application.

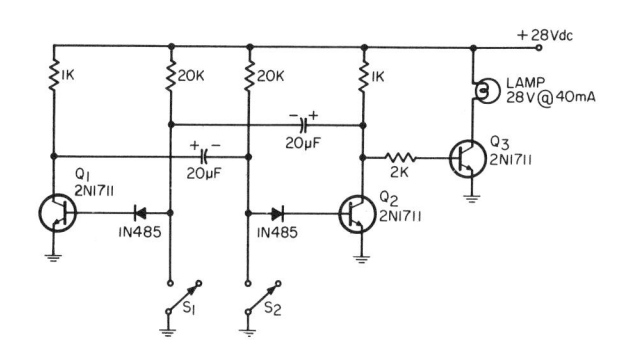

With both switches open, this indicator circuit is a free-running multivibrator.

Double-ended limit detector senses voltage with VCO

Bjorn Brandstedt,
McDonnell Douglas Corp., St. Charles, Mo.

System designs frequently call for a circuit that will indicate when a voltage goes outside some preset limits. The circuit shown in **Fig. 1** does this by monitoring the output pulse rate of a voltage controlled multivibrator to thus implement a double-ended limit detector. It consists of a 2-bit shift register, two one-shots, an inverter and two NAND gates.

The circuit compares the period of the input pulses to preset max. and min. limits. If the repetition rate of the pulses falls outside these limits then the output of the detector produces a negative going spike. In the circuit diagram, point A is normally HIGH and goes LOW coincident with the leading edge of the first input pulse. It returns to HIGH at a set time, as determined by the external RC network (R_1C_1), signifying the end of the minimum allowed period.

Point B is normally LOW; it goes HIGH coincident with the first input pulse. Its return to LOW signifies the end of the maximum allowed period.

Point C is normally LOW and goes HIGH coincident with the rising edge of the second input pulse. C goes HIGH to strobe the output NAND gate and, following a 10 nsec delay, to reset the entire circuit. The output is determined by the logic of the two NAND gates. The truth table is:

A 0 1 0 1 0 1 0 1
B 0 0 1 1 0 0 1 1
C 0 0 0 0 1 1 1 1
Y X 1 X 1 X 0 0 1

Output Y is normally HIGH and goes LOW coincident with point C if the input pulse rate is outside the set limits. The "don't cares" in the table (X) indicate combinations that cannot occur. The waveforms on the illustration explain the action which is continuous.

Output Y goes LOW whenever the input pulse rate is outside the set rate, which is determined by the R_1C_1 and R_2C_2 time constants.

Transfer parallel information without a clock

Tim O'Toole
Tektronix, Inc., Beaverton, Oregon

The design goal of this device is to take BCD from a 10-key adding machine type keyboard and shift each number, as it is entered, from left to right on the display panel. In the summation, G_8 is the first to be clocked, then G_7, then G_6, then finally G_5. The information in these latches is transferred in a parallel mode.

The secondary goal is economy and the least number of ICs possible. No internal clock was to be used in this design and there are no ac coupling devices.

The BCD is direct from the keyboard decoder, which is not debounced. The keyboard strobe is delayed 2 msec to allow time for the keyboard switches to quiet down. On the positive edge of the keyboard strobe, G_{2D} and G_{1D} send out a 250 nsec negative pulse. On the negative edge of this 250 nsec pulse G_{3A} and G_{4D} generate a 70 nsec pulse to clock G_8. On the positive edge of the 250 nsec pulse from G_{1D}, G_{2E} G_{2F} and G_{4C} send out a 70 nsec pulse to clock G_7.

On the negative edge of the keyboard strobe, G_{7A}, G_{2A}, and G_{1B} send out a 250 nsec negative pulse. On the negative edge of the pulse from G_{1D}, G_{2D} and G_{4B} combine to send out a 70 nsec pulse to clock G_6. On the positive edge of this 250 nsec pulse from G_{1B}, G_{2B} and G_{4A} combine to send a 70 nsec negative pulse to clock G_5.

The BCD outputs from G_{5-8} go directly to 7 or 10 segment decoder-drivers, such as SN7447 decoders driving RCA DR-2100 series low-voltage readouts.

All resistors in this circuit are 220Ω (±1%). Capacitors for G_{1B} and G_{1D} are 1000 pF (±5%), capacitors for G_4A, B, C and D are all 240 pF (±5%)

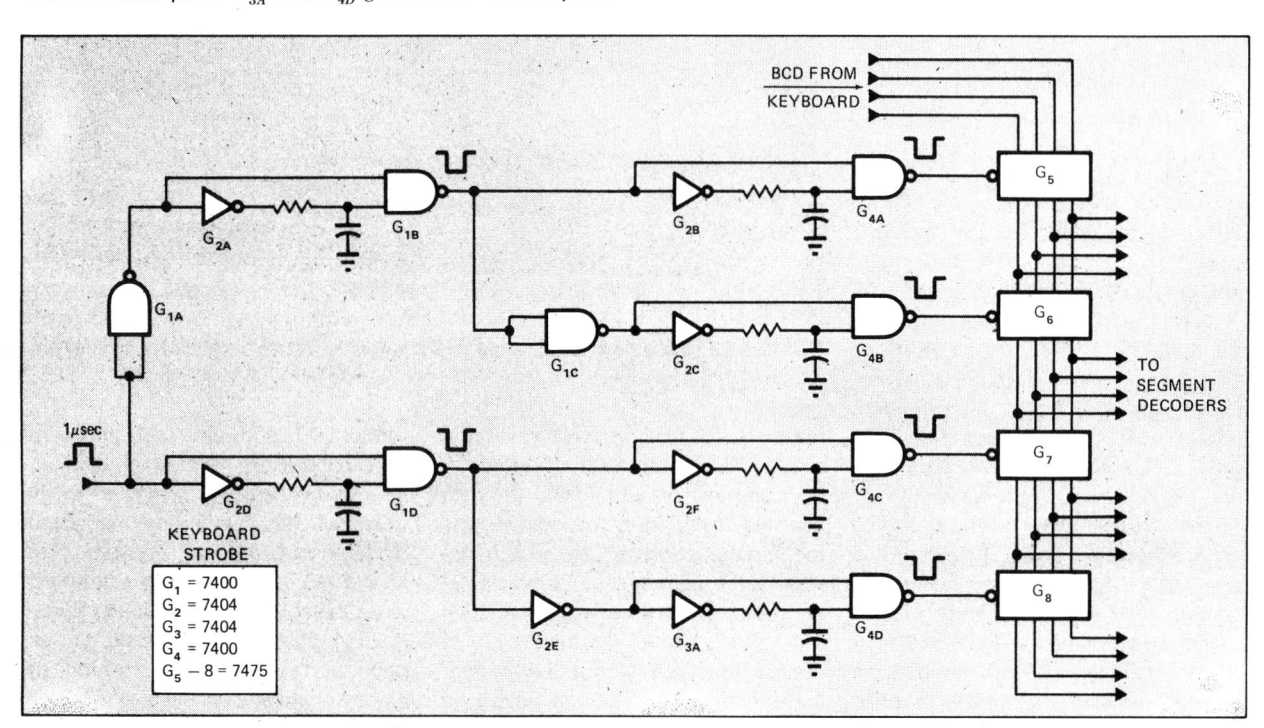

Inexpensive solution to calculator problem allows display of each digit, as entered on the keyboard, to be shifted one digit to the left without the use of a system clock circuit.

Constant-current generator speeds up wired-OR circuits

Peter Alfke
Fairchild Semiconductor Mountain View, Calif.

Open collector outputs allow easy expansion of digital integrated circuits, particularly memories, by facilitating a common bus output interconnection from many devices and allowing the activation of only one device at a time.

The output transistor of the activated device can pull the common bus LOW and a single pull-up resistor can pull it HIGH. For high-speed operation, the value of this resistor should be as low as is consistent with the fan-out (current sinking) capability of each IC, typically 16 mA as shown in Circuit "A".

While a resistor is the most economical pull-up device, it has two drawbacks: The current available to charge the stray and load capacitances decreases as the output rises; and the bus voltage rises exponentially to V_{cc}. This is much higher than necessary to insure proper TTL noise margins.

The ideal pull-up device would be a constant-current generator plus a clamp preventing the output voltage from exceeding +2.4V, as in circuit "B". The circuit shown in "C" implements this concept. Q_1 acts as a constant-current generator (16 mA) and Q_2 acts as a clamp when the output voltage reaches 2.4V. With a light capacitive load of 100 pF the response time is only slightly improved; but with a 500-pF load, rise time is decreased by 11 nsec and the fall time by 24 nsec (measured as a decreased delay at the output of the read gate), which is a significant improvement for the cost of three additional resistors and two PNP transistors.

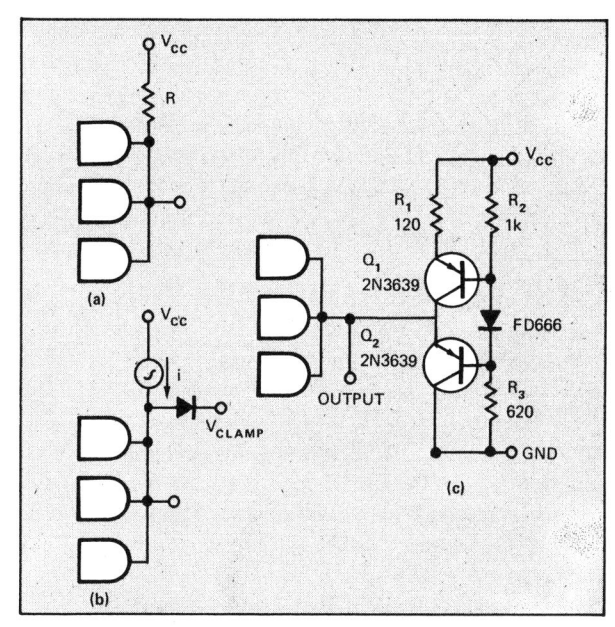

Open collector gates in a wired-OR configuration normally use a pull-up resistor, as in circuit A above, bringing the output bus to V_{cc}. Since the required input to subsequent stages is only +2.4V, the clamped constant-current scheme in circuit B will provide faster switching. Circuit C is a practical equivalent of circuit B, providing an 11 nsec improvement of risetime when driving a 500-pF load.

Non-Inverting Pulse Amplifier Uses One Power Supply

By Richard L. Sazpansky
Honeywell Inc.
St. Petersburg, Fla.

THIS PULSE AMPLIFIER increases the amplitude of 1-pps pulses from +12 v to +28 v and also decreases rise and fall times. See Fig. 1. This circuit should also work well up to 50 meg pps with component value adjustments.

Q_1 is initially held off due to the saturation of the previous stage. R_6 provides a leakage path to ground for I_{cbo} and prevents turn-on (in the absence of a pulse input) of Q_1 even at elevated temperatures. R_1 is a current-limiting resistor and C_1 is selected to minimize storage (t_s) and fall (t_f) times. A positive-going pulse impressed at the base of Q_1 turns Q_1 on and the collector of Q_1 drops from +35 v to approximately zero. This negative-going pulse is applied to the base of the pnp transistor Q_2. Q_2, which is initially held off due to the +35 v on its base, now turns on and the output at its collector (point A) is again inverted to a positive-going pulse. R_2 and R_3 are selected to give the required base drive to turn on Q_2 with a signal applied at the input.

Rise and fall times, at the output, are lowered considerably because of the speed-up capacitor C_1 and the complementary scheme of Q_1 and Q_2. Q_1, due to the presence of C_1, provides an overdriven (spiked) pulse at the trailing edge of its output. This pulse (at the base of Q_2) turns off Q_2 harder and provides a sink for the stored charge in the base of Q_2. Thus, the output pulse's trailing edge (fall time) is approved.

A 5-K potentiometer may be substituted for R_5 to give an adjustable pulse output from 0 to +28 v. An emitter-follower can be added at point A if a low impedance drive is required.

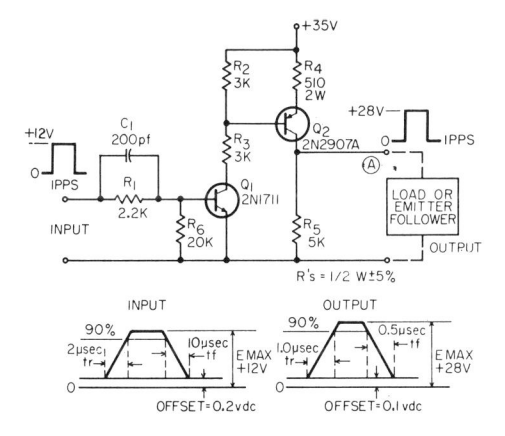

Fig. 1. Single power supply pulse amplifier.

Fig. 2. Input and output waveforms.

Poor man's LED driver is TTL compatible

Walter G. Jung
Forest Hill, MD

A simplified, unique and quite economical LED driver can be built by taking advantage of the internal current limiting and ability to voltage clamp an op amp of the 101 family.

The circuit uses an LM301A as an open-loop voltage comparator, A_1, with LED D_1 connected to receive the total source current from A_1. For positive outputs of A_1, the low forward impedance of D_1 will cause A_1's internal current-limit stage to conduct and deliver the full $I_{sc}+$ to D_1. For negative output swings of A_1, D_2 clamps the output to $-0.5V$, preventing reverse breakdown of D_1.

The circuit realizes the full available open-loop speed of A_1, since the amplifier is uncompensated and the internal voltage amplification stages are kept out of saturation by the clamping of D_2 and the current-limiting action. Therefore, D_1 may be toggled with response times in the μsec range.

The approximately 20 mA of current available for D_1 is compatible not only with the MV50 but also with a number of other LED types, including the popular LED-phototransistor isolators. The circuit, as shown, uses a TTL-compatible input, with the R_1-R_2 reference divider biasing A_1 in the center of the TTL output transition region. Other input configurations are possible, of course, as long as the input common-mode limits of A_1 are observed.

The circuit offers two-to-three orders of magnitude, better voltage and current sensitivity at the switching threshold compared with a simple transistor saturated switch. It also has the bonus of a temperature-stable threshold level for voltage monitoring applications.

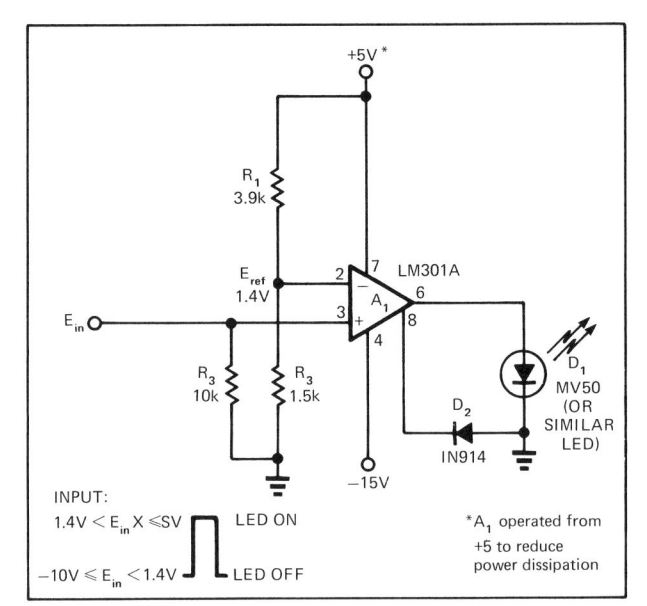

Microsecond toggling of LEDs is possible with this LED driver. The circuit is TTL compatible, and the output is current limited.

Low-pass digital filter

Thomas H. Haydon
B.F. Goodrich, Troy, OH

An SN7400 quad 2-input NAND gate, an SN7431 dual 4-input Schmitt trigger, two 1N457 diodes and two capacitors are connected as shown to form a low-pass digital filter. This arrangement is useful in retrieving a pulse train from "outside world" noise generated by contact bounce, arcing, SCR spikes and inductive surges.

A Schmitt trigger, diode and capacitor comprise a pulse-delay circuit with delay on the positive-going transition. One gate of the SN7400 is used as an inverter to drive a second pulse delay operating on the negative-going transition of the input signal. The delayed pulse trains toggle a flip-flop to generate an output frequency equal to the input, but phase delayed.

Any additional pulses occurring during the delay circuit time-out will reset the delay time with no toggle action at the output. This limits throughput frequency to $1/(2 \times \text{delay time})$. A higher frequency will latch the flip-flop, resulting in a zero output frequency. Rolloff is extremely sharp with cutoff frequency variation depending mostly on the diode thermal coefficient.

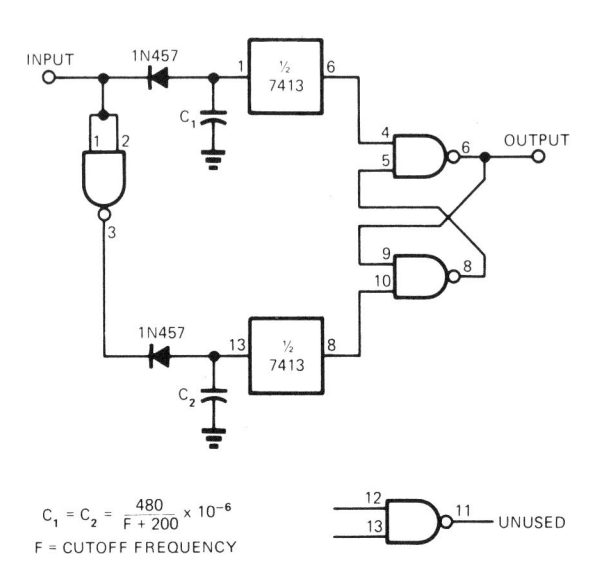

$$C_1 = C_2 = \frac{480}{F+200} \times 10^{-6}$$
F = CUTOFF FREQUENCY

This low-pass digital filter, using only two TTL ICs, can retrieve pulse train data from very noisy signal lines.

Circuit remembers random data within periodic field

by **Stephen Kreinik**
CBS Laboratories Stamford, Conn.

This simple digital circuit can detect and remember the presence of randomly-distributed information occurring within a periodic field. For example, it can produce an output as long as information appears once per field or once per line on a TV raster.

The circuit (**Fig. 1**) can be used directly in digital systems, and, with suitable modification, it can also be used in analog systems. For analog applications, the input information must first be converted into a suitable digital signal. For example, a threshold discriminator could be used to detect the presence of a signal above a certain value.

Data to be remembered by the circuit of **Fig. 1** must be in the form of a negative-true logic signal. This signal (A) is applied directly to flip-flop FF_1. A second input signal (P) occurs once per field (in the original application it was the vertical blanking pulse) and is applied to FF_2 and to a one-shot circuit (SS_1) which produces a delayed pulse (DP).

When A is true (OV), the output (X) goes to logic-1 (+4V) and remains there as long as A occurs once in each field. X will go to logic-O within one field after A does not occur. The maximum allowable repetition rates for A and P are constrained by the propagation delay through the logic circuitry plus the pulse widths of DP, P and A.

The trailing edge of pulse P generates a delayed pulse, DP, at the output of SS_1. When the A input contains a pulse within the field defined by P, flip-flop FF_1 is set, producing a low output at point B and, therefore, setting FF_2.

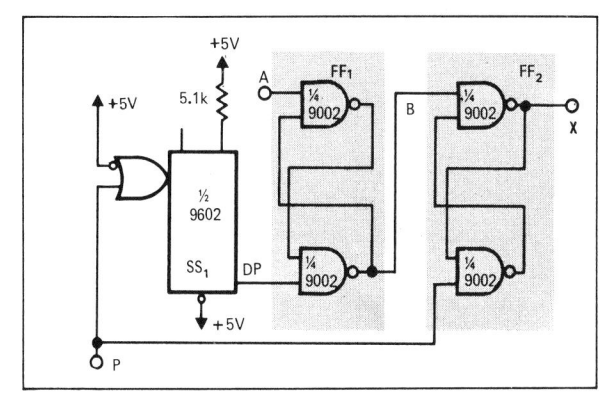

Fig. 1 – **Digital circuit** produces an output at X provided an input pulse appears at point A sometime between each two consecutive pulses at point P.

This, in turn, produces a high output at X.

When another P pulse occurs, X would return to its low value if it were not inhibited by flip-flop input B, which is still low. Therefore, X stays high.

After P has returned to its high values, DP resets FF_1 and returns B to a high value. Since P is high, X remains high. The presence of A in the next field ensures that X will remain high. If A is removed from the field, FF_1 is not set, and the next P pulse resets X to logic-O where it remains until another A input occurs.

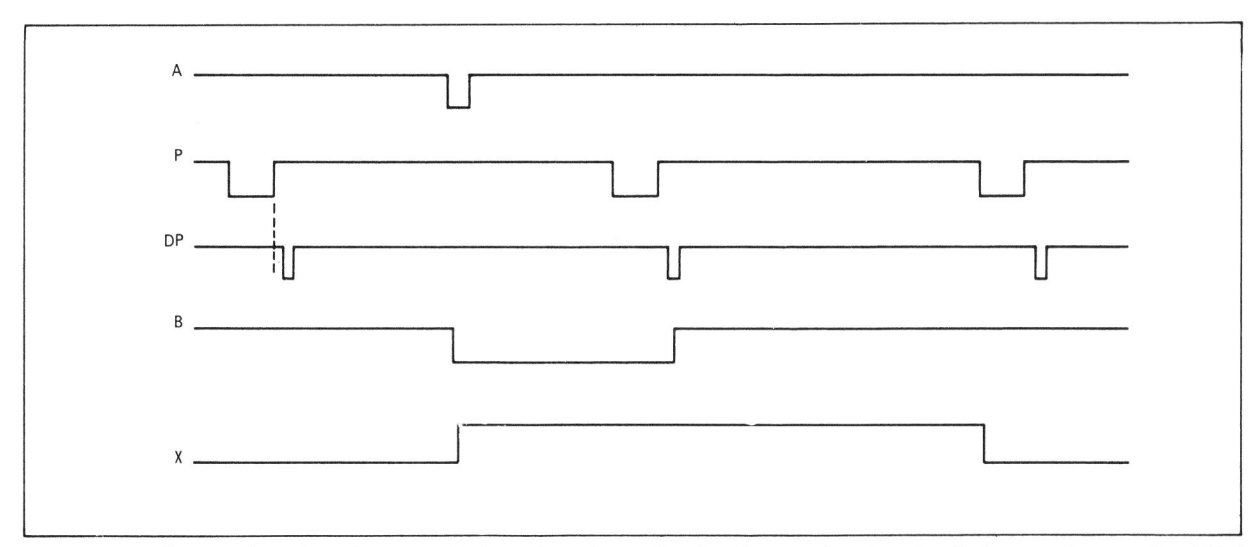

Fig. 2 – **Timing diagram** shows how the output, X, depends on the two inputs, A and P. X goes high when an A pulse appears between P pulses. Because there is no A pulse between the second and third P pulses, the third P pulse returns X to its low state.

Digital and Pulse Circuits 89

Circuit triggers one-shot on both edges of square wave

John P. Yang,
Interdata, Inc., Oceanport, N. J.

The 9602 retriggerable monostable multivibrator can be triggered either on the leading edge or on the falling edge of a square wave but not on both. This simple circuit can trigger the 9602 on both edges. It uses only two resistors and one capacitor. If a large amount of double-edge-triggering circuits are used in a system, the circuit will save 50% of 9602 multivibratiors and cut the cost in half.

The circuit works as follows: If the input is HIGH (+3.3V), capacitor C_1 has +3.3V on its B side and +2.7V on A side. It stores a charge Q = CV = 0.47 μF × (3.3V-2.7V) = 0.282μ coulomb. When the input goes LOW the B side of C_1 becomes 0V and the A side becomes −0.6V. This negative-going pulse triggers the 9602 which delivers one output pulse.

As long as the input stays LOW, capacitor C_1 will be charged to +2.7V on the A side through resistor R_1. When the input goes HIGH again, this high-going pulse is delivered to pin 12 of the 9602, thus triggering the 9602 which produces another pulse. On the rising edge of the input, the A side of C_1 jumps to +6V and then discharges to +2.7V through resistor R_3. This has no effect on the 9602.

Capacitor C_1 should be non-polarized. Its minimum value is 0.15 μF.

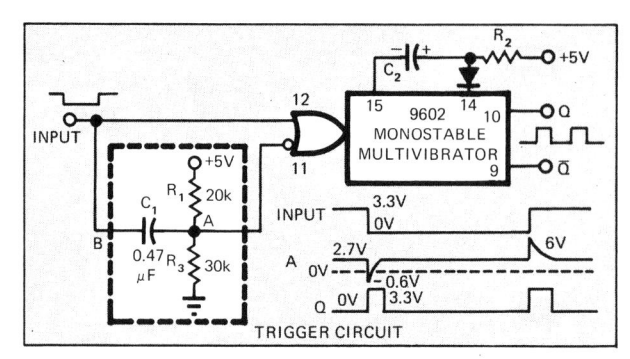

Suitable trigger pulses are delivered to the 9602 multivibrator on both the leading and trailing edges of the input square wave.

Three IC's accurately sense pulse rate

John W. Poore
General Dynamics Pomona, Calif.

Many applications arise for accurate pulse rate detection where extreme accuracy, such as that of a frequency counter, is not required. The circuit shown indicates when the input pulse rate is above (or below) a set point. The output is high for frequencies above the set point, and low for frequencies below the set point. Set point stability was measured over a 0°F to 200°F range using a mylar capacitor for C_2, and a set point of 10 kHz. The maximum shift in set point frequency over temperature was less than 0.7%. Set point frequency is the reciprocal of the one-shot delay time, i.e. $f_o = 1/(0.32 R_2 C_2)$. R_1, C_1, $G_1(A)$ and $G_1(B)$ generate a spike at the rising edge of the input signal. The one-shot (FF_2) delay time is affected by pulse width at its input terminal; adding the spike generator makes the circuit virtually insensitive to the input signal on-time or off-time. The rising edge of the input signal causes a spike at FF_2's and $G_1(D)$'s input. The leading edge of the pulse from the one-shot sets $FF_3(A)$. If another rising edge of the input signal (and thus another spike) occurs during the one-shot delay time, $G_1(D)$ gates a reset pulse to $FF_3(A)$. The trailing edge of the pulse from the one-shot clocks $FF_3(B)$ thus sampling $FF_3(A)$'s output. This provides a stable output. Two of these circuits can be used to bracket pulse rates.

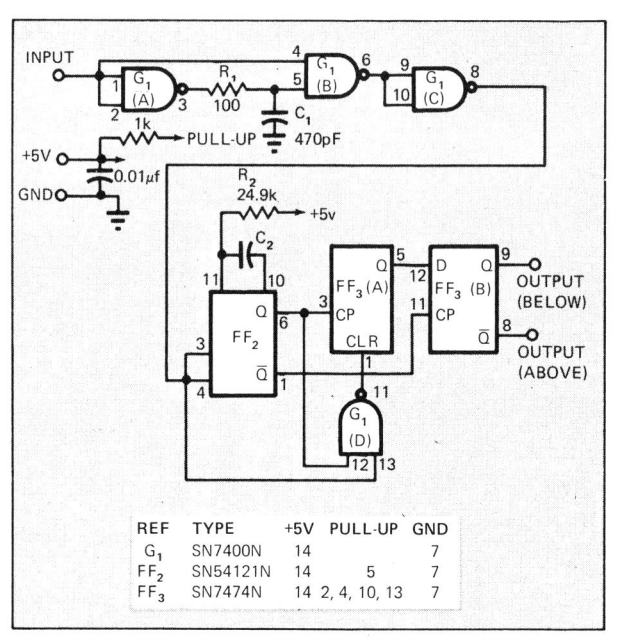

REF	TYPE	+5V	PULL-UP	GND
G_1	SN7400N	14		7
FF_2	SN54121N	14	5	7
FF_3	SN7474N	14	2, 4, 10, 13	7

Pulse rate of the input signal is accurately sensed using only 3 ICs. Two such circuits, one set to the upper limit and one to the lower can be used for constant frequency monitoring.

Frequency Comparator

By Robert Ricks
Fairchild Semiconductor
Mountain View, Calif.

THE CIRCUIT shown in in Fig. 1 can be used as a control circuit for VCOs, as a go-no-go frequency comparator, or as a frequency discriminator. There are two inputs to the circuit, a standard frequency and an unknown frequency (labeled *TACH* in the diagram). If the unknown frequency is less than the standard, then the output dc level is low. If the unknown frequency is higher than the reference frequency, then the output dc level is high.

When the two input frequencies are identical, the circuit behaves as a linear phase discriminator. Unlike frequency discriminators that use tuned circuits, this digital discriminator has no humps in its characteristic curve.

The circuit shown in Fig. 1 was originally part of a larger circuit which included oscillators and VCOs. The dual NOR gate IC_6 is not really a part of the frequency comparator circuit, so for the purpose of this analysis it can be ignored. In the original application, IC_6 gave the correct levels, rise time and fanout for driving the comparator circuit.

Assume that the inputs are at points S and T in the diagram. Assume also that the first incoming pulse is on the S line. Then, this pulse sets the A output of IC_2 to "low" and simultaneously sets the B output to "high." (Note that the gates of IC_2 and IC_5 are cross-coupled to form flip-flops.)

The next S pulse propagates through gate B of IC_4 to set the B output of IC_5 to "low," with the A output "high." These outputs are connected to Q_1 and Q_2. Therefore Q_1 collector goes "low" and Q_2 goes "high."

Because Q_2 collector is "high" it disables the B gates of IC_1 and IC_3, allowing no change for further incoming S pulses.

Now, when the first T pulse arrives after a series of S pulses, it propagates through gate A of IC_3 to set gates B and A of IC_2, "low" and "high" respectively. This drives the collector of Q_3 "low," thus enabling gates A and B of IC_4. After this occurs, the next T pulse will then propagate through gate A of IC_4 to set gate A of IC_5 "high." As the circuit condition has not been reversed, any number of T pulses in a row, without an additional S pulse, will cause no further changes in output level. This will be seen more clearly, if the reader compares Fig. 2 with the above description.

The output can be selected to be of either polarity, for a given frequency relationship, by selecting the appropriate output of IC_5.

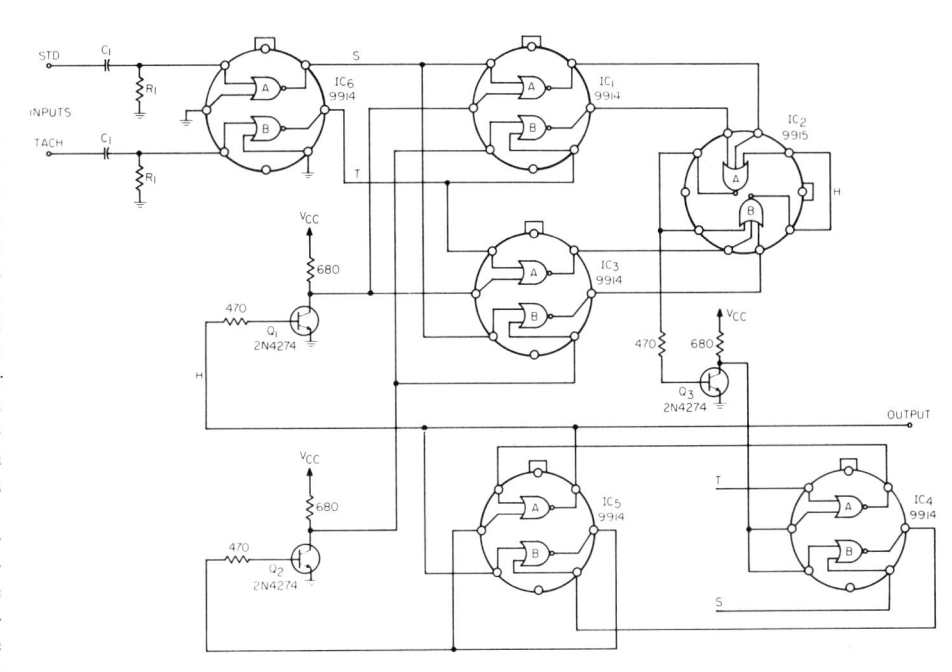

Fig. 1. Digital frequency comparator uses no tuned circuits.

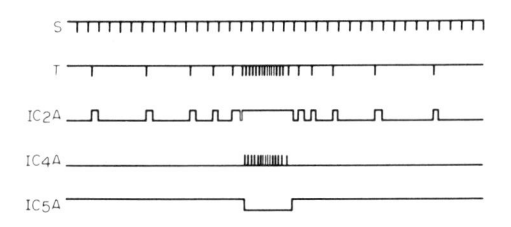

Fig. 2. Timing diagram shows the effects of T frequencies higher and lower than the S frequency.

Photo detectors can drive digital circuits directly

Dennis Berde
Grumman Aerospace, Bethpage, NY

Photo detectors, such as photodiodes and phototransistors, can deliver only small "light currents" into their respective load impedances. This current is usually in the range of a few microamperes, except for slow, high-gain devices such as photo-Darlingtons. As a result, photo detectors cannot drive logic circuitry directly. A current-gain stage, an operation amplifier or a transistor stage is therefore normally used to interface these devices with logic circuit inputs.

CMOS digital-logic families overcome this difficulty. The building blocks of these families are insulated-gate field-effect transistors. Consequently, CMOS gates have typical input impedances in the order of $10^{12}\Omega$. With such high input impedances the current-drive limitation of photo detectors is no longer a problem and direct interface is possible.

Such a circuit is shown in the illustration and its operation is as follows: under dark current conditions (no light input to the photo detector) resistor R keeps the input to the CD4050 CMOS gate at a logical ONE. When light is incident upon the photo detector, photo current I_P flows. Since all of I_P has to flow through R, due to the high input impedance of the CMOS device, a voltage drop is developed across the resistor causing a logical ZERO at the input to the CD4050. Typical values for R are 100 kΩ to 10 MΩ, depending upon the amount of incident light and the particular photo detector used. The CD4050 will drive two TTL loads directly if V_{cc} is 4.5V.

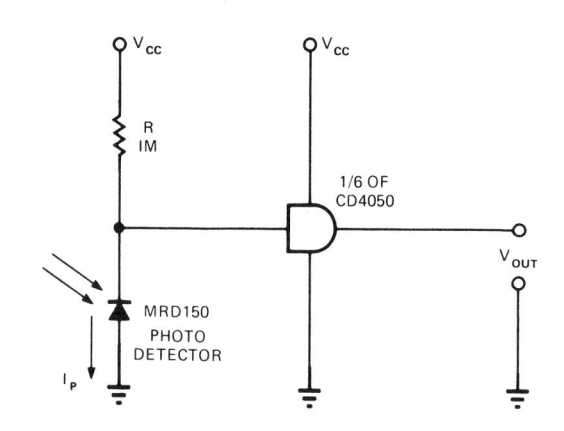

The very high input impedance of the CMOS gate makes it possible for the small photo detector "light current" to drive the gate.

Single-voltage circuit generates "power-on" reset pulse

Entry by Robert C. Snyder
GDI Inc., Melbourne, Fla.

A simple IC operating from the common +5V logic supply can be used to provide a "power-on reset" to the rest of the logic system. The output reset time is easily set with a single capacitor, C1. Fast Recovery is provided by the addition of two components (R2 and CR1).

Upon initial power supply turn-on, capacitor C1 is charged at a rate determined primarily by the input current from the SN 7413 Schmitt Trigger, and to a small degree by R1. CR1 prevents R2 from contributing to the charge current.

During the charge time, the output of gate G1 will be high and will not sink any current. When C1 is charged to 1.5V, G1's output will go low and terminate the power-on reset. R1 continues to charge C1 until it reaches the supply voltage, thus keeping G1 out of the active region and reducing its noise susceptibility.

When the power supply is turned off, CR1 forward biases and allows C1 to be discharged through R2 at virtually the same rate as the power supply.

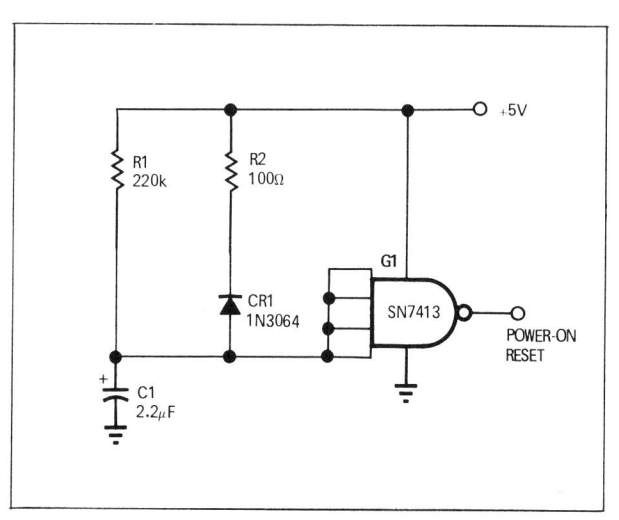

One Schmitt trigger and three discrete components insure the correct initial state of logic circuits when power is applied. The output reset time is set with capacitor C1.

92 Circuit Design Idea Handbook

Three ICs monitor pulse width

Joseph Kish, Jr.
Diebold, Inc., Canton, OH

Very often in pulse circuits, it is necessary to monitor pulse width in terms of "greater than" but "less than" given specified values. The circuit described here measures both limits constantly, and a fault indication will latch until manually reset.

The 74121 is set to coincide with the minimum pulse width spec. Should the pulse under observation fall to ground before the one-shot recovers, the J-K is clocked and the "less than" SCR is triggered.

The 8601 is set to coincide with the maximum pulse width spec. Should the pulse under observation fall to ground after the one-shot recovers, the second half of the J-K is clocked and the "greater than" SCR is triggered.

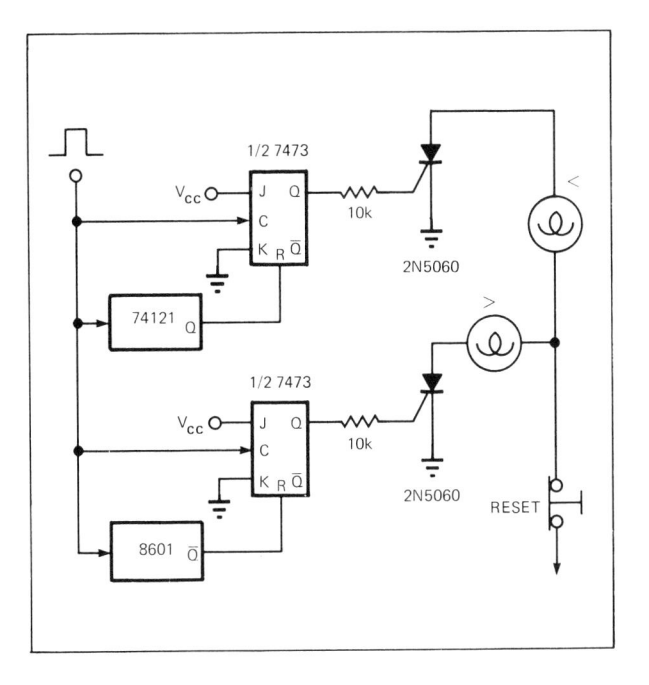

Variable-Threshold Hybrid One-Shot

By David V. Dickey
Litton Systems
Woodland Hills, Calif.

THIS ONE-SHOT is designed to take a narrow negative-going input and produce a negative-going output at least as wide as the input. The input turn-on threshold is determined by the value of R_1 (assuming $R_3 \gg R_1$). Q_2 will start to conduct when V_{B1} is about 0v due to the voltage-divider effect of R_1 in parallel with R_3, and the emitter resistor R_E.

The one-shot pulse width is determined by the RC time constant of R_3 and C, and the turn-on voltage swing of V_{out}. The integrated circuits G_1 and G_2 demonstrate the compatibility of the one-shot for use with integrated input and output circuits. Figure 2 shows test results using discrete components.

The following equations may be used to calculate component values:

$$T_{on} \approx \frac{R_3 (C - 20 \text{ pf})}{5}$$

For $T_{on} \approx 200$ nsec,

$$R_3 = 5.1 \text{ K}$$

$$V_{thr} = R_1 \left(\frac{V_{EE} - V_{BE}}{R_E} - \frac{V_{BB}}{R_3} \right) - V_{BC}$$

For $V_{in} = V_{thr} = 1.8$ v,

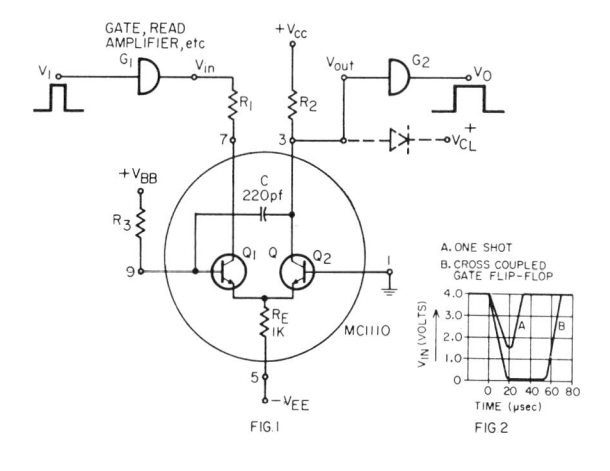

$$R_1 = 560 \text{ ohms}$$

Fig. 1. Variable threshold one-shot uses integrated circuits and discrete components.
Fig. 2. Input threshold waveforms.

$$R_2 \| R_{in (G2)} \geqslant R_E$$

For $R_E \approx 2.2$ K,

$$R_2 = 2 \text{ K}$$

$$V_{CC} = V_{BB} = +6 \text{ v}$$

$$-V_{EE} = -6 \text{ v}.$$

Divider circuit maintains pulse symmetry

Leslie A. Mann,
Radiation Inc., Melbourne, Fla.

It is often necessary, when generating clocks in a digital system, to divide the basic clock frequency by an odd number. In doing so, the 50/50 duty cycle possessed by the basic oscillator is lost. The circuit described here allows the designer to maintain a 50/50 duty cycle by the addition of only a few gates and 2 flip-flops.

The example shown in **Fig. 1** was designed as a divide-by-five counter to generate a 1-MHz, 50/50 duty cycle clock from the 5-MHz system reference. Two phases of a 1-MHz clock are generated as shown in **Fig. 2**. One phase is a decode of binary 4 from the counter while the other is a decode of 1, clocked at mid-bit time. The two phases are then recombined through gate G1 to give a 2-MHz clock. This clock is used to toggle FF-2, thus generating the desired 1-MHz, 50/50 duty cycle output.

By selecting the proper decodes off the counter, this same method may be used to generate a symmetrical square wave output through any odd number of divisions from the basic clock frequency. The only requirement is that this basic oscillator have a 50/50 duty cycle.

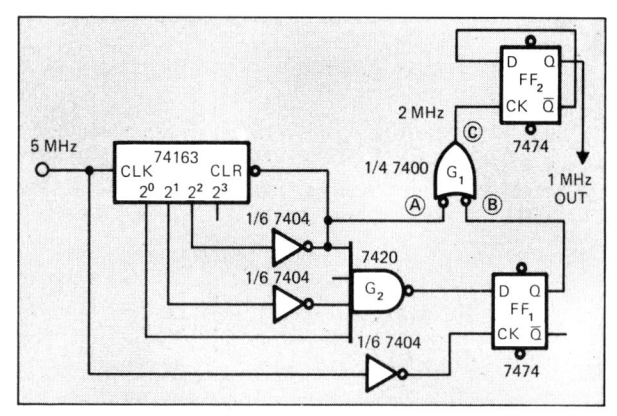

Fig. 1—**Pulse symmetry**, usually lost in odd modulo dividers, is maintained in this ÷5 circuit by first decoding through G_1, G_2 and FF_1 for a ÷2.5 count. This assymetrical pulse train is then divided by 2 at FF_2 for symmetry.

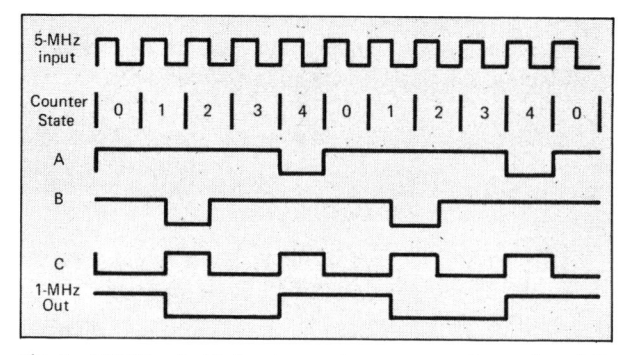

Fig. 2—**NAND gate** G_1 inputs are shown as waveforms A and B. Output (C) is then processed through FF_2. Similar techniques can be found for maintaining pulse symmetry in most odd-modulo dividers.

Single IC compares frequencies and phase

James Breese
Ampex Computer Products, Marina Del Ray, Calif.

A universal shift register, such as the 5495/7495 shown here can be connected to yield a frequency and phase determined signal as follows:

For: $f_1 > f_2$ Output = "1"
      $f_1 < f_2$ Output = "0"
      $f_1 = f_2$ Output is a square wave, the duty cycle of which varies linearly with the phase re-relationship between f_1 and f_2.

This configuration has several advantages over multiplier-type phase comparators. The carrier frequency can vary from dc to 25 MHz with no adjustment of reactive components; there are no tradeoffs of response time and acquisition range (the range is unlimited) and the frequency and phase comparisons are virtually instantaneous (requiring only two carrier cycles, worst case, for comparison).

Operation is as follows: Input f_1 shifts "1"s toward the right, and input f_2 shifts "0"s toward the left. The state of any given binary depends on its input (shift right f_1; or shift left, f_2) and the states of its neighbors. Consequently the output of binary C, for example, will be "1" if "shift-right" commands are coming along more often then "shift-left" commands. If f_1 exactly equals f_2, then one of the binaries will be toggled at the carrier rate. A closed phase locked loop, which this comparator is especially suited for, acts to ensure that the binary used for feedback, either B or C, will toggle, with output A = 1 and output D = 0.

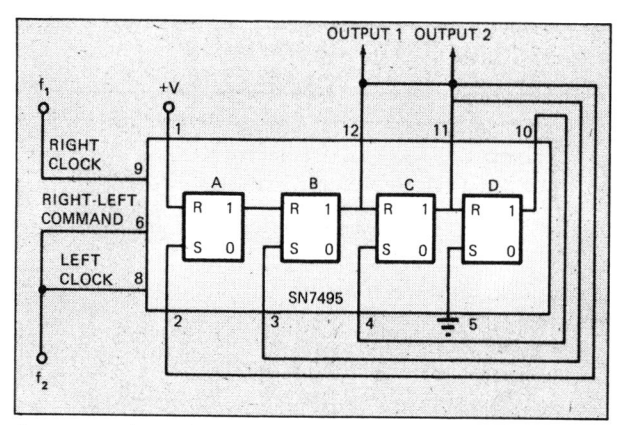

Frequency-phase comparator operates from dc to 25 MHz. Unlike multiplier-type comparators it requires no adjustment of reactive components.

Pulse-width discriminator

by Ira Spector
Sylvania Electronic Defense Labs.
Mountain View, Calif.

THE CIRCUIT described here provides an output pulse whenever the input pulse width is either less than a minimum value, or greater than a maximum value. The circuit will also function as a pulse-width error detector by providing error pulses of width equal to the amount that the input pulse width deviates from the allowable limits.

In the block diagram of Fig. 1, a pulse width T is applied to point A. The leading edge of the pulse is delayed by a delay gate (½ SG83) to avoid racing problems with the leading edge of the pulse generated by the Maximum-Pulse-Width One-Shot. This non-critical delay is usually 10 or 20% of the nominal input pulse width. The triggering pulse which drives the two one-shots is derived from the leading edge of the input pulse. The left half of Fig. 2 shows the maximum-pulse-width waveforms. Note that a pulse will appear at point F if T is greater than T_{max}, the maximum allowable pulse width. The width of this pulse at point F will equal $T - T_{max}$.

The minimum-pulse-width waveforms are shown at the right in Fig. 2. The second delay gate (½ SG83) is used to avoid leading-edge racing problems at points D and E. Note that if T is less than T_{min}, the minimum allowable pulse width, then a pulse will appear at point G. The width of this pulse will equal $T_{min} - T$.

The last NOR-gate provides a general indication any time the input pulse width is outside the test window.

An example of a circuit developed for measuring 2-ms pulse widths within a test window of ±10% is illustrated in Fig. 3. The maximum allowable pulse width is calibrated simply by applying 2.2-ms pulses at point A using a precision pulse generator and trimming R_2 to adjust the Maximum-Pulse-Width One-Shot until a pulse just appears at point H. The minimum allowable pulse width is similarly calibrated by applying 1.8-ms pulses at point A and trimming R_6 to adjust the Minimum-Pulse-Width One-Shot until a pulse appears at point H.

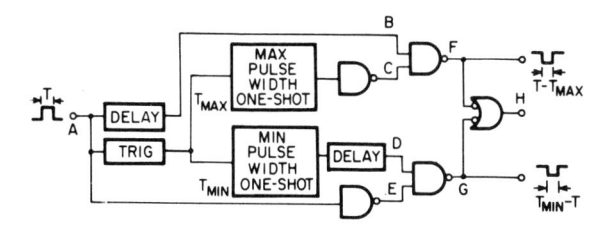

Fig. 1. Pulse-width discriminator in block-diagram form.

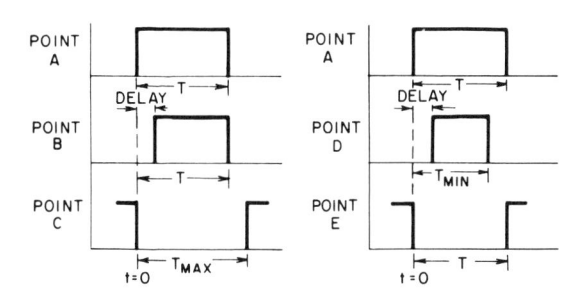

Fig. 2 Waveforms for maximum (left) and minimum (right) pulse widths.

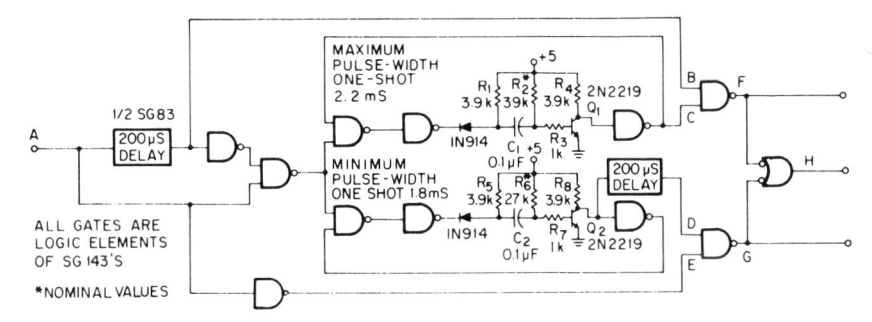

Fig. 3. Circuit for measuring 2-ms pulses with a ±10% window.

One transistor provides ECL to LED interface

George A. Altemose,
General Instrument Corp., Hicksville, N.Y.

This circuit utilizes a germanium PNP transistor to drive an LED. Only the normal ECL −5.2V dc power supply is required.

Operation of the circuit is as follows: when A' is high, A' = 0.7V and A = −1.5V. The transistor base is reverse biased, and the LED is off. When A is high, A = −0.7V, and the base is forward biased. The base current is limited by the 4.3k resistor to about 1 mA. The transitor saturates and the LED turns on. The transistor is germanium in order to insure saturation. Some types of ECL, such as Motorola's MECL II, have internal output pulldown resistors; in these cases, the 4.3k resistor is not required.

This circuit is particularly advantageous in systems consisting exclusively of ECL logic, where the addition of an ECL to saturated logic level translator would require the addition of a +5V dc power supply.

In addition to single LED's, seven-segment or other arrays can also be driven. These arrays must have the common-cathode configuration such as the Fairchild FND10, or the Monsanto MAN3. Another application is optoelectronic coupling, using a device such as Motorola's new MOC 1000 module.

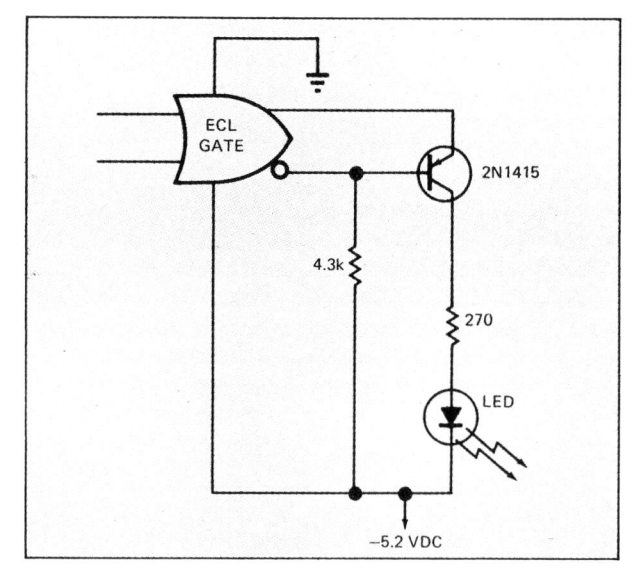

Germanium PNP transistor provides direct interface of LEDs and ECL logic systems without a second power supply. This system can be expanded to drive common-cathode LED arrays such as the MAN3 or FND10.

Time-Delayed Schmitt Sensor

By L. T. Medveson
American Photocopy Equipment Co.
Evanston, Ill.

THE SCHMITT TRIGGER CIRCUIT is widely used for industrial control sensing because the differential can be designed to be any suitable amount. However, industrial sensing circuits also often require a delay in the sensor control at start-up, until such time as the system and process are in normal operating mode.

In the example shown here, a photocell sensor command is nullified until certain fluid lines have been filled, to prevent the sensor's construing lack of fluid to be a lack of concentration.

The circuit shows a way of incorporating time delay in the Schmitt circuit itself through the use of another transistor as a switch, one that closes the bias circuit of the Schmitt. The delay is obtained with an RC network in the base of this transistor, which is designed so that the saturation current of the transistor is equal to the required bias current of the Schmitt. The output transistor is thus held inactive for the time required to charge the timing capacitor to approximately design voltage for its base.

The circuit uses economy-line transistors in a Schmitt configuration giving 0.3-V differential. With a photocell sensor, the value of R_1 will depend on the signal level when at the level for control. This sensor is connected as shown, or is interchanged with R_1 depending on the polarity of control required from the output.

For a 22-μf tantalum as C_T, the time delay is about 15 sec. The value of R_T must be small enough to allow a saturation current equal to the required bias current for the Schmitt. If less delay is wanted, then R_T should be made smaller, but for more delay the capacitor should be increased, not R_T.

The time-delay scheme can be incorporated in any existent Schmitt circuit. However, the saturation current of Q_3 must equal the existent bias current of the Schmitt, and R_X should be reduced according to what the saturation voltage drop actually is. The maximum R_T is set by the saturation current required, and so is proportional to the current gain of Q_3.

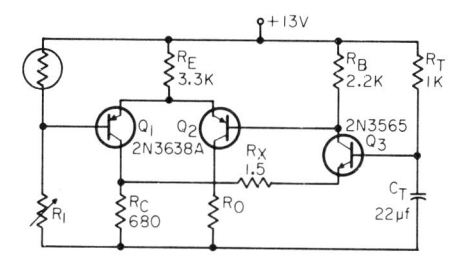

Time-delayed Schmitt trigger with photocell as sensor.

96 Circuit Design Idea Handbook

Pulse-Width Discriminator

By Phillip Cutler
Universal Electronic Controls
Garden Grove, Calif.

THE SIMPLE pulse-width discriminator shown in Fig. 1 gives linear changes in output voltage for changes in input duty-cycle. The rectangular input signal can vary in amplitude without affecting the output, because Q_1 and Q_2 are operated in the switching mode. This discriminator was used in a self-balancing servo loop in which R_2 was a potentiometer geared to a servomotor:

The time constant R_gC_c must be large relative to the period of the input waveform. This gives constant-current base drive to the "On" transistor. The current must be sufficient to cause saturation. The base of the "Off" transistor is returned to ground and reverse-biased by the emitter junction of the "On" transistor.

Since the transistors operate in the switching mode, the output voltage will swing between fixed positive and negative voltages, with the average value dependent on the duty cycle as shown in Fig. 2. The dc output voltage is given by

$$E_o = V_{oc}\left[1 - 2\,\frac{t}{T}\right]$$

where t/T is the duty cycle and V_{oc} is the open-circuit voltage seen looking into the output terminals.

$$V_{oc} = \frac{R_2\,(V_{cc})}{R_1 + 2\,R_2}$$

Neglecting the filter network, the impedance seen looking into the output terminals is simply R_2 in parallel with the series combination of R_1 and R_2.

Offset, due to unequal saturation voltages, is not a problem when the detector is used in a low-gain loop. For high-gain applications, offset can be minimized by operating the transistors in the inverted mode. The circuit will require increased drive in this mode.

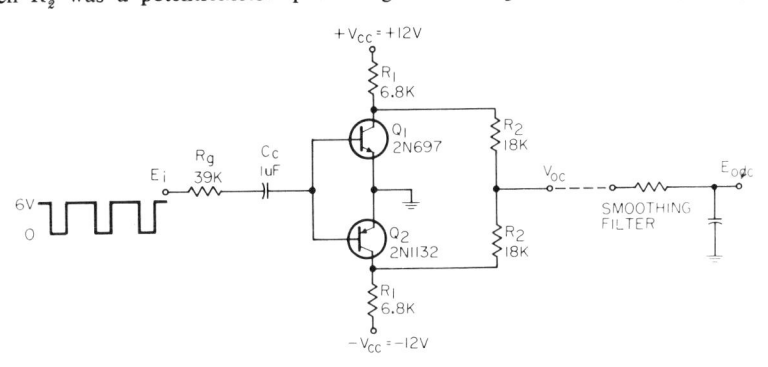

Fig. 1. Duty-cycle of input signal determines the output voltage of this discriminator.

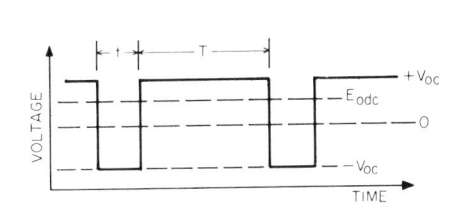

Fig. 2. Output waveform of discriminator is filtered to give the average level E_{odc}.

Digital comparator is self-adjusting

Robert A. Scher
Aerosonic Corp., Clearwater, Fla.

Magnitude comparison of two 4-bit binary coded decimal numbers will normally yield a "greater than," "less than," or "equal to" decision. But greater or less by how much?

This circuit will calculate the exact difference between two 4-bit code lines X and Y. For this case X is programmed with a BCD thumbwheel switch. Y is a count preset into the 74192 up-down counter.

A 7485, 4-bit magnitude comparator is used to compare X and Y. When the NAND gates are enabled, the comparator outputs on pins 5, 6 and 7 enable either NAND gate 1 or NAND gate 2. This enables the clock to the up or down terminal of the counter. The counter continues to count until its output (Y) equals the thumbwheel count (X).

A separate register can be used to store the up or down counts necessary to reach equality.

The "equal" signal from pin 6 of the 7485 then allows the next operation to be enabled, since the difference computation has been completed.

Comparators and counters may be cascaded indefinitely to accomodate any number of BCD digits desired.

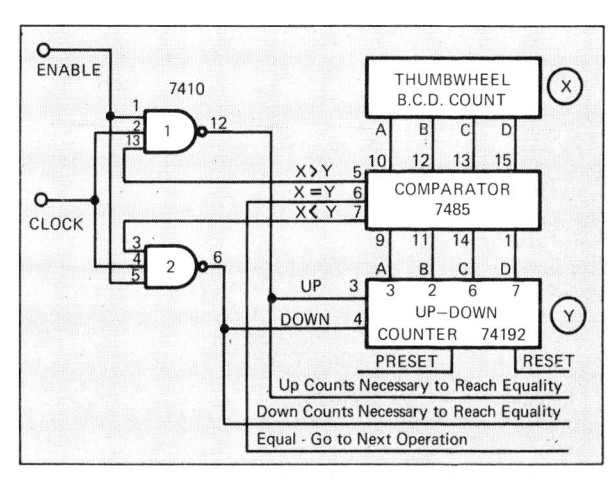

BCD comparator provides "less than," "equal to" or "greater than" comparison between thumbwheel setting and input digit, as usual. However, if equality does not exist, the circuit will then count up or down until equality is reached, thus calculating the difference between compared digits.

CMOS and LPTTL gates make low-power Schmitt trigger

Roger Cox
Hewlett-Packard Co., Loveland, Colo.

One half of a low power NAND package, and one CMOS inverter make a low-power Schmitt trigger, for applications where the power consumption of a 7413 IC Schmitt trigger is acceptable.

The two NAND gates form an $\overline{SR}$ flip-flop. Q goes high when Vin > 2.1V, the CMOS threshold $\overline{Q}$ will not go low until Vin < 1.2, the low-power TTL threshold.

Both polarities of the output signal are available. The power savings of this circuit over standard TTL methods is quite impressive, as shown in Table I.

Table I

Power Comparisons: (for 2 Schmitt triggers)

STANDARD TTL	LOW POWER TTL/CMOS
1 SN7413 32mA	1 SN74L00 1.4mA
	1 CD4001AE .5mA
	(for resistors)
Worst case current: 32mA	1.9mA
Worst case power: 160mW	9.5mW

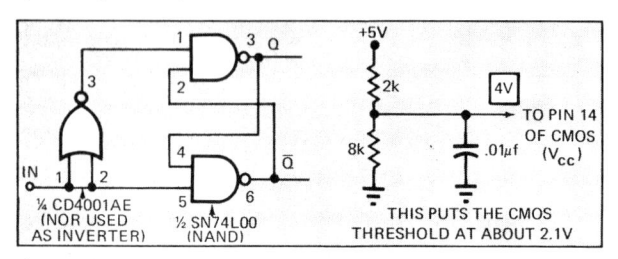

Fig. 1—**One CMOS and 2 LPTTL gates** are all that are needed for a very low power Schmitt trigger. Threshold for the CMOS gate is dependent upon its Vcc. Using 4V for V_{cc} as shown, will result in a threshold of approx. 2.1V.

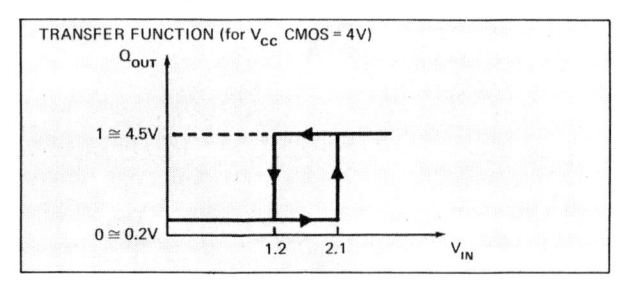

Fig. 2—**Switching thresholds** for the low-power Schmitt circuit, when Vcc for the CMOS gates is 4V, provide about 900 mV of hysterises. The 1.2V threshold is fixed, but the upper threshold can be varied from 2.1 to 2.5V by increasing CMOS V_{cc}.

Clock pulse generator has addressable output

Bjorn Branstedt,
McDonnell-Douglas Corp., St. Louis, MO

This clock pulse generator (**Fig. 1**) provides an expandable 3-bit binary output and four overlapping clock pulses for each of the unique binary outputs. The A-output of the binary counter is used along with the clock input to form a 4-phase overlapping clock function. The B, C and D outputs are used to address multiplexers, ROMs and other units. For each selected address, the four clock pulses are provided to control the data selected during that time. The timing diagram (**Fig. 2**) shows the sequence of the output pulses.

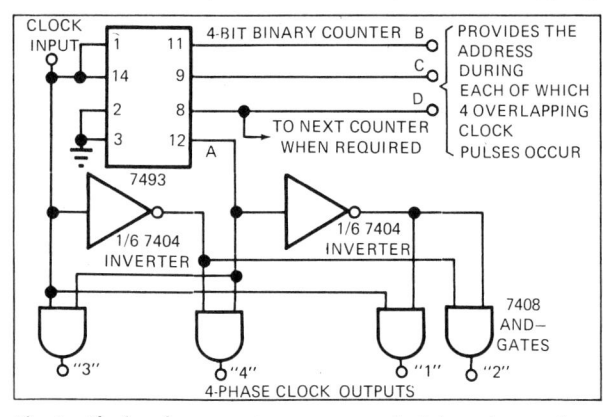

Fig. 1—**Clock pulse generator** converts a clock input into 4-phase clock outputs.

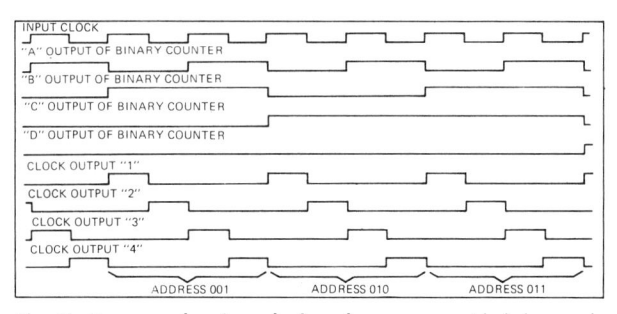

Fig. 2—**Four overlapping clock pulses** are provided for each unique binary address output from the binary counter (outputs B, C and D).

SECTION III
Signal Sources

Triangular and square wave generator has wide range

R.S. Burwen,
Analog Devices, Norwood, MA

This oscillator circuit delivers ± 13V square waves and ± 10V triangular waves simultaneously. The values shown are for operation at 100 Hz. By simply scaling R_1, R_2 and C_1, a wide range of frequencies can be covered, even to below 0.1 Hz. The square-wave rise time is typically 1.5 μsec and the fall time 0.5 μsec.

The oscillator consists of an integrator, A_1, feeding a second amplifier, A_2, used as a comparator. Regenerative feedback through R_4 converts A_2 into a Schmitt trigger having ±10V hysteresis levels at the input. When the output of A_2 is pos., A_1 integrates in the neg. direction toward −10V. At this point, the input to A_2 reverses polarity, causing a neg. output from A_2. A_1 integrates in the pos. direction toward +10V, where the polarity reverses again.

The frequency is somewhat determined by the saturation voltages of A_2 and by the power supply voltages. The pos. supply has little effect, and it requires a 20% change in the neg. supply to produce a 1% frequency change. By using stable components, a frequency stability of 0.02%/°C is readily attainable. Capacitor C_1 is preferably a polycarbonate type for temperature stability. It is possible

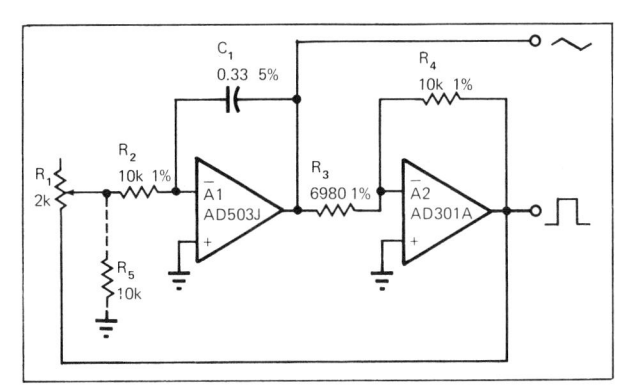

A_1 **operates as an integrator** and A_2 as a Schmitt trigger in this triangular- and square-wave generator.

to change the duty factor by feeding a bias current to the input of A_1 and, in fact, this is a means of producing pulse-width modulation.

Operation at 0.1 Hz can be produced by changing R_2 to 10 MΩ. Tuning can be accomplished by potentiometer R_1 to as low as 2 kΩ by loading the arm with a 10 kΩ resistance, R_5. Best performance with such high resistance values is attained by using a lower-input-current operational amplifier for A_1, such as the AD503K, which is rated at 5 pA max. at 25°C. Alternatively, lower values of resistance can be used if C_1 is increased to produce a 3.3 sec time constant.

Inexpensive UJT-SCR Intervalometer

By E. L. Dewig
University of California
Medical Center
Los Angeles, Calif.

OFTEN A CIRCUIT is required that operates at the end of a pre-determined period for a second pre-determined period (unlike the time delay relay, which merely switches on or off). A monostable multivibrator will do this, but for time periods more than about 4 sec the timing capacitor becomes bulky, since it is not feasible circuit-wise to have the timing resistor large in transistor MV circuits.

A transistor MV, for 5 sec, requires about 500 μf. The circuit shown, however, works with small values of timing capacitor, here 3 μf, low standby current (5 mA) and with the values of R_t shown, gives time periods from below 5 to near 10 sec. Longer periods are easily obtainable.

Circuit operation is based on the familiar SCR with commutator capacitor for switching "other off" connected anode-to-anode. When power is first applied, SCR_2 is switched to conduction by the voltage-divider action of the relay coil through the 2.2-K resistors to ground. The voltage at the gate is about 0.75 V. The unijunction base 1 is at nearly the same potential as base 2, since point A is near supply potential. Thus, the unijunction does not conduct.

When a trigger pulse is fed to the SCR_1 gate, it is turned on and commutates SCR_2 off. The unijunction base-1 circuitry is now near ground potential. SCR_2 loses its 0.75-V gate signal since point A is near ground. The unijunction now functions as a familiar pulse generator.

The first pulse triggers SCR_2, which commutates SCR_1 off, and the circuit is once again in standby, awaiting the next signal pulse. Other uses can be obtained by replacing the relay coil with a different type of load. Total cost of the SCRs and UJT shown is $4.95.

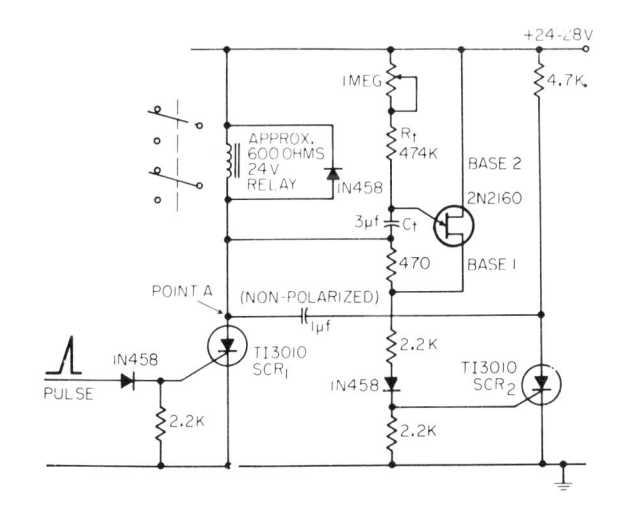

Low-cost intervalometer using unijunction transistor and two SCRs.

Linear bidirectional ramp generator

by Richard W. Currell
Newell Research
Saratoga, Calif.

A simple bidirectional linear ramp generator can be constructed using a dual op amp (such as the µA747) and a few external components. The circuits described here were originally designed to vary the reference voltage for a dc motor servo, and thus control acceleration and deceleration rates.

In the basic circuit (**Fig. 1**), ramp rate is determined by R_2 and C_1. Potentiometer R_1 determines the ramp amplitude by controlling the magnitude of V_{in}.

If R_1 is adjusted so that V_{in} is ±10V (for SW_1 in positions 1 and 3), then the circuit will behave as follows:

With SW_1 in position 2, V_{in} is zero and V_{out} will ramp to zero from its previous level. With SW_1 in position 1, V_{in} is now 10V, and V_{out} will ramp to this level. Similarly, with SW_1 in position 3, V_{in} is −10V, and V_{out} ramps to −10V.

Typical input and output waveforms are shown in **Fig. 2**. If the values of R_2 and C_1 are as indicated in **Fig. 1**, then V_{out} will take one second to ramp from zero to 10V.

The circuit can be easily modified as shown in **Fig. 3**, so that the ramp rate can be externally controlled. A FET switch, Q_2, introduces a shunt resistor which diverts current to ground, and thus slows the ramp. With the component values shown, the fast ramp (FET off) has a duration of 100 msec and the slow ramp (FET on), 1 sec. Transistor Q_1 allows the FET to be conveniently controlled by a 5V logic signal.

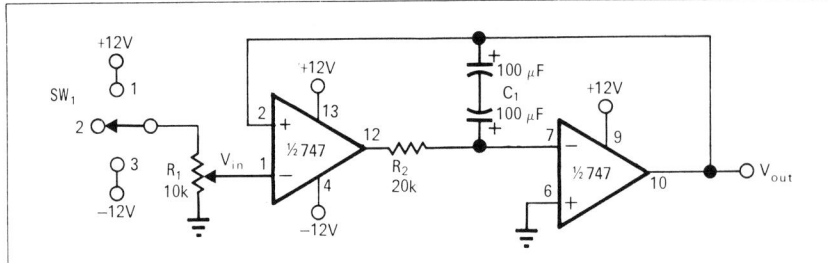

Fig. 1 – **Ramp generator** integrates the stepped input voltage V_{in}.

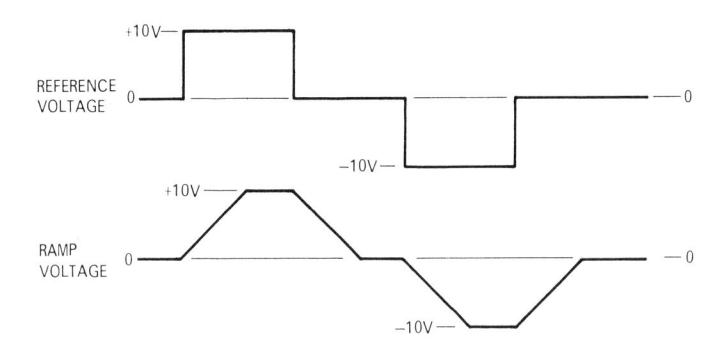

Fig. 2 – **Typical input** and output waveforms for the circuit of Fig. 1.

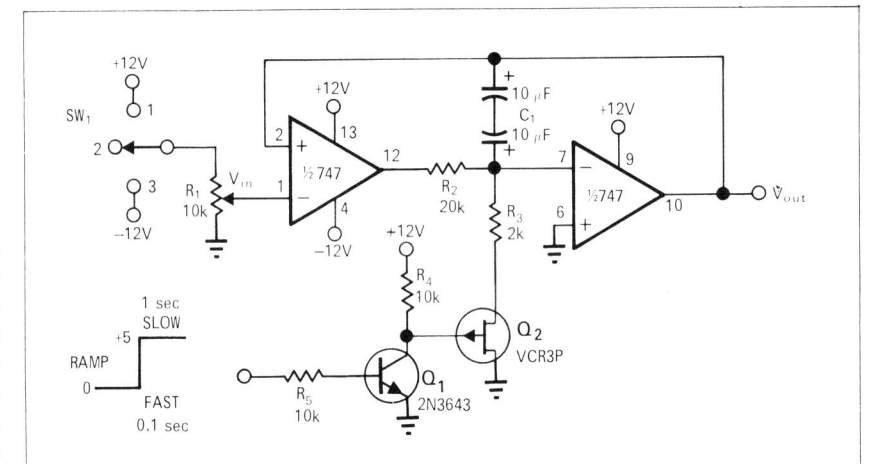

Fig. 3 – **Modified circuit** includes a FET switch which allows the ramp rate to be controlled by a logic signal.

Op amp and one transistor produce ramp function

Larry Wing,
Motorola Semiconductor, Phoenix, Ariz.

It is well known that an integrating op amp configuration can be used to generate a triangle wave from a square wave input. The circuit shown in the schematic offers the designer an economical method of carrying this one step further to convert this triangular wave to a ramp function with a predetermined negative starting level. This waveform change is affected by the addition of only one transistor and a few associated components.

When the input signal changes in polarity from positive to negative, the output begins to go positive as a ramp function, like a conventional inverting integrator. This allows capacitor C_4 to charge with the output voltage.

When the square wave input changes polarity from negative to positive and the output begins to go negative, the voltage at point A also starts to go negative. The resultant negative voltage swing at point A forward biases the emmitter-base junction of transistor Q_1 causing it to conduct. This drives the non-inverting input of the op amp negative. Since the input square wave is in a positive voltage condition at this time, the op amp output is forced to go negative at the slew rate of the MC1709C. The output remains negative until the input square wave switches in the negative direction. At this time the output starts positive and the cycle repeats itself.

The purpose of the clamping network, R_4, R_5, and D_2 is to limit the reverse bias on of Q_1 below breakdown potential.

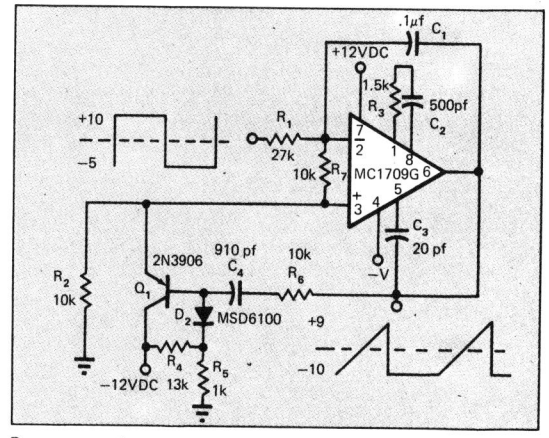

Ramp generator uses one op amp and one transistor. When the input goes positive, Q_1 conducts and drives the op amp negative at it's max. slew rate.

Wide-range ramp generator has programmable outputs

Charles F. Wojslaw and Warren A. Buschmann,
National Semiconductor, Santa Clara, Calif.

This generator, although implemented with only three op amps, offers the flexibility of programmable start and stop voltages and slope, with a wide range for each control.

In the circuit, op amp A_3 functions as a summing amplifier, A_1 as an integrator and A_2 as a stop-limit comparator for the JFET switch S_1. Amplifier A_2, through switch S_1, regulates the generator output at the stop voltage set by V_4.

The generator output voltage, V_o, is the sum of the integrator output voltage, V_3, and the initial or start voltage, V_2.

$$V_o = -\left[\frac{R_4}{R_2} V_3 + \frac{R_4}{R_3} V_2\right],$$

where $V_3 = \frac{I_1}{C_1} \Delta T = -\frac{V_1}{R_1 C_1} \Delta T = -M\Delta T$.

If $R_2 = R_3 = R_4$

then $V_o = M\Delta T - V_2$.

The ramp period, ΔT, is a function of the ramp magnitude, V_4 and the ramp's slope, M.

For a high-voltage ramp generator, only A_3 is required to be a high-voltage amplifier. Resistors R_2, R_3 and R_4 can be used to scale the lower voltages of V_2 and V_3. The only limitation for the start and stop voltages is the max. voltage limitation of A_3. With the proper selection of V_1, R_1 and C_1, the range of slope value can be set from 10^{-3} to 10^6V per second. Polarity may be controlled through the signs of V_1, V_2, V_4 and the inputs of A_2. Controls V_1, V_2, and S_2 may be externally or computer controlled.

Applications for the generator include timing and sweep circuits as well as use as a function generator.

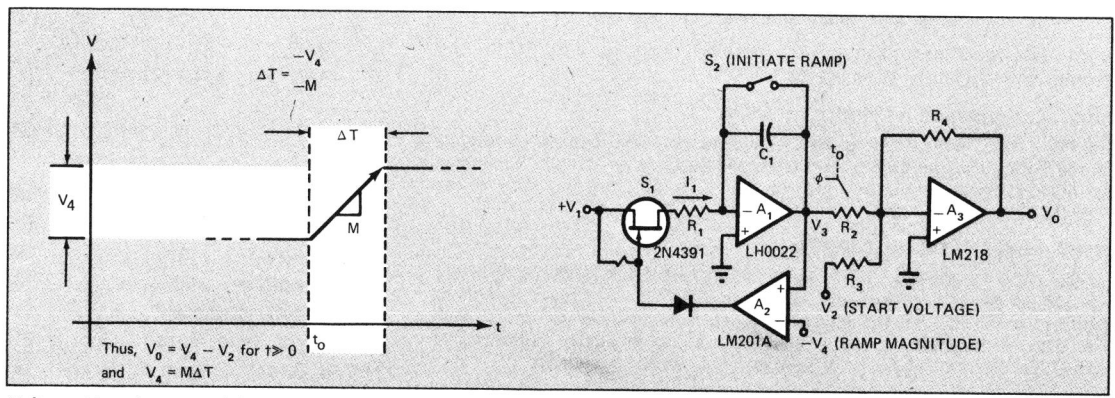

Voltages V_2 and V_4 control the outputs of this wide-range ramp generator.

FET stabilizes sine-wave oscillator

Frederick Macli
Goldmark Communications, Stamford, CT

This FET stabilized sine-wave oscillator consists of a peak detector and a FET, operated in the voltage variable-resistance mode, added to the standard double-integration circuit with regenerative feedback. In the standard configuration, the resistor, R/3, and the FET are replaced by a potentiometer such that the sum of the two resistances is equal to R/2. The pot is then adjusted until oscillation is barely sustained. That configuration, however, is highly susceptible to power supply and temperature variation. The improved version of the circuit peak detects the sine-wave output to provide a dc voltage to control the resistance of the FET, thereby automatically adjusting the condition for oscillation and improving stability. The peak detector is referenced to an adjustable dc voltage to allow for variation in FET resistance from unit to unit. This also provides a dc bias point for the feedback loop.

For the component values shown in the diagram, this circuit provides a 10V p-p, 1460-Hz sine wave into a 500Ω load with power supply voltage ranges of 8 to 18V without any variation in output amplitude or frequency. Over a temperature range of +10 to +65°C, the circuit varies less than 1.5% in frequency and 6% in amplitude. Under all conditions the second harmonic was 25 dB below the fundamental.

The circuit can be modified and extended to other frequencies of operation.

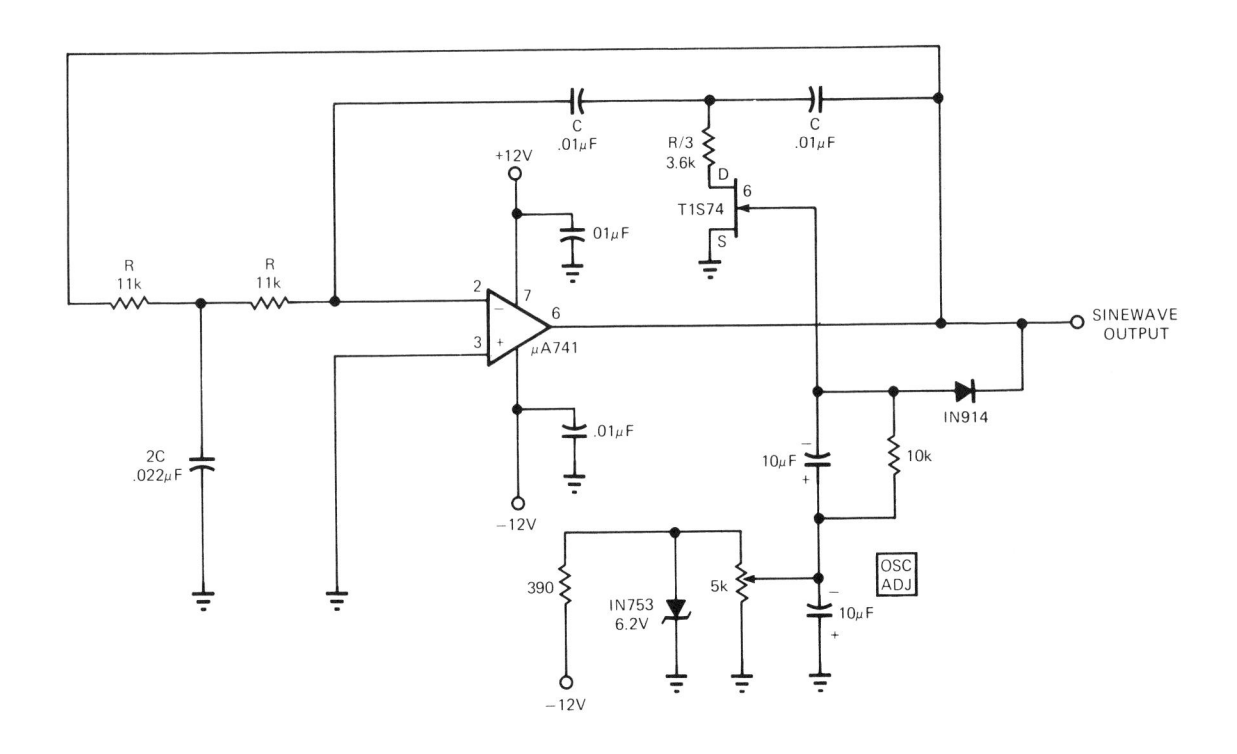

Voltage controlled resistance FET replaces the potentiometer in this otherwise classic oscillator to provide improved stability.

Zener-diode controls Wien-bridge oscillator

W. B. Crittenden and E. J. Owings, Jr.
Westinghouse Electric Corp., Baltimore, Maryland

The usual method employed to control the amplitude of a wien-bridge oscillator such as the one shown in **Fig. 1** is to employ a light bulb or an FET as a variable resistance element to maintain a gain of 3 from point one to point two.

The gain and phase from two to three at $F_{osc} = 1/(2\pi RC)$ is equal to $1/3 < 0°$, thus the overall loop gain $= 3 \times 1/3 < 0°$.

The circuit shown in **Fig. 2** operates to maintain the amplitude symetrical about ground by using a single zener (D_5) and a bridge of diodes (D_1, D_2, D_3 and D_4). As the output e_o approaches the soft knee threshold of conduction of D_5, the zener impedance decreases and shunts R_2. This violates the requirement for oscillation that $R_2 = 2R_1$. The output then starts to decrease sinusoidally, and as the swing decreases the gain increases until e_o reaches the negative

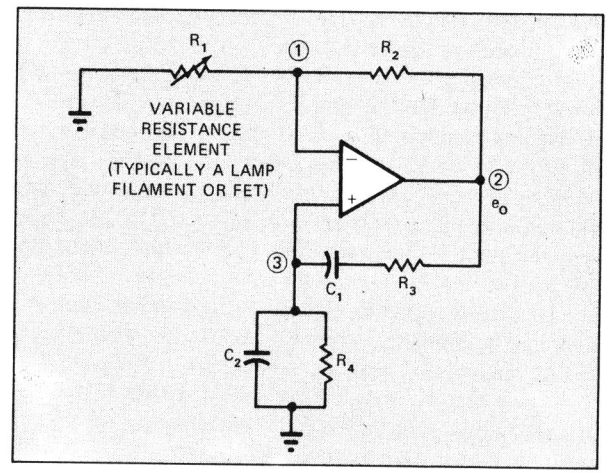

Fig. 1—Conventional Wien bridge circuit employs a variable resistance element, typically a lamp filament or an FET to maintain the required op-amp gain.

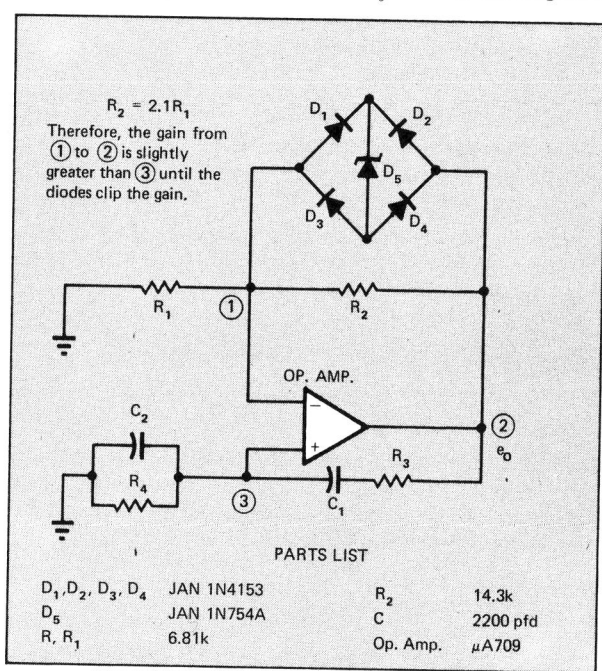

Fig. 2—Diode clipping circuit shown here uses the "soft" knee of a low-voltage zener to provide amplitude control for the oscillation circuit. The diode bridge allows the use of a single zener to control both positive and negative amplitudes.

threshold. The signal reverses then and again starts going positive.

Fig. 3a is the output waveform for $f_o = 10.5$ KC and **Fig. 3b** is an expanded view of the peaks. Clipping does not distort the waveform appreciably.

The results of thermal tests from $-55°$ C to $+95°$ C are:
Fundamental $F_o = 10.5$ KHz $\pm .15$ Kz
2nd Harmonic $\leq 0.2\%$
3rd Harmonic $\leq 1\%$
$e_o = 5 \pm 0.15$ VRMS

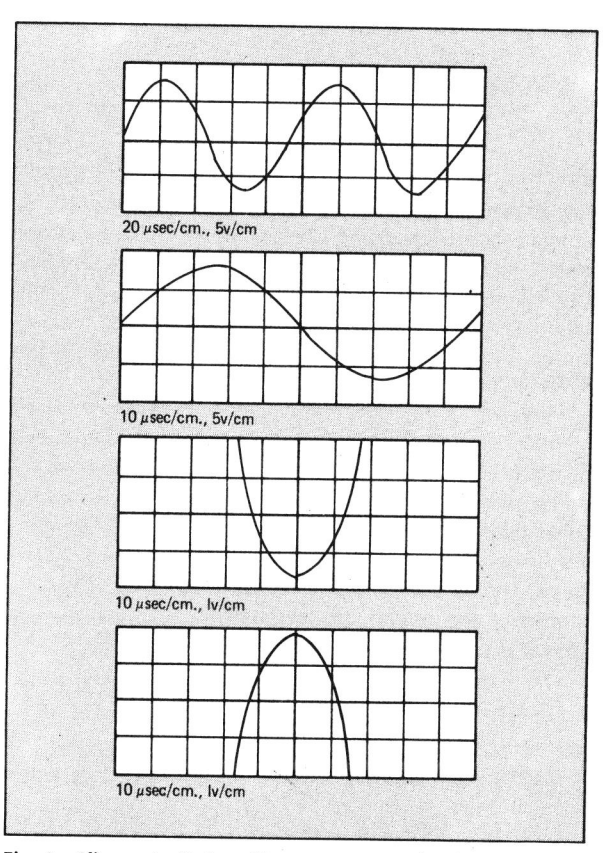

Fig. 3—Clip controlled oscillator output using a 709 op amp, 1N4153 diodes and a 1N754A zener shows that very little distortion was introduced by clipping.

Start of oscillation was positive over the entire temperature range.

Staircase generator uses current-regulating diode

by Robert G. Warsinski
Ford Motor Co.
Detroit, Mich.

The output of a pulse-generating position transducer must sometimes produce an analog input for the horizontal axis of an X-Y recorder or storage scope. Unfortunately, many pulse-to-analog converters depend on the repetition frequency of the pulses (transducer velocity) as well as on the actual number of pulses (transducer position).

A well-designed staircase generator can produce an analog output which is proportional to the number of input pulses, regardless of repetition frequency. The key to the transducer-interface circuit described here is the use of a current-regulating diode (CRD) as a constant-current source in a staircase generator circuit. The rest of the circuitry is fairly conventional.

Basically, the staircase generator operates as a charge-exchange integrator. Each input pulse causes ceramic capacitor C_5 to become charged to approximately 7 mV. Between pulses these charges drain off, resulting in current flow through diode CR_6, and thus causing the LH201 op amp to charge low-loss capacitor C_6 by an equal amount. This transferred charge is held by the integrator capacitor, and is added to all subsequent charges received by charge exchange from C_5. The integrator's output voltage continues to increase up to the saturation level of the op amp in a series of 7-mV staircase steps.

To insure that the same amount of charge goes to C_5 with every pulse, regardless of pulse rate, a monostable multivibrator is employed to switch on Q_3 and the CRD for a constant time duration with every pulse.

Relay AR automatically resets the integrator to zero when the output voltage reaches almost 8.7V. This prevents data from being lost if the op amp should saturate before the data run is completed.

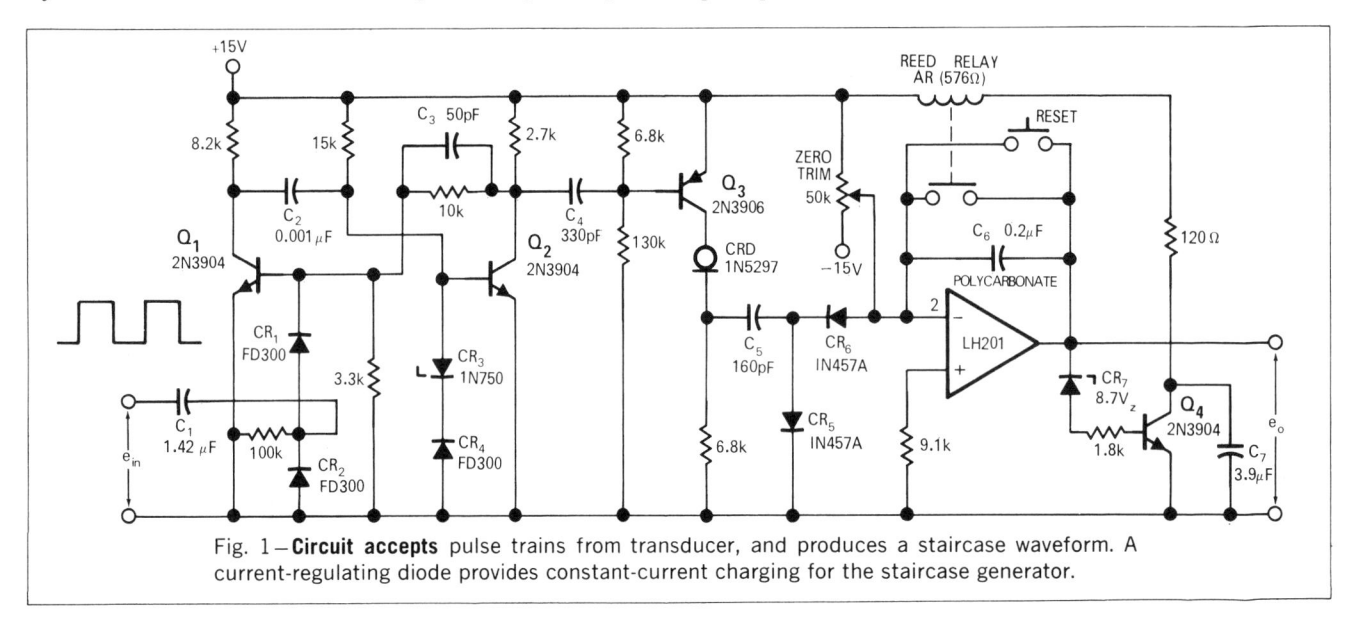

Fig. 1—**Circuit accepts** pulse trains from transducer, and produces a staircase waveform. A current-regulating diode provides constant-current charging for the staircase generator.

Simple stair-step generator uses 1 IC and 3 transistors

Edwin E. Morris
General Electric Corp., Utica, N.Y.

Synchronized stair-step generators are needed for, among other things, a gray-scale test signal for television equipment.

The illustrated circuit is a synchronized LC oscillator driving the stair-step generator, both of which are reset by the horizontal-blanking input signal.

The synched oscillator is a simple, positive-feedback LC oscillator composed of two TTL gates biased at their linear range by negative feedback resistors R_1 and R_2. With a LOW input on pin 10 of G_1, pin 8 goes to its HIGH output level regardless of the input on pin 9. During this OFF condition, C_1, C_2 and L_1 assume their steady state conditions. As a result, the oscillator always starts in the same condition.

The circuit generates a stair-step by integrating a train of equally spaced pulses. The voltage across C_4 is −10V at the start of the stair-step due to the clamp Q_2. As a function of time the voltage across C_4 is expressed by:

$$V_{C_4} = -10 + \frac{1}{C_4} \int_o^t i \, dt.$$

The only currents into C_4 are a small amount of leakage from Q_2 and the gate of source follower Q_3 and the input current from Q_1.

The current from Q_1 is a series of pulses due to input current to this common base stage. The input current pulses result from applying a series of step voltages across the capacitor C_3. The current into C_3 is expressed by:

$$i_{C3} = C_3 \frac{dv}{dt}$$

Since the step voltage of the low output impedance TTL gate is applied directly to both C_3 and the low impedance input to the common base stage, Q_1, the current approximates one pulse.

C_3 charges rapidly into the emitter of Q_1 on the positive going transition of the square wave input. On the negative going transition Q_1 is turned off and C_3 slowly discharges through R_3.

The magnitude of each step is determined by:

$$[(G_1 \text{ Step Voltage Change}) - (V_{EB})] \times [C_3 \div C_4]$$

The number of steps is determined by the frequency of the oscillator.

Q_3 is simply a source follower buffer stage.

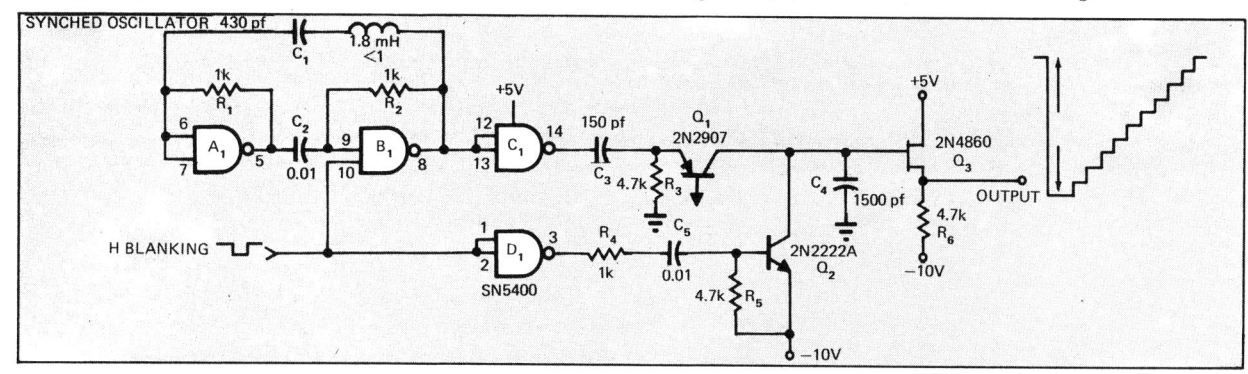

Synchronized stair-step generator, designed as a gray-scale test generator for video equipment, is reset by the horizontal blanking signal.

Op amp makes variable-frequency triangular wave generator

Geroge R. Begault
Harris-Intertype Corp., Melbourne, Fla.

The circuit in **Fig. 1** is a new twist to an old theme. If the diode/FET bridge were replaced with a single feedback resistor, the circuit would be the classic textbook op-amp squarewave generator. The output of the op amp would toggle between + and ÷ Vcc whenever C_1 charged to the voltage at the junction of R_3 and R_4. The operation of the circuit as shown is exactly the same, with the exception that the diode/FET arrangement causes C_1 to charge from a constant current source (Q_1), thereby forcing the voltage across C_1 to change at a linear rate. The linearity of the triangular waveform is surprising, considering the simplicity of the circuit.

For a more detailed look at circuit operation, assume power has been applied, C_1 has acquired no charge, and the op amp has been driven into its upper bound by the positive feedback. Pin-6 now sits at about +12V and pin-3 because of the divider action of R_3 and R_4 is at +1V. Current flows through C_1 to pin-6. Note that the arrangement of the diodes allows current to flow through the FET in only the source-to-drain direction, and current path would be through D_3, R_2, R_1, Q_1 and D_2 into the op amp. When the voltage at pin-2 exceeds the +1V level of pin-3, the amp output is driven to its negative bound (about −12V). This puts pin-3 at −1V, the current flow through C_1 reverses (this time through D_4 and D_1) and the voltage across C_1 (pin-2) now charges toward −12V.

Then the voltage at pin-2 drops below the −1V level of pin 3, the op amp output switches again to +12V and the process repeats.

With the component values shown, the frequency can be adjusted from 500 Hz to greater than 20 kHz at constant output amplitude. The short term stability is better than ±1 part in 10,000. Also since the same R and C are used to generate both sections of the waveform then the positive slope must be the same as the negative slope (assuming diodes are matched and the amplifier + and − bounds are the same.)

At higher frequencies (above 25 kHz approx.) the waveform will increase in amplitude because the time it takes the amp to slew to the new bound becomes an appreciable portion of the ramp. This can be compensated by C_2, which will reduce the voltage to which C_1 must charge by decreasing the rise time of the voltage at the junction of R_3 and R_4. The value of C_2 will depend upon ramp frequency, amplifier slew rate and ramp amplitude.

To generate a sawtooth, simply remove C_2 and replace a diode pair with a short. For positive going ramps replace D_1 and D_4; for negative ramps short D_2 and D_3. The same of linearity is obtained since C_1 must still charge through a constant current source in forming the ramp.

The circuit output (C_1) should work into an impedance of at least 200 kΩ.

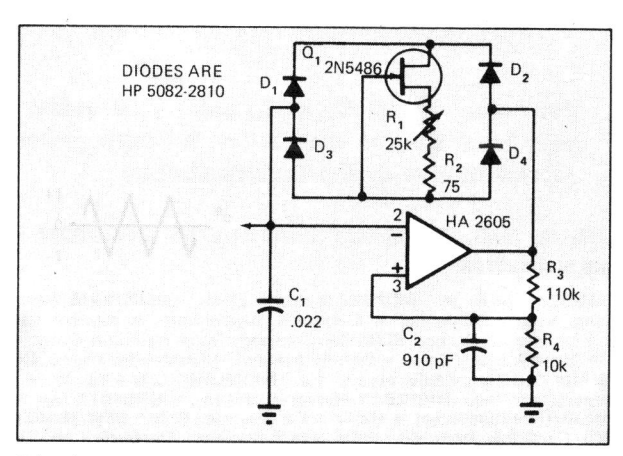

Triangle wave generator is adjustable from 500 Hz to 25 kHz. By removing C_2 and shorting D_1 and D_4, the circuit can be changed to a ramp generator.

Triangle wave circuit has wide range controls

Jerry Graeme, Burr-Brown Research Corporation
Tucson, Arizona

With only two op amps, a precise triangle wave generator can be formed which has wide range controls for every characteristic of the waveform. Precision is provided by high gain feedback and buffered control loading of the op amps. Although there is some control interaction, the characteristics set by the controls are accurately predictable and highly stable.

The triangle wave generator, as shown in **Fig. 1**, is basically an integrator with a feedback comparator that switches the reference voltage to be integrated. Forming the integrator are A_1, R_f, and C. The voltage integrated is the difference between those supplied by the comparator output and the symmetry control, of $e_1 - V_s$. Since e_1 switches polarities, V_s alternately increases and decreases the voltage integrated. This increases one integration rate and decreases the other to make the triangle wave asymmetrical. Symmetry is affected by control potentiometer R_s. The accuracy of this control is primarily set by the accuracies of R_s and its biases, while the errors produced by the input bias current and offset voltage of A_1 are negligible in most cases. Control range is limited by the maximum integration rate set by either the op amp slew rate or the charging rate limit of C by the output current of A_1.

As the output of A_1 traces out the triangle wave, the waveform peaks are set by the trip points of the comparator, composed of amplifier A_2, the zener diode output limiter and hysteresis feedback through R_A. The amplitude is controlled by the setting of amplitude control R_A. Both the accuracy and the stability of the amplitude are set by those of R_A and V_z. The range of amplitudes attainable is limited below the output swing limits of A_1 and above the comparator input offset voltage.

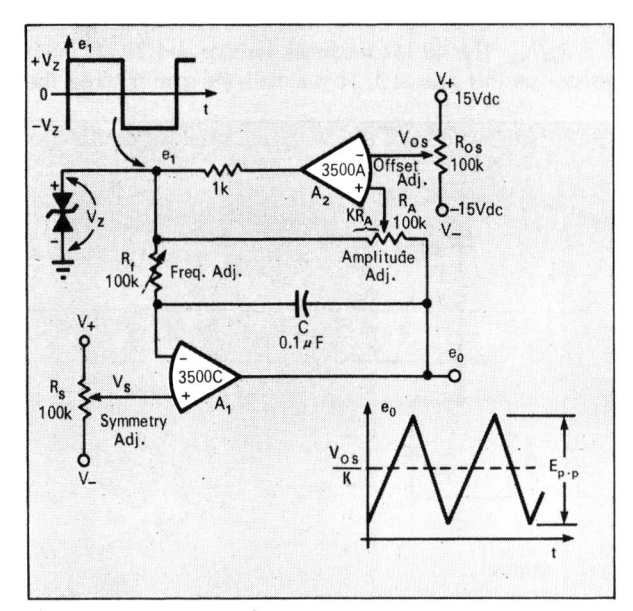

Fig. 1—Comparator feedback around an integrator produces a linear triangle wave. The input characteristics of the two op amps permit accurate control with potentiometers.

Depending on the actual components used, this circuit can provide linear triangle waves from 0.1 Hz to 10 kHz with amplitudes of 10 mV to 20V and with varying degrees of offset and symmetry. Excellent linearity is provided by the op amp integrator with only 0.01% nonlinearity at low frequencies. Control errors will typically be 0.1% to 1% as determined by the potentiometer and supply voltage accuracies.

IC op amp makes gated oscillator

Frederick Macli
North American Philips, Briarcliff Manor, N.Y.

This circuit is a gated 5-MHz relaxation oscillator with several unique features. Its output always starts in the same phase with respect to the gating signal. The major circuit components are a gate controlled wideband amplifier, MC 1545G, and a frequency selective network comprised of R_1, R_2, and C. This network provides positive feedback around the amplifier. The conditions for oscillation are controlled by varying the amplifier gain via the gate input. For example, a gate signal supplied from T²L logic can vary the amplifier gain from −70 db at logic "0" to +20 db at logic "1".

The selection of resistors R_1 and R_2 should be limited by:

$$[(R_1 + R_2) \div R_1] > K$$

where K is the amplifier gain with a high gate voltage. R_1 is also limited in value by bias current considerations of the amplifier. The period of oscillation has been found to be approximately:

$$T \approx 4(R_1 + R_2) C.$$

The MC1545G has a 3 db frequency of 75 MHz with a single ended voltage gain of 20 db. Thus, extending the operating frequencies well beyond the 1-10 MHz range should not be any problem.

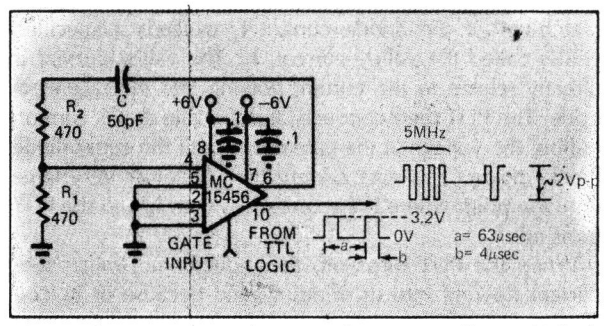

This single IC op amp makes a gated 5 MHz oscillator which always starts with a positive pulse. Adaptation for frequencies above 10 MHz should present no problems.

108 Circuit Design Idea Handbook

High Efficiency Relaxation Oscillator

By Carl F. Andren
The John Hopkins University
Applied Physics Laboratory
Silver Spring, Md.

THIS CIRCUIT was developed for use as a voltage-controlled oscillator in a high efficiency switching regulator. It is a relaxation-type oscillator which provides short, fast pulses to trigger a multivibrator. It has many advantages over the common unijunction relaxation oscillator, such as: higher power output, lower power consumption, faster pulse risetime, higher maximum operating frequency, higher reliability, and, with the new plastic-case transistors, lower cost.

The oscillator consists of a simple R-C ramp generator coupled to a high efficiency trigger circuit. In the circuit, Q_1 is initially off if the voltage at its base is less than:

$$V_{B1} = \frac{V R_3}{R_3 + R_4 + R_5} + V_{BE1} + V_{BE2}$$

Timing capacitor C_1 will charge up via control resistor R_1 until its voltage reaches the point at which Q_1 begins to conduct. If resistor R_1 can supply enough current (about 1/2 μa) Q_1 turns on. As Q_1 turns on its collector voltage falls, driving the base of Q_2 negatively, causing Q_2 and thus Q_1 to conduct more. Both Q_1 and Q_2 will be driven into saturation by the charge in C_1 and C_2. The base current of Q_1 in saturation will discharge C_1 through limiting resistor R_2; and if resistor R_1 cannot supply enough current to keep Q_1 in saturation (about 0.5 ma), the trigger will turn back off by the same regenerative action that turned it on.

Using the parts values shown, the permissible range of values for R_1 is found to be between 50 K and 25 meg. With R_1 set at 1 meg the oscillator operates at about 75 kHz putting out 10-v negative-going pulses with risetimes of 10 nsec and fall times of 1 μsec. Power consumption is 1.5 mw.

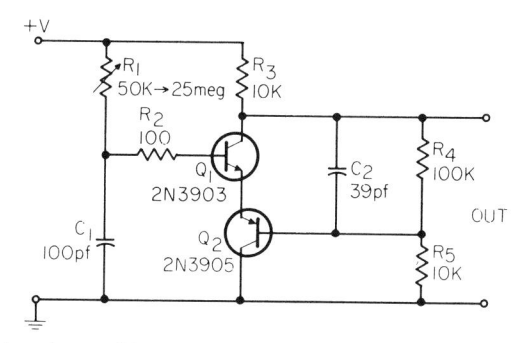

Relaxation oscillator composed of ramp generator driving a trigger circuit.

With R_3 at 10 K the upper operating frequency is about 1 MHz, limited by the recharge time of C_2 through R_3 and R_5. Pulse width is about 0.5 μsec but can be lengthened by increasing C_1 and/or R_2. For example, with $C_1 = 100$ μf, $R_1 = 1$ meg, the pulse width is 25 msec and the frequency is 0.25 Hz.

Supply voltage and temperature affect the frequency since they set the charging rate of C_1 and the triggering voltage. In a closed loop feedback system, with R_1 replaced by a control transistor, the effects of supply voltage and temperature on the frequency will be compensated for by the feedback.

Costs might be estimated as follows: Q_1 and Q_2: $0.50 each; R_1 to R_5: $0.05 each; and C_1 and C_2: $0.23 each, for a total component cost of $1.71.

CMOS linear-ramp generator has amplitude control

Kenneth Bower
ESL Inc., Sunnyvale, CA

Most ramp generators are constructed using an RC time constant. Using an RC circuit to drive a VCO however, creates problems with temperature stability and poor linearity. In order to eliminate the temperature and linearity problems a digital ramp was generated using a digital-to-analog technique with a single CD4024A.

The ramp itself was generated from a 50 kHz clock and is was stopped by a reset pulse to the counters. Normally a digital ramp has the problem that the amplitude cannot be changed without changing the clock frequency. This causes the discrete steps, or discontinuities to become larger. Using CMOS logic solves this problem. By using a stable but variable supply to the IC, the ramp output amplitude may be adjusted. This is accomplished by using an IC adjustable voltage regulator.

The output from the ramp generator consists of many very small steps. If these steps are too large, a second CD4024A can be added and the clock frequency increased. Also, if the 741 op amp does not provide an adequate frequency response for very steep ramps, a higher slew rate op amp may be used.

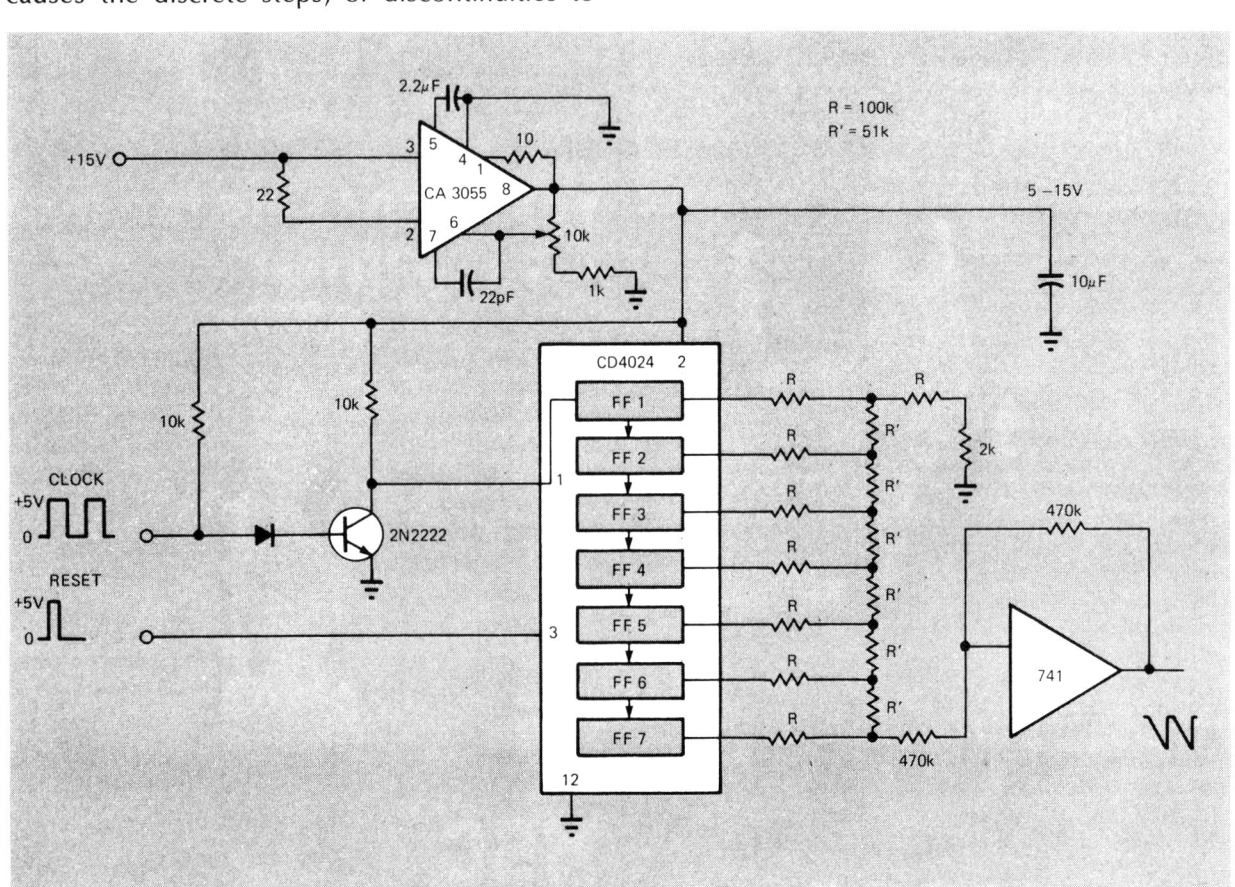

A single CMOS shift register forms the heart of this DAC/ramp generator and allows adjustment of amplitude not possible with other logic families.

Stepped-sawtooth tone generator

by Heinrich Krabbe
Analog Devices
Norwood, Mass.

THE CIRCUIT shown uses a digital approach in producing a sawtooth waveform for tonal color generation in an electronics organ. A sawtooth waveform is useful since it contains all odd and even harmonics on a monotonically decreasing basis.

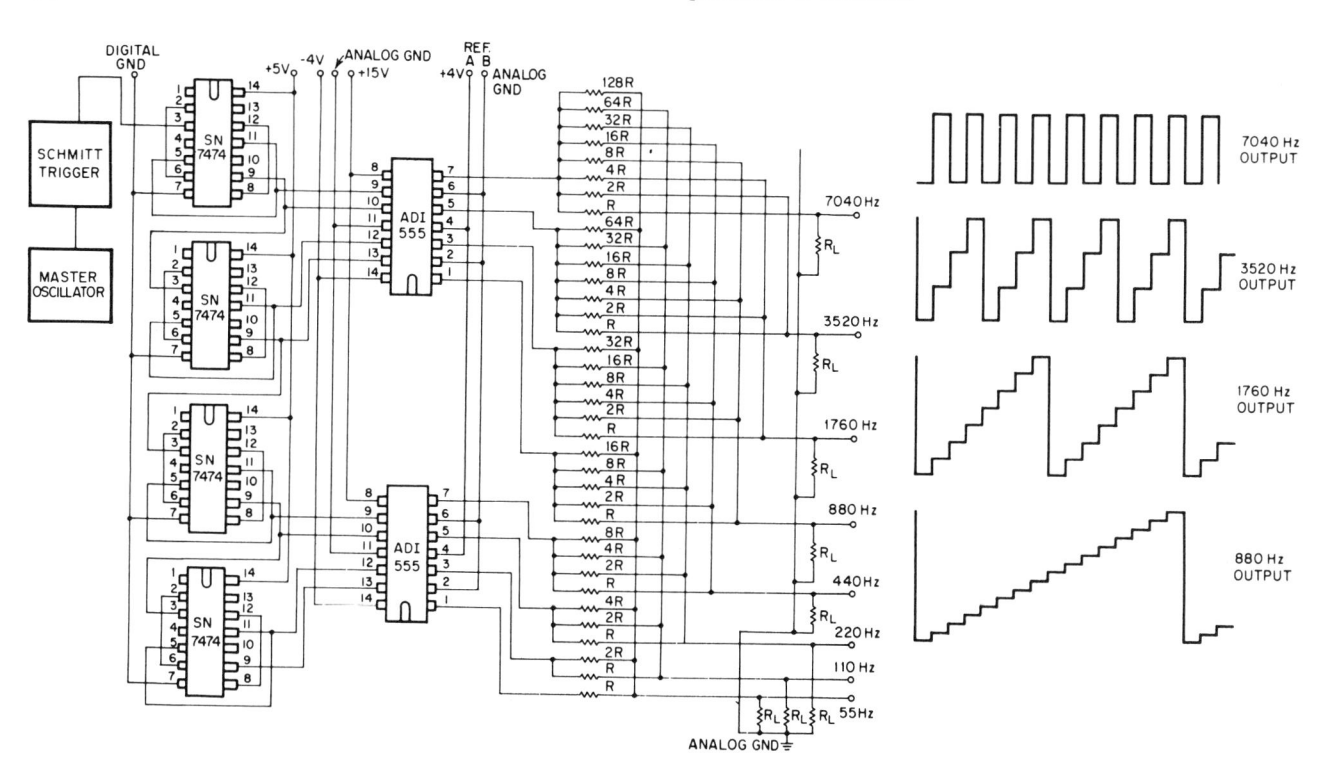

This tone generator circuit uses a digital approach to generate approximate sawtooth waves.

The sawtooth is generated from binary divider outputs, two μDAC voltage switches and a handful of resistors.

The figure shows a complete circuit diagram for the 8-tone generator divider strip. A highly stable L-C oscillator provides a frequency (14,080 Hz) an octave above the highest frequency desired. A Schmitt trigger squares the oscillator output to provide proper drive for the TTL flip-flops. The output of the first flip-flop provides the highest tone output (7040 Hz). All other tones are derived by the binary counters formed by the string of SN-7474s.

Each flip-flop drives one switch of the AD 1555 quad switch. The switch outputs are then resistively mixed through binary-scaled resistors so that a staircase waveshape results. This waveshape contains all odd and even harmonics up to a point, because the fundamental is mixed with ½ of the second harmonic, ¼ of the fourth harmonic, ⅛ of the eighth harmonic and so on. Since all odd harmonics are present in the fundamental, only even harmonics must be added. The best sawtooth approximation is obtained at the lowest frequency tone. Wave forms become progressively coarser at higher tones. The highest tone is simply a square wave.

Typical stepped outputs from the tone generator. Note that the 7040 Hz output is a square wave while all other outputs approximate a sawtooth.

Positive or Negative Slope Generator

By Gilbert Marosi
Friden Inc.
San Leandro, Calif.

THIS CIRCUIT GENERATES linear ramps, either negative or positive, by switching on and off two current sources charging a capacitor.

A negative gate into the base of Q_1 turns on Q_1, which turns on Q_2, and the emitter of Q_3 therefore sits at −12 v, thus turning on Q_5 and turning off Q_4. The constant current furnished by Q_5 will charge C_3 in the negative direction until clamped by the base-collector junction of Q_5. When Q_2 turns off, the emitter of Q_3 will be at about +12 v. Q_4 turns on and Q_5 turns off. The capacitor charges in the positive direction until clamped by the base-collector junction of Q_4. The purpose of Q_3 is to present an equal low impedance either at +12 v or at −12 v to each of the current sources.

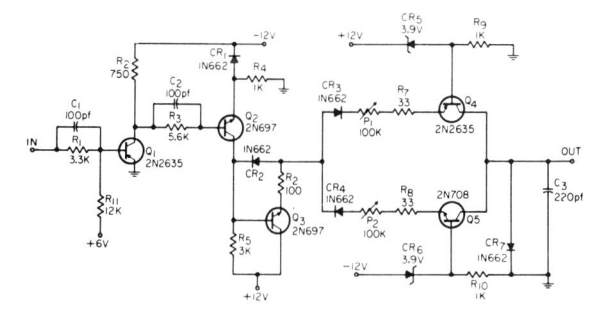

Positive or negative slope generator.

Signal Sources 111

Cascade UJT oscillator generates linear frequency sweeps

by Raldon Smith, Jr.
Boeing Co.
Seattle, Wash.

This sweep oscillator consists of two stages, shown separated by the dotted line in Fig. 1. The first stage generates a low-frequency sawtooth signal that controls the higher-frequency sawtooth generator of stage two. Both stages are simple relaxation oscillators using unijunction transistors.

The first stage determines the repetition rate and frequency range of the swept output. The second stage generates the output sawtooth waveform and determines the amplitude.

Figure 2 shows the emitter-voltage waveform of Q_1. The amplitude of this signal determines the frequency range of the sweep, while the frequency determines the repetition rate of the sweep.

When the first-stage output voltage is applied to the second stage via buffer-amplifier Q_2 the second stage oscillation frequency increases as the control voltage rises. Output frequency can be calculated from the following equation:

Figure 3 shows the relationship between V_{E2} and the output frequency. Fig. 4 lists the effects of the various components that control the performance of the generator.

Of course, the maximum operating frequencies are restricted by the characteristics of the unijunction transistors. Also, the frequency of the second stage must be higher than the frequency of the first stage. Usually, a ratio of ten or more is acceptable.

Both oscillators in this circuit have better linearity than standard unijunction oscillators. This is because the capacitor charging voltages are greater than the corresponding inter-base voltages of the unijunctions. Sweep linearity can be further improved by replacing R_3 with a constant-current source.

In the original application, the sweep generator output was passed through a narrow bandpass filter to produce a short-duration signal. Figs. 1, 2 and 3 show component values and performance for the circuit that was built and tested for this application.

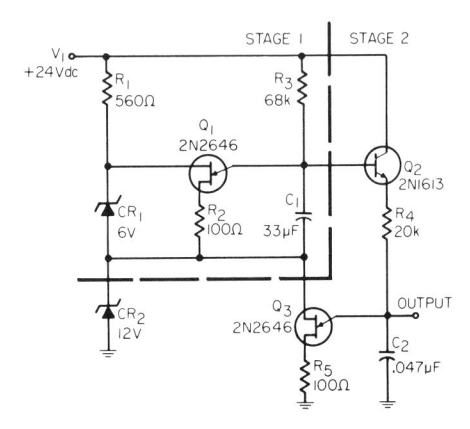

Fig. 1. Simple sweep-frequency generator consists of two cascaded unijunction oscillators.

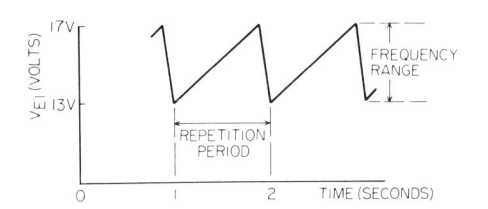

Fig. 2. Waveform of the first unijunction oscillator determines the repetition rate and frequency range of the swept output.

Fig. 3. Output frequency of the second stage is a linear function of the emitter voltage of Q_2.

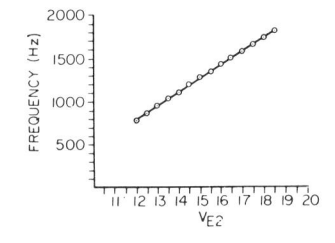

Fig. 4. Circuit parameters can be varied independently by changing component values as shown in this table.

$$f = \frac{1}{R_4 C_2 \ln\left[\dfrac{V_{E2} - V_{E3}(\min)}{V_{E2} - V_{E3}(\max)}\right] + t_f} \qquad (1)$$

where $V_{E3}(\max) = V_P$ = emitter peak-point voltage and t_f = emitter-voltage fall time

	Adjust	To Increase	To Decrease
Freq range	CR_1	Increase V_{CR1}	Decrease V_{CR1}
Rep rate	$T_{C1} = R_3 C_1$	Decrease T_{C1}	Increase T_{C1}
Freq output	$T_{C2} = R_4 C_2$	Decrease T_{C2}	Increase T_{C2}

Recycling Timing With Variable Duty-Cycle

By Paul Gheorghiu
Hi-G, Inc.
Windsor Locks, Conn.

THIS CIRCUIT was designed to offer, at low temperatures, a variable recycling time delay with adjustable time-on, time-off.

In the circuit, the 100-K variable resistors control on and off times. Some refinements are necessary to make the circuit operational down to −65°C (a minimum beta of 20 at this temperature is required). For −50°C, an unselected transistor will perform well with 300 ma load.

If a small current is used, 50 ma or so, the second transistor could be eliminated, as could SM 72 as long as the emitter current of the trigger is increased, by changing the bias, to about 3.5 ma.

The circuit as shown has a time delay of 0.300 sec to 6 sec. Care should be taken when the 300 msec level is set because the base bias, i.e., 100 K, may now be almost as low in value as the load resistance, and the multi will not start.

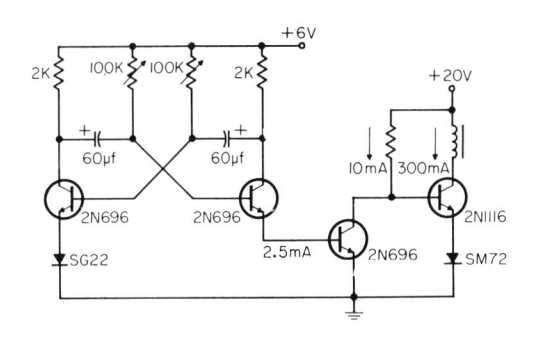

Recycling timer with variable duty-cycle.

PUT delivers ultra-low-power, high-energy pulses

Jeffrey P. Stein
Honeywell Inc., Ft. Washington, Penn.

This circuit shows how a programmable unijunction transistor (PUT) can be used to generate an ultra-low-power, very-low-duty cycle pulse. The circuit has been applied in a battery back-up for an MOS memory system to periodically refresh the internal data-node capacitors by switching the power supply on for a very short time.

The circuit is basically a standard PUT oscillator with the voltage divider R_2/R_3 setting a threshold or trigger level. This voltage is such that the voltage across R_2 is much less than the breakdown of D_1. The zener diode, D_1, should be about 2 volts less than the battery or supply voltage. Capacitor C_1 charges through R_1 until V_A equals $V_G + 0.7V$. The PUT fires at this point and capacitor C_1 discharges into the base of Q_1 via R_4 and into inductor L_1. This turns on the Darligton driver pair very hard, sinking large currents from the load, and also stores some energy in the field of inductor L_1.

In order to complete one cycle and start the next oscillator cycle, the PUT must be turned off. This is done by the zener diode. When the PUT conducts, the device will "latch-up" if the anode current I_A exceeds a specified value, called the valley current, I_V. The valley current is directly related to the current flowing into the gate electrode. The PUT then conducts and the gate draws current, pulling the voltage at the gate down until the zener diode breaks down. The gate current then becomes very large until the point where I_A becomes less than I_V and the PUT turns off.

When the PUT turns off, the inductor no longer sees current flowing into its winding, and because of its collapsing field, generates a negative transient at the base of the Darlington, cutting it off rapidly. This generates a high-power pulse with very sharp rise and fall times. The charging sequence then starts again to sustain the oscillation.

This circuit generates a very high energy pulse of short duration and low duty cycle. Typical values are:

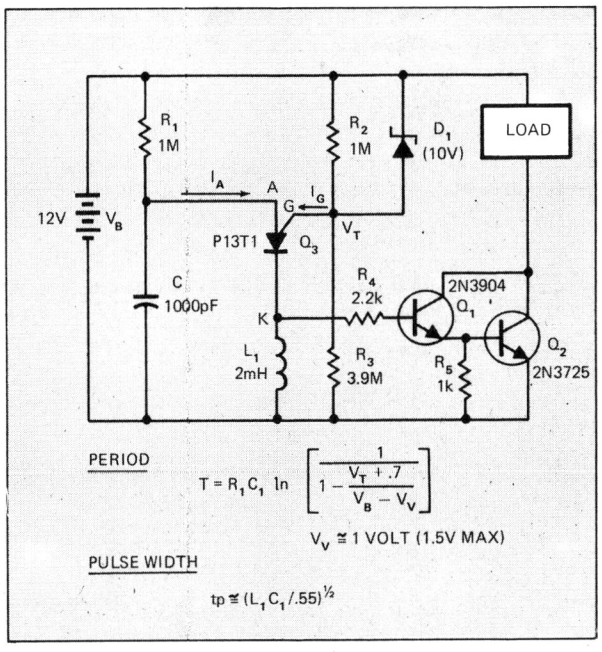

PERIOD
$$T = R_1 C_1 \ln \left[\frac{1}{1 - \frac{V_T + .7}{V_B - V_V}} \right]$$

$V_V \cong 1 \text{ VOLT } (1.5V \text{ MAX})$

PULSE WIDTH
$$t_P \cong (L_1 C_1 / .55)^{\frac{1}{2}}$$

PUT pulse generator delivers very-high-energy pulses in an extremely-low-duty-cycle mode. This circuit was developed to provide memory-refresh pulses from a standby battery to an MOS memory system.

t_p = 1.5 μsec
T = 1.5 msec
I_L = 1A Peak
$t_r \leq$ 200 nsec
$t_f \leq$ 200 nsec

Standby power is consumed only in the resistor divider R_2/R_3 and the charging resistor R_1 which can be very large. The only other currents are small leakage currents across the device junctions. The real key to low-power operation is the elimination of the interbase channel resistance (6k-10k) of the more common UJT used in this type of circuit. Also, the zener diode allows R_2/R_3 to be made very large while still maintaining proper operation without "latch-up."

Sequential bipolar multivibrator

By Gerard T. Flynn
MIT Lincoln Lab
Lexington, Mass.

A SINGLE circuit can replace cascaded one-shot multivibrators that are often employed to generate a delayed pulse or a pair of sequential pulses. The circuit, shown in Fig. 1, provides two sequential pulses of opposite polarities, with the duration of each pulse independently adjustable over a wide range.

Other versions of the circuit can provide two short pulses separated by a long pulse of the opposite polarity, or allow independent triggering of the positive and negative pulses (*i.e.*, with the pulses not automatically sequential but mutually exclusive in time).

The basic circuit of Fig. 1 works as follows: Initially, I_1 and I_2 are both zero. When a positive trigger pulse is applied, the output swings negative. The negative pulse is coupled through capacitor C_1 causing D_1 to conduct. We choose $R_1 < R_f$ so that, at first, I_2 is greater than I_1. Then, as C_1 charges, $|I_2|$ decreases until it is equal to $|I_1|$ and the circuit switches back to the quiescent zero state. This switching transient is coupled by the capacitance of diode D_2 and causes the output to swing positive, which turns on diode D_2. The same process is then repeated for the positive pulse.

If the first time-constant is short compared to the second, capacitor C_1 charges to the output voltage during time τ_{2A}. Therefore, when the output returns to zero, the voltage across C_1 is sufficient to turn on diode D_1. Thus a second negative pulse is generated. This pulse is longer than τ_1 because of the higher initial charge across C_1. Capacitor C_2 does not fully discharge during τ_3 and, when the out-

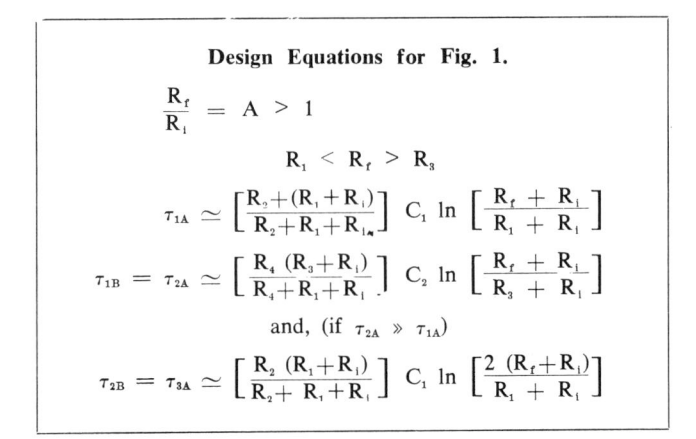

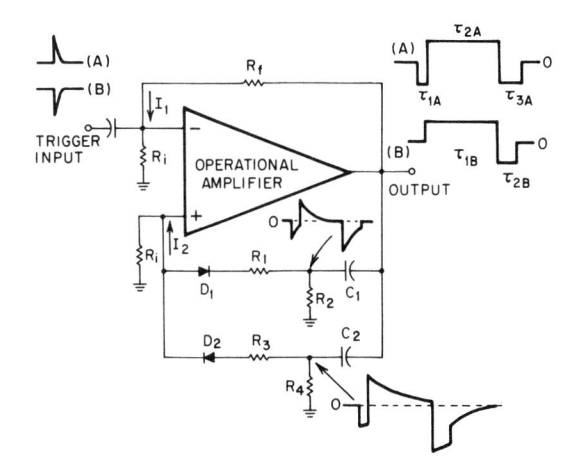

Design Equations for Fig. 1.

$$\frac{R_f}{R_i} = A > 1$$

$$R_1 < R_f > R_3$$

$$\tau_{1A} \simeq \left[\frac{R_2 + (R_1 + R_i)}{R_2 + R_1 + R_i}\right] C_1 \ln\left[\frac{R_f + R_i}{R_1 + R_i}\right]$$

$$\tau_{1B} = \tau_{2A} \simeq \left[\frac{R_4 (R_3 + R_i)}{R_4 + R_1 + R_i}\right] C_2 \ln\left[\frac{R_f + R_i}{R_3 + R_i}\right]$$

and, (if $\tau_{2A} \gg \tau_{1A}$)

$$\tau_{2B} = \tau_{3A} \simeq \left[\frac{R_2 (R_1 + R_i)}{R_2 + R_1 + R_i}\right] C_1 \ln\left[\frac{2 (R_f + R_i)}{R_1 + R_i}\right]$$

Fig. 1. Bipolar multivibrator is similar to a one-shot, but it generates a sequential pair of pulses of opposing polarities.

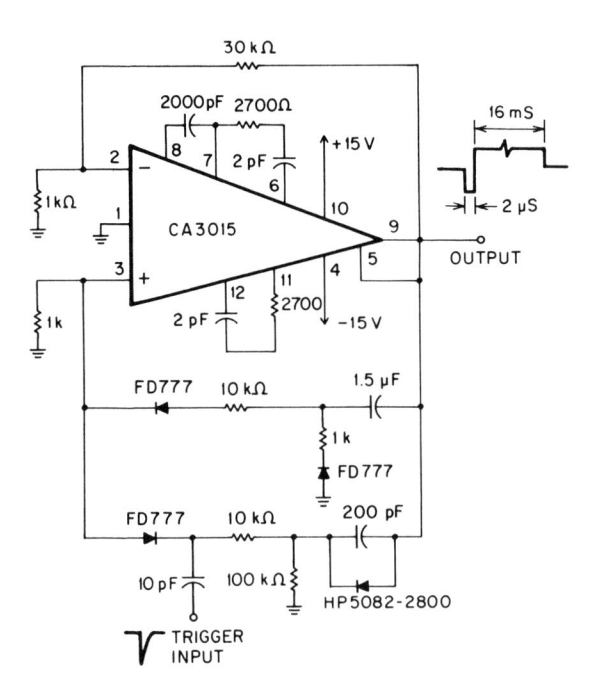

Fig. 2. This practical version of the bipolar multivibrator gives a 2-µs negative pulse and a 16 ms positive pulse.

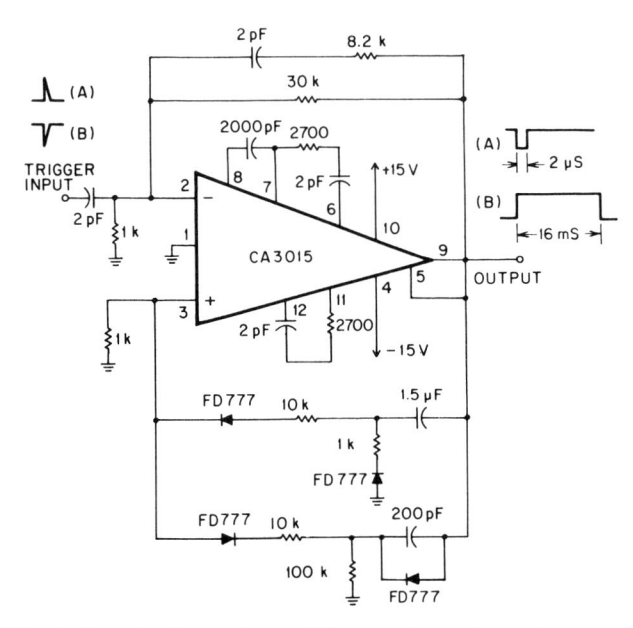

Fig. 3. The circuit can be modified, as shown here, to suppress the sequential action. Output is either a positive or a negative pulse, depending on the polarity of the input pulse.

put returns to zero, diode D_2 remains sufficiently back-biased to keep the circuit from switching positive.

If a negative trigger pulse is applied (situation (B) in the diagram), the sequence starts with the positive pulse. Subsequent operation is then the same as described for a positive input. The diode, of course, can be reversed to give opposite-polarity pulses.

Figure 2 shows a practical circuit that generates a 2-µs negative pulse followed by a 16-ms positive pulse. The diode across the 200-pF capacitor prevents reverse charging of the capacitor during the positive portion of the cycle. Note that the triggering point has been relocated to allow negative triggering. The 10-pF coupling capacitor couples the trigger pulse and suppresses the transient during turnoff of the positive cycle. With this circuit there is no second negative pulse generated after the positive pulse.

The 1kΩ resistor, in series with the diode connected to the 1.5-µF capacitor, provides a fast discharge path for the capacitor after the positive cycle. This gives the longest possible time-constant during the positive cycle (when the diode is reverse-biased) and the shortest possible recovery time. The circuit can be triggered as often as once every 20 ms. Amplitude of the output pulse is −7 and +7 V.

Figure 3 shows how the circuit can be modified to supress the sequential action. With the trigger applied to the negative input of the operational amplifier, a positive trigger will yield a negative pulse and a negative trigger a positive pulse, with both pulse durations independently adjustable and mutually exclusive in time.

Series approximation sine-wave generation

Don Aldridge and **Karl Huehne**
Motorola Semiconductor, Phoenix, AZ

Generation of waveforms in a function generator requires generating one waveform, then shaping this waveform into the other desired waveforms.

One of the most difficult waveforms to produce is the sine wave. This approach for sine-wave generation, uses a series approximation of the sine wave:

$$\sin X = X - \frac{X^3}{3!} + \frac{X^5}{5!} - \frac{X^7}{7!} + \ldots$$

From this equation, it is evident that the sin X

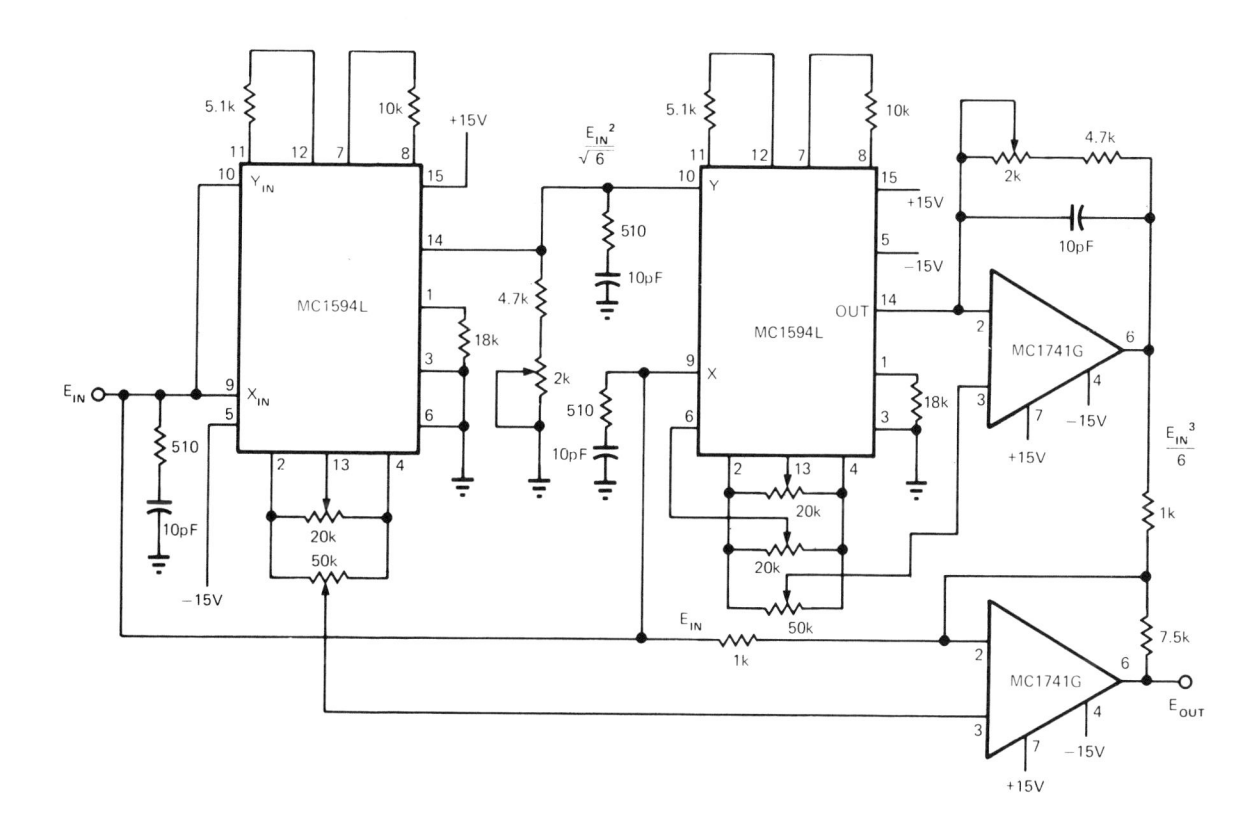

Fig. 2—Detailed schematic of the "infinite" series sine-wave shaper which requires only four ICs.

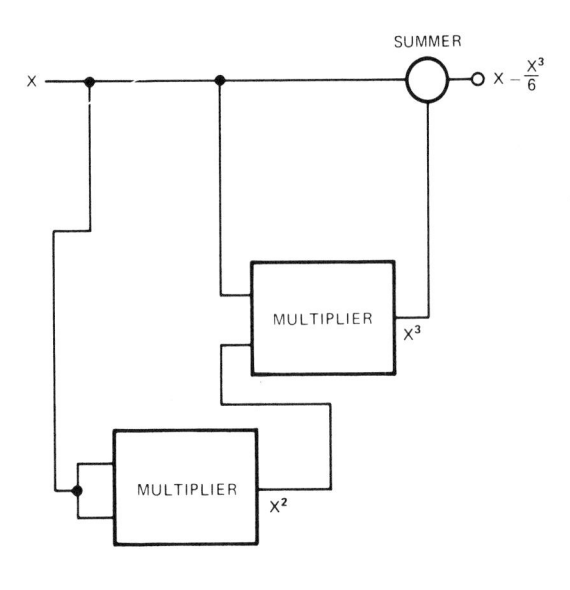

Fig. 1—Using IC multipliers to approximate sine-wave series
(sin X = X − $\frac{X^3}{3!}$ + $\frac{X^5}{5!}$ − $\frac{X^7}{7!}$ ...) by generating X − $\frac{X^3}{6}$

can be approximated from a linearly increasing value of X, which is a triangular waveform.

Using the first two terms of the series, a rather good approximation of the waveform to be produced is:

$$\sin X = X - \frac{X^3}{6}$$

This function can be easily implemented with a pair of analog multipliers. **Fig. 1** shows the block diagram of this technique with the complete schematic using the Motorola MC1594 monolithic multiplier shown in **Fig. 2**. The two multipliers generate the X^3 term while an operational amplifier is used to sum the X and the X^3 terms.

The other op amp is a current-to-voltage converter for the multiplier. A constant amplitude of 3.14V p-p is required for the bipolar triangle-wave input. The output sine wave has an amplitude of approximately 15V p-p.

A sine wave of less than 0.3% total harmonic distortion can be generated with an upper-frequency limit of about 1 kHz. The circuit can be operated to an upper-frequency limit of around 15 kHz, above which the distortion becomes noticeable.

A sine function look-up table can also be generated from this method. Here the input would be the angle, with the output being the sine of this angle.

2 gates make quartz oscillator

Edward G. Olson
Naval Air Development Center, Warminster, PA

A type SN7400 IC quad-2 input NAND gate, a quartz crystal of the desired frequency, and a resistor may be arranged as shown to provide a square-wave output of approximately 3.5V. One of the two remaining gates may be used to gate the generator output.

Resistor R_1 biases the emitters of this NAND such that the output terminal (lead 6) is approximately halfway between V_{cc} and ground which in turn, biases the emitters of the upper NAND at a suitable point. The crystal provides a resonant path for feedback. With the crystal removed from its socket, the circuit will oscillate momentarily at a much higher frequency due to the capacitance between terminals of the crystal socket providing feedback. With the crystal installed, this momentary oscillation shocks the crystal into oscillation at its resonant frequency.

The circuit has operated reliably at frequencies from 120 kHz to 4 MHz. The upper-frequency limit is not known although the output waveform is approaching a clipped sine wave around an operating frequency of 4 MHz. This waveshape triggers type SN7490 decade counters reliably. The output terminal (3) is capable of normal fan out.

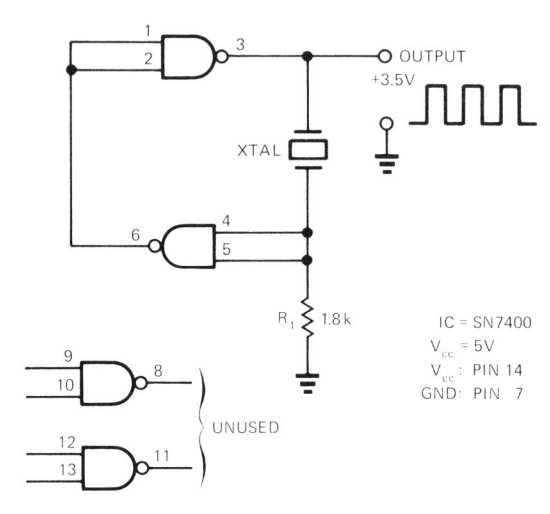

Half a TTL package and a quartz crystal combine to make stable clock sources up to 4 MHz.

Monolithic timer generates 2-phase clock pulses

Gerd Schlitt
Signetics Corp., Sunnyvale, Calif.

The Signetics 555 Timer can be used as an oscillator to generate non-overlapping clocks, which are required for most 2-phase dynamic MOS memories and shift registers. The features that make this device appealing for the application shown in **Fig. 1** are its accuracy and adjustable duty cycle. The 555 has a temperature stability of 0.005% per °C and is insensitive to supply-voltage variations.

As shown in **Fig. 2**, the duty cycle can be programmed by two external resistors which, together with the timing capacitor, determine the frequency of oscillation.

The pulse-width of clock phases ϕ_1 and ϕ_2 is readily adjustable by varying R_A and R_B. The 7473 flip-flop controls the phase that is switched on through the 7402 NOR-Gates. The timing waveforms shown in **Fig. 3** give the response of the circuit for the timing component values shown in **Fig. 1**.

The maximum operating frequency of this configuration is 1 MHz, limited by the timing circuit.

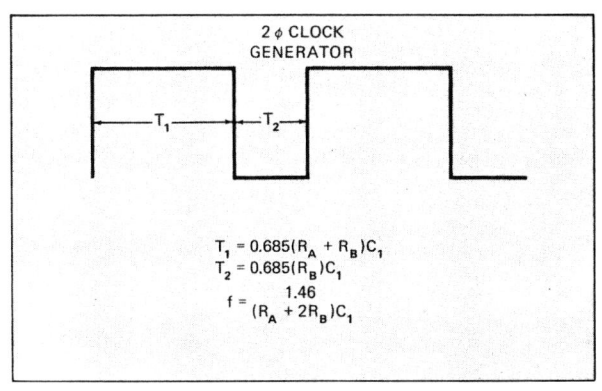

$$T_1 = 0.685(R_A + R_B)C_1$$
$$T_2 = 0.685(R_B)C_1$$
$$f = \frac{1.46}{(R_A + 2R_B)C_1}$$

Fig. 2—**Timing equations** given here determine component values required to achieve the duty cycle and desired frequency of the pulse circuit shown in **Fig. 1**.

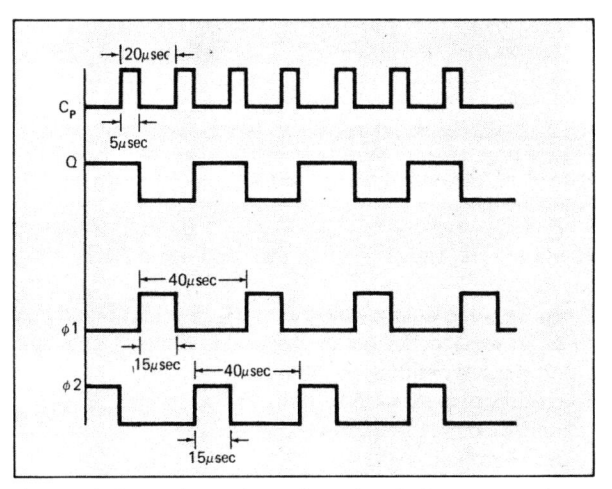

Fig. 3—**Timing and waveforms** from the clock generator shown here are those produced by the values of R_A, R_B and C_1 given in **Fig. 1**.

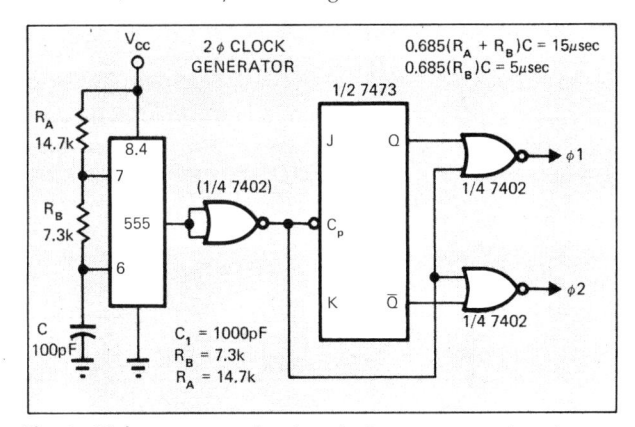

Fig. 1—**Biphase non-overlapping clock generator** is based on a monolithic timer IC and two TTL packages to provide the timing pulses required for MOS memory operation.

Current-controlled triangular/square-wave generator

Sergio Franco
Univ. of Illinois, Urbana, IL

This circuit is centered around an operational-transconductance amplifier (CA3080), i.e., an amplifier whose output signal is a current rather than a voltage. If the differential input voltage $|V^+ - V^-|$ of the CA3080 exceeds a silicon-junction voltage drop (about 0.6V), the output current assumes a saturation value and becomes simply equal to $\text{sign}(V^+ - V^-) \times I_{BIAS}$. Hence, by controlling the value of I_{BIAS} and the polarity of $V^+ - V^-$ one can easily turn a CA3080 into a current-controlled integrator of both polarities and use it, for instance, in the implementation of a current-controlled triangle oscillator.

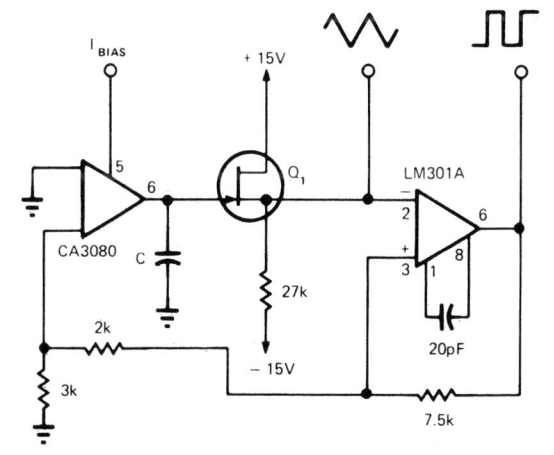

Transconductance amplifier CA3080 operates as a controllable integrator and the LM301A op amp as a Schmitt trigger in this current-controlled triangular and square-wave generator.

Signal Sources 117

The LM301A op amp, due to the regenerative feedback, acts as a Schmitt trigger with about ±5V of hysteresis, and Q_1, which is a high input-impedance voltage follower, prevents loading of the integrator output. Whenever the output of Q_1 reaches one of the threshold levels of the Schmitt trigger, the direction of integration is reversed, owing to the fact that the op amp output changes polarity, thereby causing the polarity of $V^+ - V^-$ to change as well.

The amount of overdrive $|V^+ - V^-|$ is about 3V, which is within the maximum rating specified by the manufacturer. The voltage offset, V_{GS}, introduced by Q_1 is irrelevant, since it is compensated for by causing the CA3080 output to swing about V_{GS} rather than about ground. Of course, if desired, one can always replace Q_1 with a low input-bias-current op amp connected as a voltage follower.

The circuit works very satisfactorily over at least three decades of I_{BIAS} control current (1 μA to 1 mA). The output frequency is $v = 0.025\, I_{BIAS}/C$, where v is in Hz, I_{BIAS} in amperes and C in farads. By deriving I_{BIAS} from a voltage-to-current antilog converter, and by sending the triangular waveshape through an appropriate diode-shaping network, a voltage-controlled sine-wave oscillator can be built that works over the whole audio range (20 Hz to 20 kHz).

Gated clock generates pulse train or single pulse

by J. V. Sastry
Westinghouse Electric
Pittsburgh, Pa.

The circuit shown in **Fig. 1** either gates a train of full clock pulses through or it generates a single output pulse, as required.

When the control line is held at logic 0, the circuit transmits a train of complete clock pulses, beginning with the first clock pulse that starts to rise after application of the gate signal and ending with the last clock pulse that starts to rise before the gate signal falls.

In other words, regardless of the relative timing of the gate signal, full clock pulses are always gated through.

When the control line is held at logic 1, the circuit transmits one complete clock pulse after the gate signal rises. Continuance of the gate signal thereafter is immaterial. To send another signal pulse, the gate signal must be removed and reapplied. (A logic-1 gate signal is assumed.)

In **Fig. 2**, timing diagrams show the outputs for various combinations of clock and gate signals.

The circuit uses a total of seven gates and can be constructed with only two ICs. A working model has been built using a Fairchild LPDTμL 9047 (triple 3-input NAND) and a 9046 (quad 2-input NAND). Other compatible DTL or TTL NAND gates can be used.

For the circuit mode in which the control line is held at logic 0, gate G_5 provides an additional output line (with a train of full clock pulses).

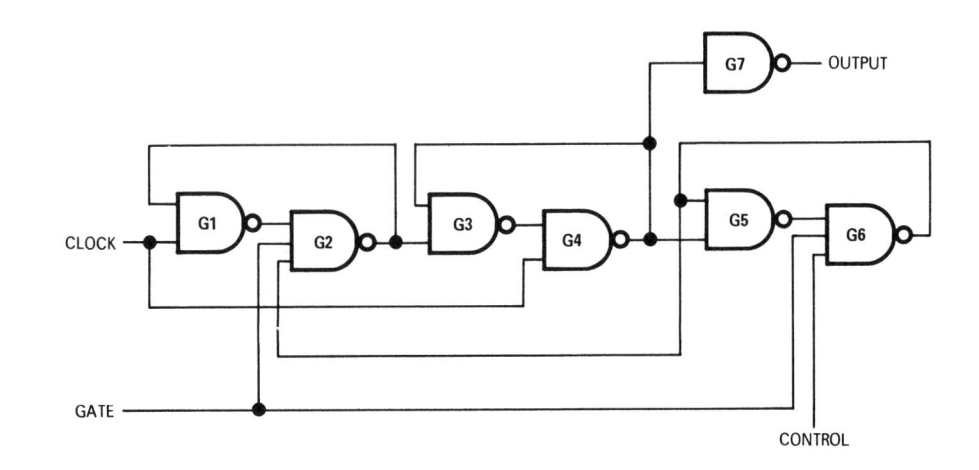

Fig. 1 – With logic 0 on the control line, this gate circuit transmits a train of clock pulses; with logic 1 it transmits a single pulse.

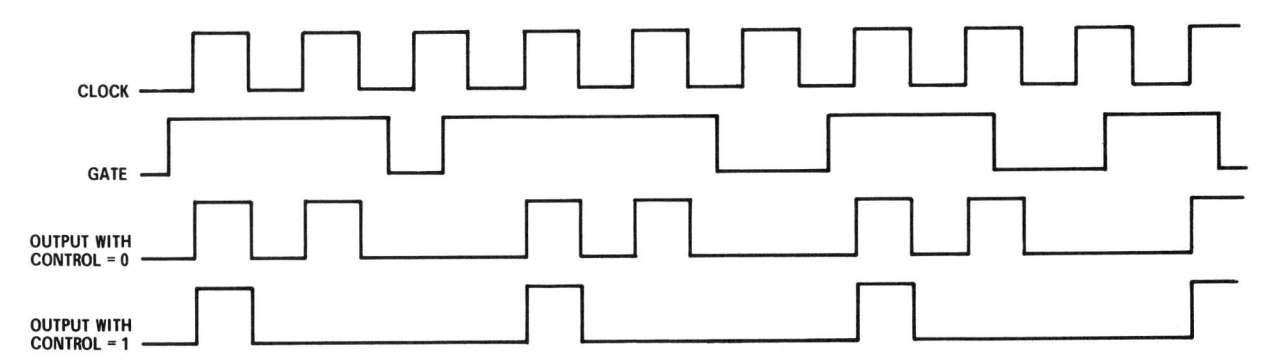

Fig. 2 – As can be seen in this timing diagram, full clock pulses are transmitted; if the gate signal falls to logic 0 during a clock pulse, the remainder of the pulse is still transmitted.

Crystal-controlled relaxation oscillator

by Robert D. Clement
and Ronald L. Starliper
Western Electric Co.
Burlington, N.C.

In a conventional relaxation oscillator, operating frequency is determined primarily by the circuit's RC time constant. But any variations in resistance, capacitance, ambient temperature or power-supply voltage will cause the oscillator s frequency to change.

Frequency stability of a relaxation oscillator can be improved, however, by adding a crystal in the frequency-determining circuit. Since the crystal controls oscillator frequency, the circuitry is therefore less susceptible to power-supply variations and component drift.

The circuit shown in **Fig. 1** is a fairly conventional unijunction-transistor relaxation oscillator in which the charging capacitor has been replaced by a 100-kHz quartz crystal (Texas Crystals Type FT243 or equivalent.) The output waveform at base 2 of the UJT is a distorted sine wave which is produced when the crystal capacitance discharges through the UJT. The signal at the emitter of the UJT is a damping-type discharge as shown. The circuit shown yielded a measured output frequency of 99,925 ±1 Hz. A version with a 1-MHz crystal had an output frequency of 999,663 ±1 Hz.

Another type of crystal-controlled relaxation oscillator is shown in **Fig. 2**. This version uses a four-layer diode as the active element.

Capacitor C_1 is needed across the crystal because the crystal capacitance alone cannot supply enough energy to sustain oscillation. Variable resistor R_1 allows adjustment of the RC time constant so that it is close to the crystal's resonant frequency, thus allowing the oscillator to lock to the crystal frequency. Note that this resistor can be adjusted so that oscillation occurs either at the crystal's fundamental frequency or at half the fundamental frequency.

For the circuit shown in **Fig. 2**, measured fundamental- and subharmonic-mode output frequencies were 99,934 ±1 Hz and 49,966 ±1 Hz, respectively. The output pulse appears across R_3 each time C_1 discharge through the four-layer diode.

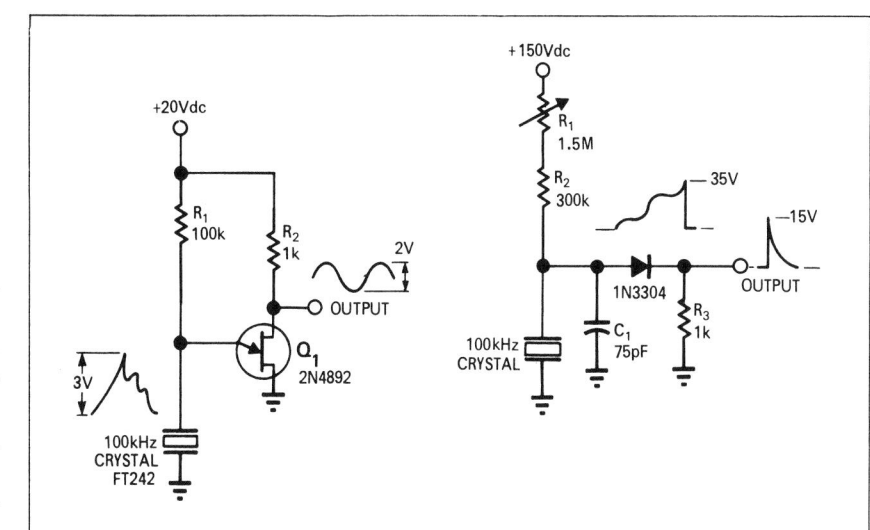

Fig. 1—**Crystal-controlled** relaxation oscillator is similar to RC oscillator except that a crystal replaces the capacitor.

Fig. 2—**Alternative** type of relaxation oscillator uses four-layer diode. Resistor R_1 controls oscillator frequency which can be the fundamental frequency or a subharmonic of the crystal.

TTL inverter makes stable Colpitts oscillator

by Charles A. Herbst
Dumont, N. J.

MOST IC-clock circuits use either RC or crystal-controlled multivibrators as their oscillators. The crystal clock is very stable but expensive, while the RC clock is relatively unstable but inexpensive. This circuit provides a happy medium between the two clocks by changing the RC oscillator to a more stable LC type.

G_1 operates as a Colpitts oscillator with C_1 setting the feedback level and $L_1 C_2$ setting frequency. R_1 and R_2 provide isolation for the resonant circuit which improves the circuit frequency stability. Low dc resistance through R_1, R_2 and L_1 provides high negative dc feedback around G_1 and biases it into its linear region. G_2 squares up the output of G_1 to appropriate TTL-logic levels.

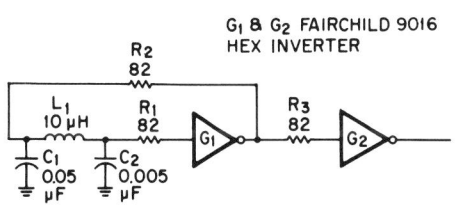

A stable LC oscillator is built from two TTL inverters. For the values shown, output period is 1.2 µs and rise and fall times are less than 20 ns.

Simple op amp relaxation oscillator generates linear ramp output

Jerry Graeme,
Burr-Brown Research Corporation, Phoenix, Arizona

The common relaxation oscillator shown in (a) employs a unijunction transistor (UJT), and provides a timing pulse train and a crude ramp output. A fairly precise timing pulse rate can be achieved by appropriate bias of the UJT to remove temperature sensitivity. However, the zero temperature-coefficient bias varies widely between UJTs of the same type, making consistent low-drift bias impossible without trimming. In addition, the ramp generated by the relaxation oscillator generally has serious limitations.

Poor ramp linearity and output offset are two of these limitations which can be overcome with the op amp circuit shown in (b). The discharging action provided by the UJT is now provided by Q_1 and Q_2, and the capacitor charging current through R_1 is now controlled by the op amp. To insure a constant charging current to C_1, the op amp feedback holds one end of R_1 at ground level. The voltage on R_1 is then constant, and the charging current is independent of the capacitor voltage. This results in a linear rise in output voltage.

The output-voltage rise continues until Q_1 and Q_2 turn on, as initiated by the emitter-base breakdown of Q_2. Breakdown results from the inverted connection of Q_2 and provides a low output-voltage limit. Since the collector-base junction of inverted Q_2 is forward biased, its voltage drop is that of a forward biased junction.

This peak voltage is temperature stable because the typical 2.6 mV/C sensitivity of $-BV_{EB}$ is largely cancelled by the -2mV/C variation in V_{BE}. As a result, the temperature coefficient of the output peak V_P is around 0.01%/°C. Because the signal is a linear ramp, frequency drift is also 0.01%/°C Frequency, as set by the integration time, is:

$$f = \frac{V-}{V_p} \cdot \frac{1}{R_1 C_1}$$

Once the emitter-base junction of Q_2 breaks down, a base current is supplied to Q_1. Then Q_1, in turn, supplies base current to Q_2 and discharging current to the capacitor. As the capacitor voltage drops below the breakdown point, Q_1 and Q_2 (because of their positive feedback) remain on, to continue discharging C_1. Discharging continues until the capacitor voltage will no longer sustain V_{BE}. At this point the capacitor voltage equals V_{BE} and the output er_{min} is zero. The cancelling voltage provided by V_f removes any output offset.

With this op amp circuit, the ramp-train frequency is limited only by the slewing rate of the op amp selected. Ramp linearity is limited by leakage from Q_1 and Q_2, the sensitivity of V_f to e_r, and the op amp input-overload-recovery following discharge. Leakage from Q_1 and Q_2 is limited by R_3, which also prevents this leakage from turning the transistors on. A stray turn-on current is also created by capacitance coupling from the output, which means that the selected value of R_3 is also frequency dependent.

Linearity errors from variations in V_f are a result of the change in the diode current as e_r changes the drop on diode bias resistor R_2. Nonlinearity from all of these effects is rarely more than 3%.

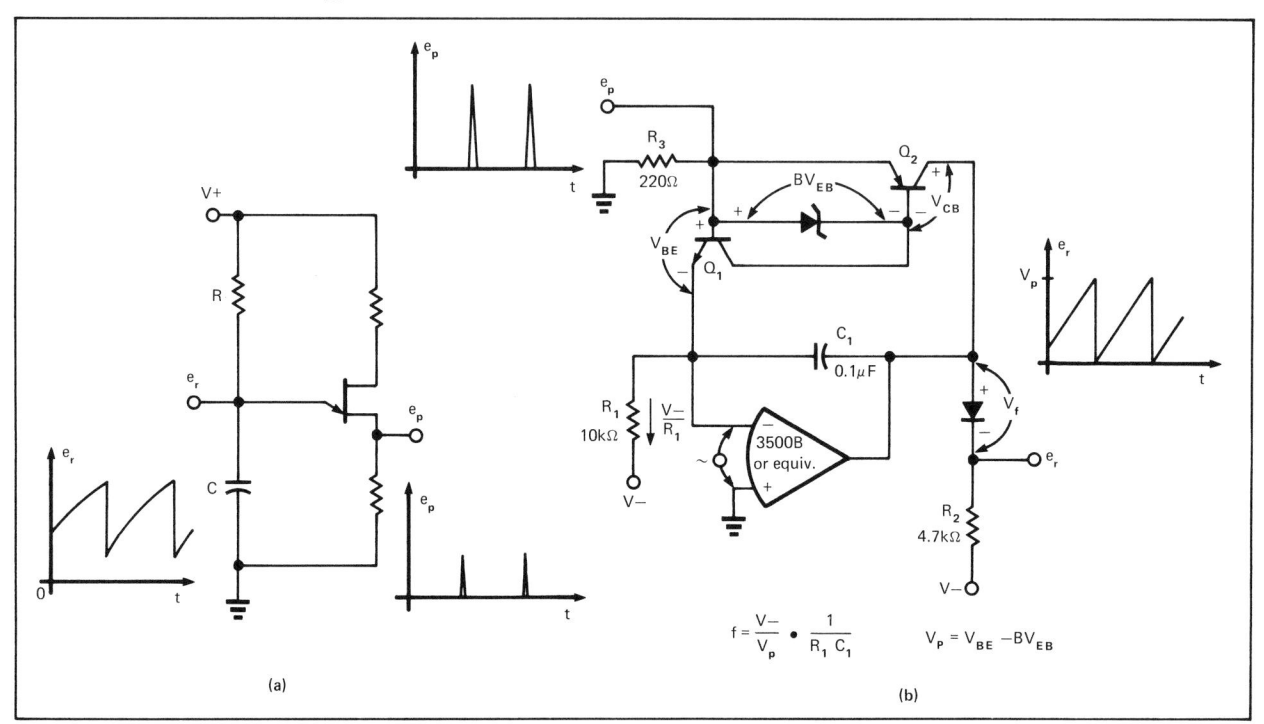

Frequency drift and nonlinearity of conventional relaxation oscillator (a) are removed by compensating junctions and linearizing of the operational amplifier feedback (b).

Astable operation of IC timers can be improved

John P. Carter
Southwest Research Institute, San Antonio, TX

The 555 IC timer has an almost limitless number of possible uses. The astable oscillator mode, however, as shown in **Fig. 1** from a manufacturer's applications note, leaves quite a bit to be desired. The charge time of C_1 to HIGH (approx. 2/3 V_{cc}) is given by:

Output ON time = $t_1 = 0.69 (R_A + R_B)C$
and the discharge time for the same capacitor is:
Output OFF time = $t_2 = 0.69 (R_B) C_1$.
Solving for a duty cycle equation gives:

$$D = \frac{t_2}{t_1 + t_2} = \frac{R_B}{R_A + 2R_B}$$

This result is rather discouraging because it is immediately apparent that a square wave (50% duty cycle) can never be achieved, and the ON-OFF timing controls are not independent.

The addition of a few components to the same circuit, as shown in **Fig. 2** can yield an extremely versatile astable oscillator with totally independent ON and OFF times. The modified circuit employs a second RC network which is controlled by the output terminal of the timer.

During the charging time of C_2 through R_C, the output is held very close to V_{cc}. When C_2 reaches 2/3 V_{cc}, the internal threshold detector clamps C_2 and forces the output to a low state. Now, C_3, which has been held at approximately V_{cc} by the output terminal, begins discharging through R_D. When the voltage on C_3 discharges to 1/3 V_{cc}, the circuit triggers and the cycle repeats. The time constant for each mode is given by:
$t_1 = 1.1 R_C C_2$
$t_2 \simeq 1.1 R_C C_3$
and the oscillator duty cycle, if $C_2 = C_3 = C$, is given by:

$$D = \frac{R_C}{R_D}$$

The free-running period is simply the sum of the two time constants and, therefore, the frequency is:

$$F = \frac{1}{T} = \frac{0.90}{C(R_C + R_D)}$$

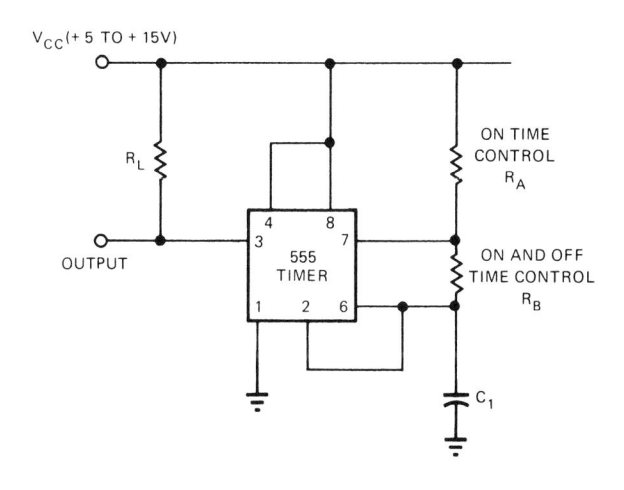

Fig. 1—Astable operation of IC timer as shown in application notes has limited duty-cycle range. Output OFF time can never equal or exceed ON time.

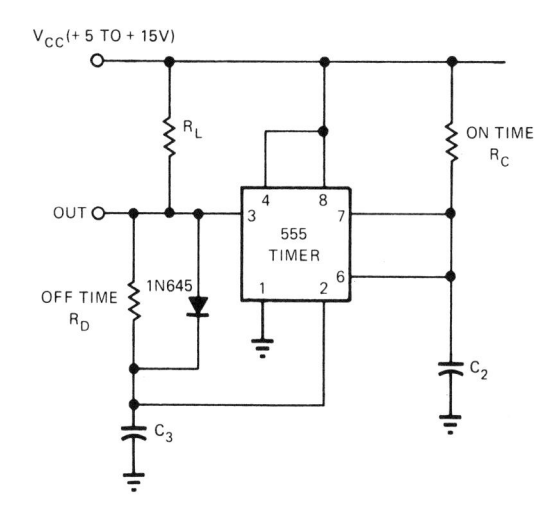

Fig. 2—Alternate circuit for astable operation allows independent control of ON and OFF time and gives a full range of duty-cycle coverage.

Ultra low distortion oscillator

by Richard Burwen
Analog Devices
Cambridge, Mass.

THIS AMPLITUDE-stabilized, 1-kHz oscillator delivers 7 Vrms with only 0.01% typical total harmonic distortion. The oscillator consists of an amplifier A_1, whose closed-loop gain of 3 is set by a negative feedback network (R_2, R_3 and R_4). Regenerative feedback through a band-pass filter, consisting of C_1, C_2, R_1 and R_5, determines the frequency of oscillation.

Output amplitude is stabilized by a multiplier which increases negative feedback via R_3 when the output signal reaches 7 Vrms. The control voltage of the multiplier is derived from an integrator A_2. This amplifies the voltage difference between a 15-V reference and the rectified output of A_1. Amplifier A_2 acts as an integrator or low-pass filter for feeding a smoothed dc into the multiplier. CR_2 allows only positive outputs from A_2 so that polarized capacitors can be used.

The multiplier is used only for vernier gain adjustment on the oscillator, so its distortion has little effect on the output. Distortion is limited primarily by the characteristics of op amp A_1 which is an AD503J with a high slewing rate. A type μA741C can be used if 0.04% distortion is tolerable.

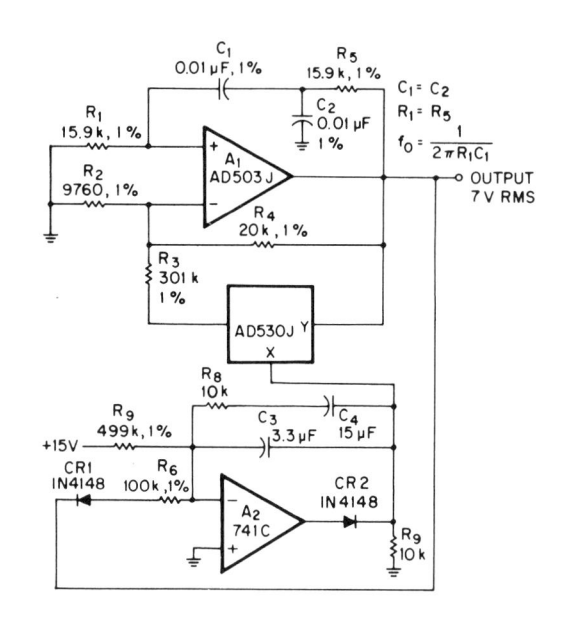

This oscillator circuit uses a monolithic IC multiplier as part of an amplitude-stabilizing loop.

CMOS circuits generate arbitrary periodic waveforms

John R. Tracy
Litton Systems, Van Nuys, CA

In the testing of analog systems it is often desirable to drive the system with a certain arbitrarily-shaped periodic waveform. Such waveforms can be effectively synthesized using CMOS analog multiplexers. The hardware complexity of the synthesis depends on the spectral content of the desired waveshape and the accuracy desired. For example purposes, the periodic waveshape shown in **Fig. 1a** will be synthesized.

The synthesizer circuit is illustrated in **Fig. 2**. The two CD4051 CMOS analog multiplexer chips form a 16-channel analog data selector, which is continuously sequenced at a constant rate through all its states by a binary counter (CD4024). The binary counter is clocked at a rate 16 times the fundamental frequency of the desired waveform.

At each multiplexer input there is a weighting resistor connected to a fixed source. As the data selector is sequenced through its states, the output of the selector forms a step-approximation of the desired waveform as a function of the input weighting resistors (**Fig. 1b**). The data selector is followed by a 2-pole active-filter stage to remove components of the sampling frequency. The period of the analog waveform is 1.67 msec. The sampling rate is 9.6 kHz, which is 16 times the 600-Hz fundamental frequency. The cutoff of the output filter is 4.8 kHz.

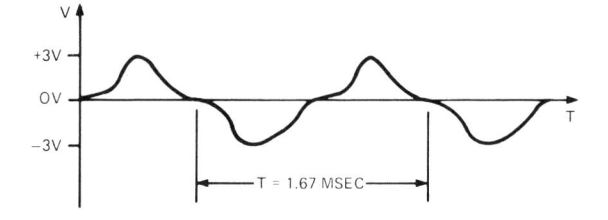

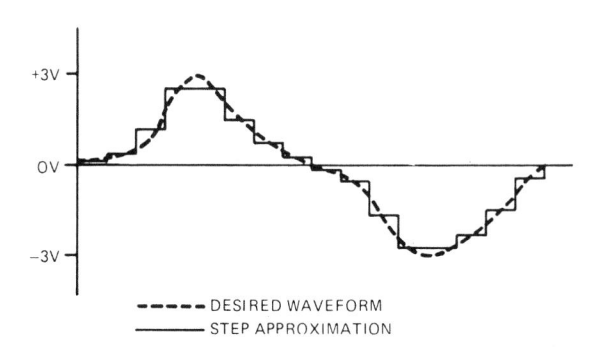

Fig. 1—The desired waveform shown in (a) can be synthesized by the step-approximation of (b).

If greater accuracy is required in the output waveform or if there are sharp transitions in the output waveform causing increased spectral content at higher frequencies, the sampling rate can be increased. This requires additional multiplexers and a correspondingly higher-cutoff output filter.

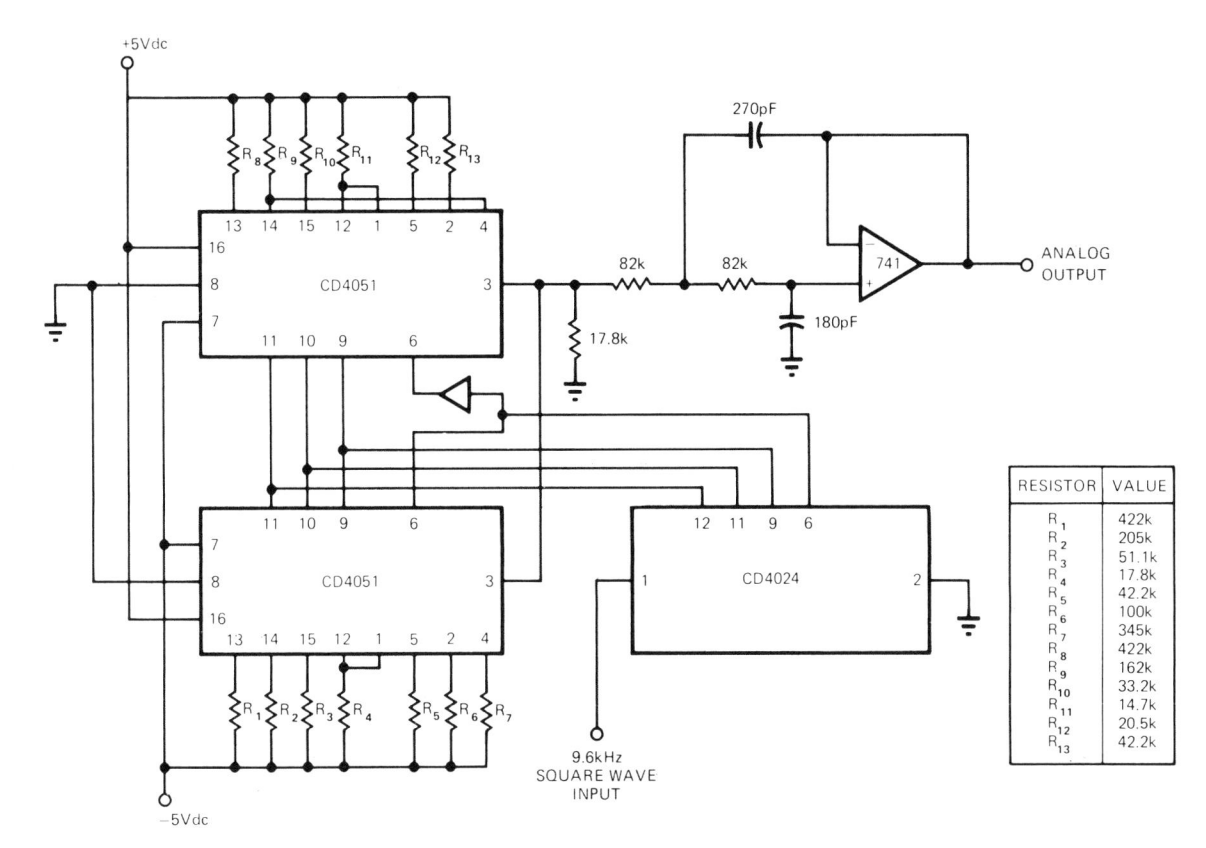

Fig. 2—Shape of the output waveform is determined by the values of the weighting resistors connected to the inputs of the multiplexer chips.

Improved Multi With Continuously Variable Rep Rate

By Joseph H. Bayne, Jr.
and Robert J. Haislmaier
U. S. Naval Ordnance Laboratory
Silver Spring, Md.

THE BASIC MULTIVIBRATOR shown has a square wave output across the collector resistor and a triangular wave output across the capacitor. (See Circuit Design No. 56, EEE, Oct., 1964 by Peter Lefferts.) When this circuit is biased from a constant-current source, the tops of the square wave become flat, and the triangular wave becomes linear. When the current source is made variable, the repetition rate becomes variable.

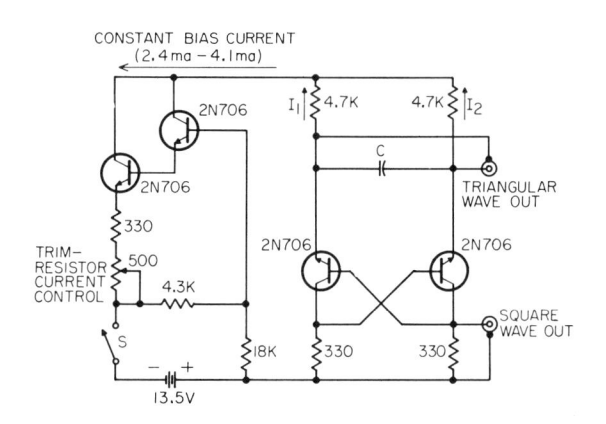

Improved multi with continuously variable rep rate.

The expression describing this variation of the period with current is:

$$T \propto \frac{C(I_1 + I_2)}{I_1 I_2}$$

In a symmetrical arrangement $I_1 = I_2$.

$$T \propto \frac{2C}{I_1}$$

The repetition rate thus varies directly with the magnitude of the constant biasing current.

Capacitance Value-C	Repetition Rate—Continuous Variation
100 µf	5.6 cps to 10 cps
10 µf	56 cps to 100 cps
1 µf	560 cps to 1 kc
0.1 µf	5.6 kc to 10 kc
0.01 µf	56 kc to 100 kc
0.001 µf	560 kc to 1 mc
330 pf	1.55 mc to 2.68 mc

Therefore, by using a variable current control resistor, one multivibrator can be used over a very wide range of repetition rates with few components, as is shown in the table. For this circuit the repetition rate may be varied by more than 70 percent. If an analog waveform is impressed on the current biasing transistor, voltage-to-frequency conversion results.

Because the amplitudes of both the square wave and the triangular wave are fixed by the transistor internal voltage drops, these amplitudes remain fairly constant over the full frequency range.

Signal Sources 123

Precision oscillator is versatile

Leonard Accardi, Kollsman Instrument Corp.,
Elmhurst, New York

This design features a low-distortion amplitude-stable oscillator. It does not require the usual trimmers or trial-and-error selection of components to determine the output level. Also of interest are the psuedo full-wave rectification by a single op amp, and the use of an op amp in the "pure" integrator configuration. Usually "pure" integrators can not be built since bias current would cause the amplifier to drift into saturation. The circuit incorporates a Wein Network as the basic resonator.

Normally, response time and linear operation are mutually exclusive properties, and therefore, lamps or similar devices, which introduce a time constant of their own and require circuit trimming are used as gain control elements in order to achieve linearity. A good rule of thumb for the distortion introduced by a FET in the voltage controlled resistance mode is that a drain-source voltage of 1% of pinchoff results in 1% distortion. Knowing that the A_1 circuit gain must be exactly three, this circuit is designed so that the FET resistance merely provides a correction term to the nominal gain, thus minimizing the (drain-source) voltage across the FET and at the same time producing a high level, low distortion output. For a 20V pk-pk output, V_{ds} is nominally 8 mV peak or 1/6% of the minimum pinchoff voltage of Q_1 indicating a distortion of about 1/6%. A_1 is in the non-inverting configuration, therefore its gain is $1 + R_F/R_{in}$. The 40 kΩ feedback resistor and 20 kΩ input resistor set this gain at 3. To maintain the gain of three, the 20 kΩ and 40 kΩ resistors and using a silicon resistor for the 100Ω resistor so that its temperature coefficient matches that of the drain-source resistance of the FET. Frequency and frequency stability are determined solely by quality of the RC components used in the Wein network, if the closed loop amplifier A_1 introduces no phase shift or no changes in phase shift at the oscillation frequency. Most general purpose op amps are adequate for oscillation frequencies up to 1 kHz.

The remainder of the circuit is for amplitude determination and control. A_2 is connected as a standard precision half-wave rectifier; however, the addition of Resistor R_2 effectively provides a full wave rectified current input to the inverting input of A_3. In the steady state, the dc value of the full wave rectified current is equal and opposite to the current in R_3 which is precisely determined by a reference zener diode. A_3 is designed to give a dc gain equal to the open loop gain of the op amp (3×10^5, typical) and a low ac gain so that ac components of the current input are sufficiently attenuated. A_3 is thus connected as a "pure" integrator (no dc feedback path) which, in open loop circuitry, is forbidden. In this closed loop configuration, however, dc offsets introduce only small errors in the output amplitude of oscillation. Only the values of the integrating capacitor and input resistors to A_3 determine the bandwidth of the regulating loop and thus the transient response of the system, since there is no time constant associated with the FET. Unlike a lamp, the FET's response is essentially instantaneous and thus entirely predictable.

The high dc gain of the integrator essentially means that its dc output can adjust itself to whatever control voltage is

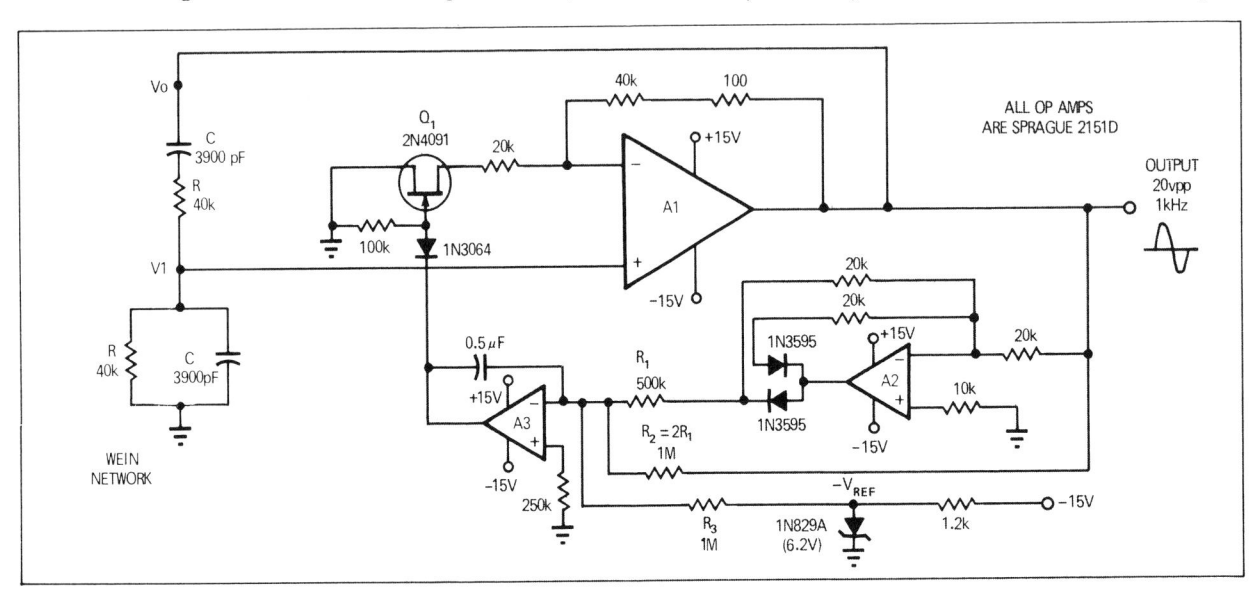

100Ω resistor forces the FET to operate at a drain source resistance of 50Ω, a highly linear region near the R_{ds} ON maximum of 30Ω for this FET. Distortion over the temperature range is kept to a minimum by specifying tracking of required by the FET to produce the drain-source resistance necessary to maintain the gain of A_1. The voltage output remains constant at 20V peak-to-peak despite variations of the gate-source voltage requirement of the FET.

A wideband, linear VCO

by Gil Bank
Westinghouse Ocean Research
San Diego, Calif.

A voltage-controlled oscillator should have good stability, excellent linearity and a wide range of operation. In addition, a sinusoidal output is frequently desired. This circuit does not have a sine-wave output, but its triangular waveform has only about one third as much third harmonic as a square wave of the same fundamental frequency. This factor makes it relatively easy to convert the triangular wave to a sine wave by filtering.

In the circuit, A1 inverts the positive input signal while Q1 serves as a switch to select the direction of integration. Amplifier A2, with feedback capacitor C1, forms an integrator. Op amp A3, a 709, is used as a comparator. Zener diodes D1 and D2 produce a plus or minus reference voltage with which the output of the integrator is compared.

To follow the circuit through one cycle of operation, neglect R6 and assume Q1 is OFF, which means the output of A3 is negative. The input signal, which must be positive, forces a current through R4 into the summing junction of the integrator, A2, whose output starts slewing negatively at a rate proportional to E_{in}. When the output of A2 reaches approximately −7V, the comparator output goes positive. This action switches Q1 ON, changes the reference voltage of the comparator from −7 to +7V and locks the comparator into a positive output state.

With Q1 ON, a negative current is drawn through R5 that is double the current supplied through R4. The net current to the integrator is equal but opposite to the previous integration current. The integrator output voltage now slews in a positive direction and rises toward +7V. At a 7V input, the comparator output is driven negative which turns Q1 OFF and causes the cycle to repeat.

Resistor R6 attenuates the drain voltage of Q1 so that it will remain completely off (during the appropriate half cycle) for high input voltages. R6 does not affect the magnitude of current injected into the summing junction of A2. Resistors (R7 and R8) and diodes (D3 and D4) are used to prevent latch-up during turn-on transients.

For the values shown and a +15V power supply, the circuit transfer function will be about 1 kHz/V. The circuit will easily operate over a 100:1 frequency range with a linearity error of less than ±0.5%.

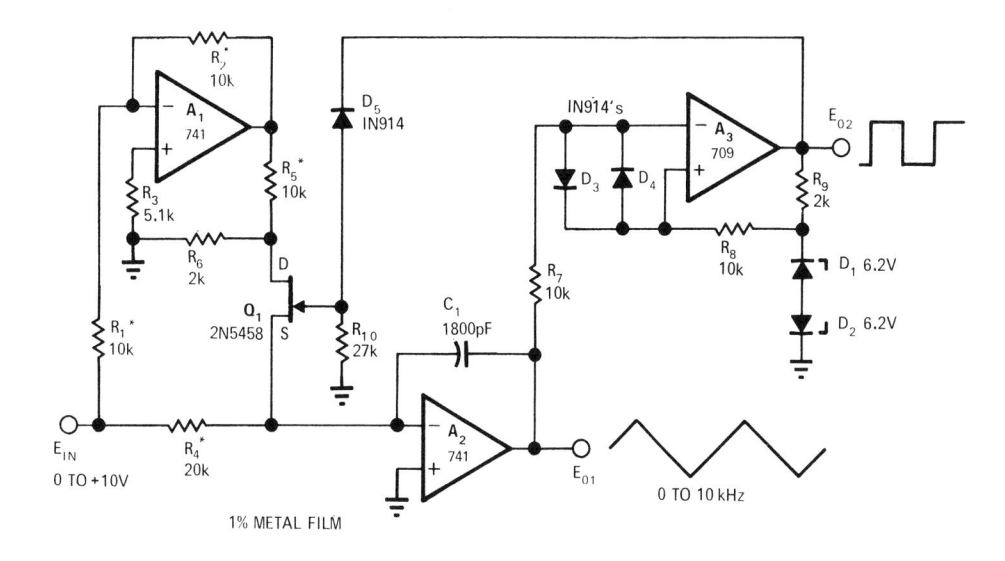

Three IC op amps and a junction FET form a VCO with a 100:1 range and ±0.5% linearity.

PUT oscillator has 4-decade frequency range

Herb Cohen
Electret Corp New York N.Y. 10023

A simple PUT relaxation oscillator, as shown in (a), has a tendency to "latch up" when the current in P_1 exceeds the valley current of the PUT. This severely limits the component values and voltages for the design of wide range oscillators.

The addition of a single transistor, as shown in (b), will prevent latching. When the PUT fires, Q_1 goes into saturation, causing C_1 to be shorted out. Q_1 acts as a shunt path for the PUT and reverse biases the anode and cathode of the PUT momentarily, thereby assuring turn-off of the PUT.

An added benefit of C_1 discharging through Q_1 is that Q_1 has a lower saturation voltage than the PUT. The difference varies from a few millivolts for some transistors to 1.5 volts for many PUTs. When considering temperature stability, this difference allows much greater design flexibility. The frequency range of this circuit is 7 Hz to 23 kHz.

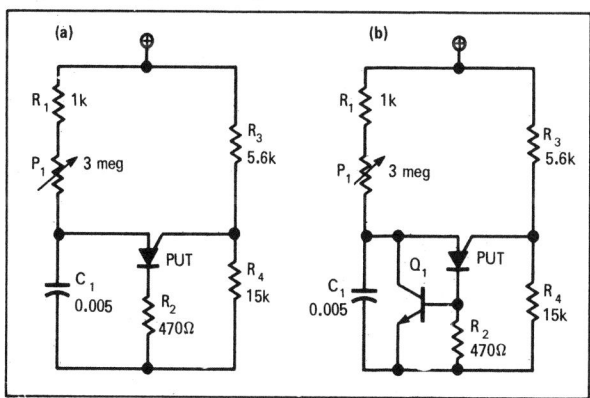

The usual approach to PUT oscillator design (a) offers limited frequency range, and often suffers from "latch-up." **The addition of** Q_1 across C_1 (b) prevents latch-up and permits an operating frequency range of 4 decades.

Simple sinewave oscillator

by John C. Freeborn
Honeywell Inc.
West Covina, Calif.

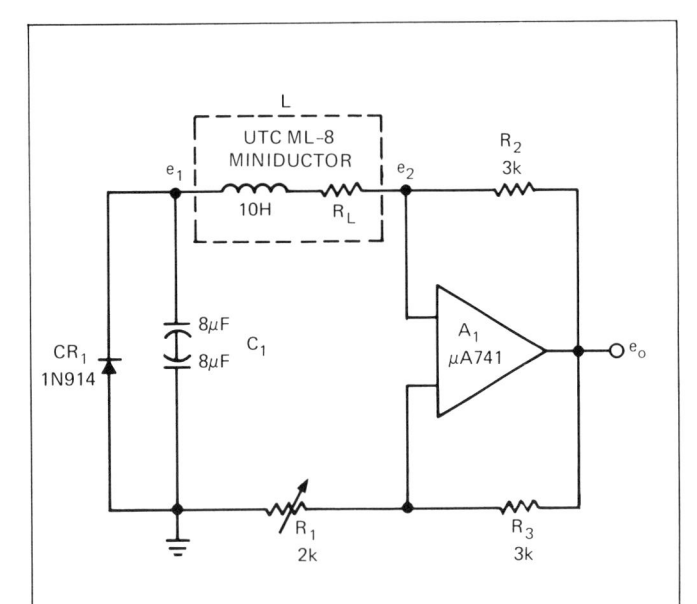

The upper frequency limit of this oscillator is determined by the response of the op amp. With the values shown, the frequency is around 25 Hz.

With the component values shown, this circuit produces a low-distortion sinewave with a frequency of approximately 25 Hz. The output voltage e_0 is around 8V pk-pk. Any combination of inductor and capacitor can be used. Output frequency can be calculated from the following well-known equation:

$$f_0 = \frac{1}{2\pi\sqrt{LC_1}}$$

Potentiometer R_1 allows adjustment of the amount of regeneration applied to the series-tuned circuit. To cause oscillation, the value must be roughly equal to the series resistance R_L of the choke plus a resistance value that represents core losses at the resonant frequency.

Diode CR_1 clamps the signal voltage e_1 and limits amplitude regeneration to avoid saturation of the inductor or op amp. Signal voltage e_1 will be an accurate cosine version of the output sine function. The amount of distortion can be made quite small by keeping R_1 at the minimum value needed to sustain oscillation.

Voltage e_2 is an accurate representation of voltage loss across the inductor (core losses plus IR drop), since it is just equal to the voltage required to overcome these losses when R_1 is adjusted to the minimum value for oscillation. The output voltage e_0 is merely e_2 multiplied by the amplifier gain R_2/R_1. (Resistors R_2 and R_3 have the same value.)

Using other component values, the basic oscillator circuit has been operated at frequencies from 15 Hz to 100 kHz, with total harmonic distortion less than 0.5%. The upper frequency limit appears to be set by the response of the op amp.

Modified UJT Oscillator Has No Timing Error

By Joseph V. Crowling
U. S. Army Training Center
Warrenton, Va.

WHEN POWER is applied to a conventional unijunction oscillator circuit (Fig. 1), the capacitor C must charge from zero volts to the peak-point voltage V_P before the first pulse occurs at B_1. For succeeding pulses, C charges from $V_{E(SAT)}$ to V_P. This causes a difference between the period of the first cycle and that for succeeding cycles, as shown in Fig. 1b. The error may be serious in some applications. It can be eliminated by a simple circuit modification.

Fig. 2a shows a method requiring one extra transistor and two resistors. When power is applied, Q_2 saturates and rapidly charges C to $V_{E(SAT)}$. The voltage at point A is equal to $V_{E(SAT)} + V_{BE}$. Q_2 is cut off and no current can flow into C except through the timing resistor R_T.

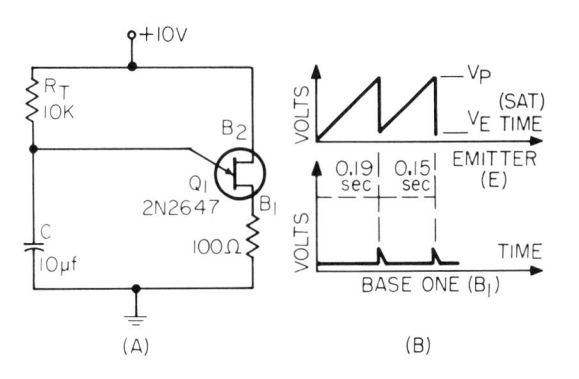

Fig. 1. Basic UJT oscillator (A) with waveforms (B). This circuit has timing error, due to time taken to charge capacitor to V_p during first cycle.

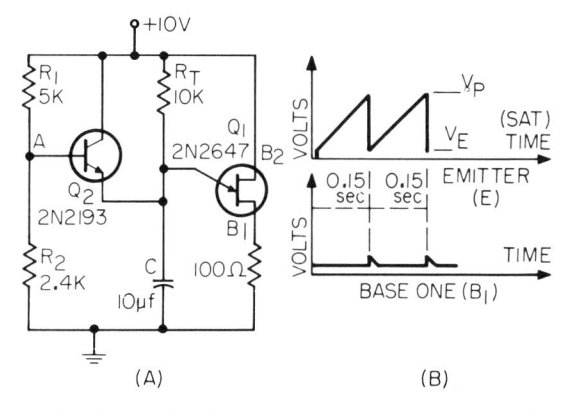

Fig. 2. Modified circuit has no error because capacitor is initially charged by transistor Q_2.

126 Circuit Design Idea Handbook

Zero V = Zero-frequency VCO

I. A. Glibbery
Rohr Industries, Chula Vista, CA

This circuit provides a linear ramp and a pulse output proportional to input voltage e_1, when operating from equal positive and negative supplies of 5 to 15V.

The most interesting application for this circuit is to use it in conjunction with a counter on the 1-kHz range to read out directly a dc voltage applied to the input. (Where 1 Hz = 1 mV.) Use of input attenuators will permit higher ranges.

Reversal of the input voltage causes the circuit to cease oscillating. This condition can be indicated with the LED shown between the ramp output and zero or ref. voltage.

The first op amp, in conjunction with R_1 and C_1, is the familiar integrator circuit, resettable with the FET gate CD4016A. The second op amp operates as a voltage comparator, with positive feedback applied via R_3, which causes latching of the circuit when the output swings positive.

Initiation of a cycle is with the integrator output at zero and the comparator output saturated negative. An input e_1 causes the integrator output to move positive until the voltage comparator inputs are such that its output swings positive. Owing to its latching action, the comparator now holds its output positive. Under the action of the positive voltage on its gate, the FET switch is now closed and consequently, shorts capacitor C_1.

The integrator output is rapidly returned to zero, at which time the comparator reverts to its original condition of negatively-saturated output.

The FET gate, now being negative, is turned OFF and allows the cycle to restart.

The linearity of the VCO is essentially determined by the ratio of the reset time to the integrate time, which at 1 kHz is 2 parts per 1000. Zero stability is a function of the op-amp offset and typically not more than 2 mV or 2 Hz at the output. The freq.-out vs input (e_1) stability is effected mainly by temperature effects on the capacitor C_1, and the op-amp characteristics; consequently, a capacitor of good temperature stability should be chosen. (The prototype circuit changed output frequency by 3 Hz for 75°F.)

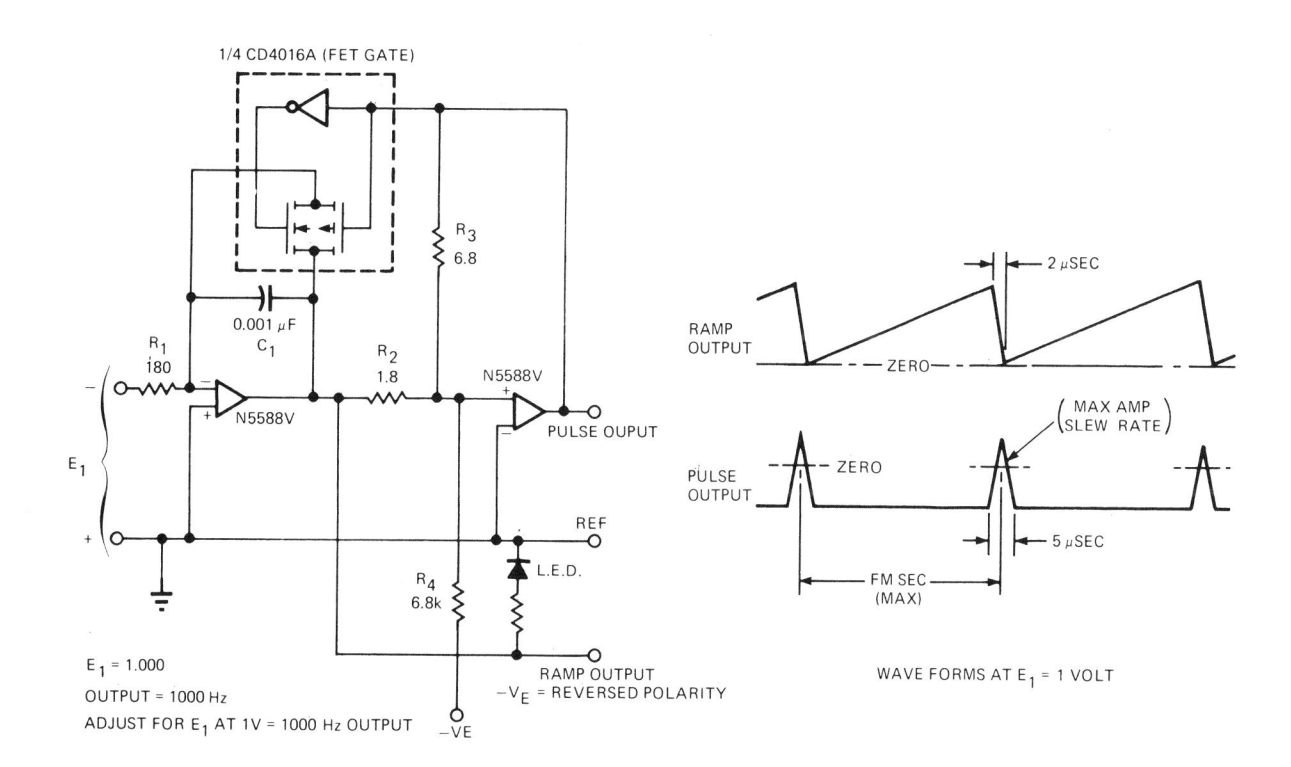

Linear VCO will not oscillate with 0V input and maintains 1 Hz/mV frequency throughout its range of operation.

Ramp generator has adjustable nonlinearity

Hank Olson
Stanford Research Institute, Menlo Park, CA

The engineering world has devoted considerable effort to the attainment of a perfectly linear ramp function. However, once the linear sweep has been attained, it is then used to sweep some other device (like a voltage controlled oscillator) which usually has its own type of nonlinearity. The result is that, with a ramp input voltage that is linear with time, the output of the circuit being swept is generally not linear with time.

The ramp generator described here will enable one to pre-distort the sweep with either concavity or convexity to compensate for the nonlinearity of the circuit being driven. The amount of distortion is continuously adjustable by means of a potentiometer.

The ramp generator is shown in **Fig. 1**. Q_1 operates as a constant-current source. The amount of constant current is proportional to the voltage difference between the ±15V supply and the base voltage of Q_1.

$$i \approx \frac{15 - e_b}{R_1} \text{ (where } e_b = \text{base voltage of } Q_1\text{).}$$

If the wiper of the "curvature" pot is set to minimum position (ground), then e_b is equal to twice the voltage on the wiper of the "starting current" pot. In this setting, the ramp output is linear. SCR_1 is periodically pulsed ON by the gate trigger impulses shown. The period of the pulses determines the period of the ramp function, because capacitor C is discharged by SCR_1 each time a positive gate trigger pulse is present.

The noninverting follower U_2 assures that the ramp voltage on C is sampled at a high enough impedance ($10^9\Omega$), so as not to cause unintended distortion of the ramp by partially discharging C. The output of U_2 provides a ramp output at reasonably low impedance for operating other circuits.

U_3 is operated as a multiply-by-two amplifier, in

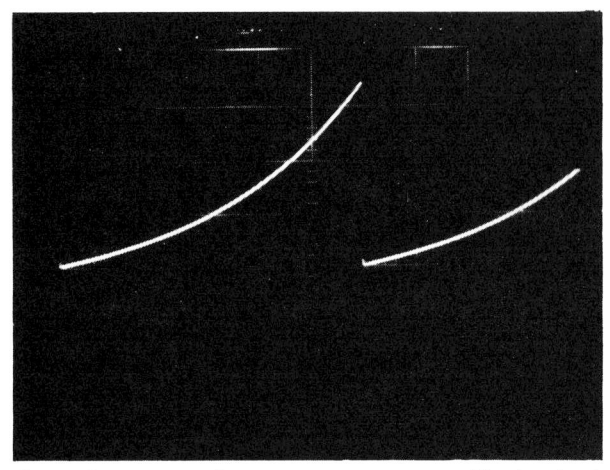

Fig. 2—Scope trace of a concave ramp generated by the circuit in **Fig. 1**. The value of C is 0.22 µF and the period is 6 msec.

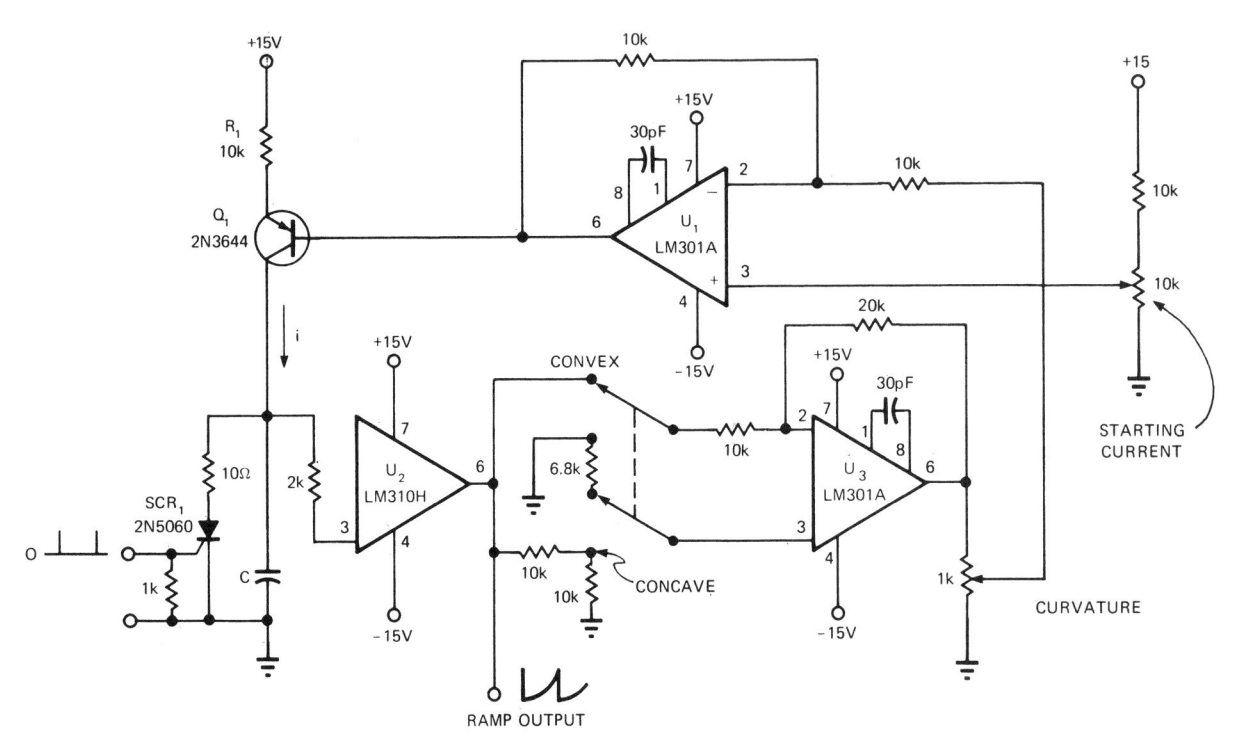

Fig. 1—Adjustable linearity/nonlinearity of this ramp generator makes it ideal for driving circuits which are nonlinear. When the final output must be linear, this circuit can be programmed to cancel the nonlinearity of VCOs and other similar circuits.

either the inverting or non inverting mode—depending on the input switch setting. In the inverting mode (switch in the "convex" position) the output has the effect of decreasing the voltage (15V − e_b) as the capacitor charges. Decreasing this voltage decreases the current that charges C and produces a ramp that departs from linear in a convex fashion.

If the switch is in the "concave" position, the rising voltage on C causes the voltage (15V − e_b) to increase, which increases the charging current into C. This causes the ramp to be distorted from linear in the direction of concave.

Unlike some earlier circuits which used capacitive coupling of the "bootstrap" feedback, this circuit uses dc coupling. This frees the designer from using huge coupling capacitors which must be considered when any change in the ramp length is made. The photograph in **Fig. 2** is an actual waveform generated by the ramp circuit; the value of C is 0.22 µF and the period is 6 msec.

Pulse generator offers wide range of duty cycles

by Jerald Graeme
Burr-Brown Research Corp.
Tucson, Ariz.

Sampling time control and pulse-width modulation are commonly achieved with asymmetrical pulse trains. Unfortunately, if one needs sampling times which are small compared to the test period or if one needs large dynamic ranges of pulse width, a conventional astable multivibrator won't provide the necessary precision. However, a single op amp can readily provide precise duty cycle control over a wide dynamic range. With suitable component values, the circuit described can generate pulse trains with duty cycles from 0.1% to 99.9%.

Essentially, the op amp forms the gain element of an astable multivibrator with the two characteristic time intervals determined by feedback elements. Astable operation is insured by positive feedback via R_3 and R_4. From this feedback, the switching threshold voltage is set at the noninverting input. Switching occurs when the inverting input voltage is brought to this threshold level by the charging of capacitor C.

Capacitor charging current is supplied from e_o by one of the feedback resistors, R_1 or R_2. A positive e_o forward biases CR_1, and R_1 controls the rate of charging with a resultant time constant of R_1C. When e_o reaches e_t, the output switches, thus reversing the polarities of e_t and the charging current. The negative output then discharges C through R_2 with a time constant that is now R_2C. Since separate resistors control the time intervals of the two output states, the duty cycle is set by simply choosing the resistors. Duty cycle is defined by the following equation:

$$\text{Duty cycle} = \frac{R_1}{R_1 + R_2}$$

For the resistor values shown, the duty cycle is equal to 0.001.

Period depends on the values of R_1 and R_2 and on the size of the capacitor. The complete equation is as follows:

Period =

$$(R_1 + R_2)C \ln\left[\frac{V_z - V_f + e_t}{V_z - V_f - e_t}\right]$$

where, $e_t = \dfrac{R_4 e_o}{R_3 + R_4}$

For the component values indicated in the schematic, the period is one second.

In a practical circuit, the range of duty cycles attainable is limited by the slewing rate, output current and input bias current of the op amp. Slew rate controls the large-signal rise time, and thereby limits the minimum time interval of either state. This minimum is also limited by the slewing rate of the capacitor as set by the amplifier output current available for charging. With the minimum interval determined by the preceding two factors, maximum interval length is limited by input bias current. This input current must be small compared to the charging currents supplied by R_1 and R_2. For this reason, high input resistance must be maintained by the op amp under input overload. A good general-purpose op amp, such as Burr-Brown's Model 3500, can provide adequate performance and most FET op amps are satisfactory for this application.

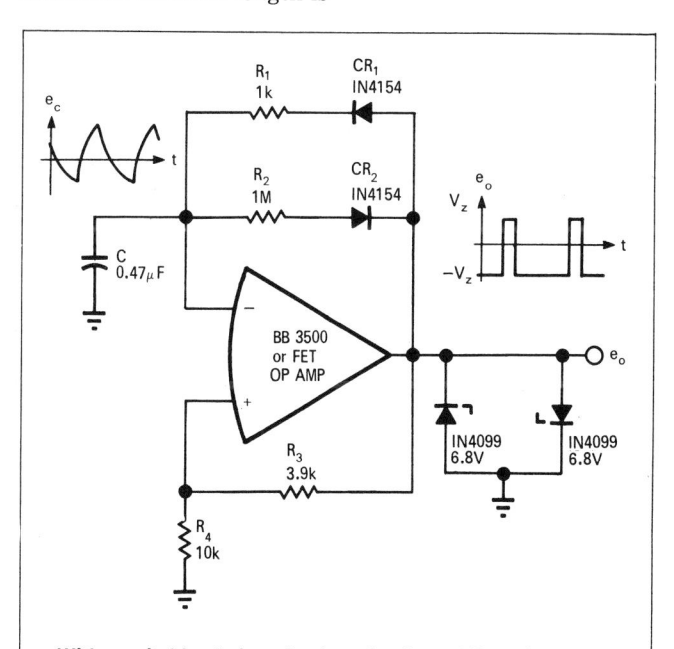

With a suitable choice of values for R_1 and R_2, this circuit can produce pulse trains with duty cycles from 0.1% to 99.9%.

Signal Sources

Self-Completing Gated Astable

By Michael Neidich
Sanders Associates, Inc.
Plainview, N. Y.

This circuit, using standard logic circuits, produces a pulse train containing any integral number of pulses. The first pulse starts when switch S_1 is closed and the last pulse is completed when S_1 is opened.

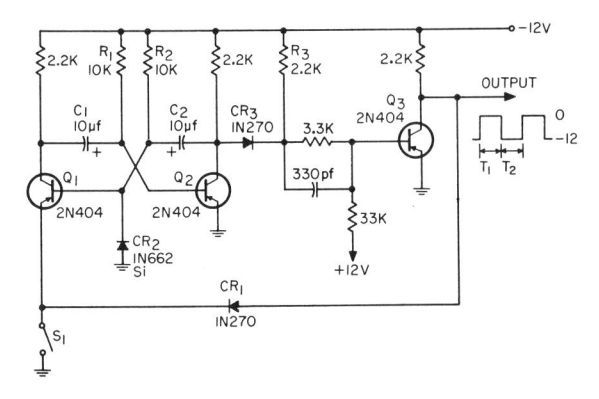

Self-completing gated astable.

The self-completing action is produced by the "OR" function of S_1 and CR_1. When S_1 is closed, Q_1 turns on and pulses are produced by astable $Q_1 - Q_2$. If Q_1 is on when S_1 is opened, it will remain on by obtaining emitter-current through CR_1 for the duration of the last pulse. When the last pulse is completed, Q_1 and Q_3 turn off.

CR_2 prevents the base voltage V_B of Q_1 from going to -12 v when S_1 is open by clamping it to -0.7 v. This prevents the first pulse from being wider than those following. Since CR_2 is silicon, it will not rob Q_1 of base drive since V_B of Q_1 never exceeds -0.7 v when Q_1 is on. If silicon transistors are used, simply use two silicon diodes in series for CR_2.

By incorporating network $CR_3 - R_3$, Q_2 is unloaded when off, and the standard timing equations may be used.

ON Time $\quad T_1 = 0.69\, R_1 C_1$
OFF Time $\quad T_2 = 0.69\, R_2 C_2$

R_1 and R_2 should be adjusted only as a last resort as their values are determined by worst-case analysis. For the values shown, the output is a square wave at 7 cps.

Clock-pulse generator

Robert Bolvin
Signetics, Inc., Sunnyvale, CA

In most older automatic IC test equipment, clock-pulse generators were not an integral part of the system. For binary and decade counters and shift registers, a clock pulse or train of clock pulses may be required to set outputs in the proper state for testing. This requires that an internal supply be programmed as a pulse generator, adding considerably to programming and test time.

The circuits shown in the figure were designed to supply up to 16 clock pulses to the DUT with a minimum of programming. A 74123B 1-shot sets a 7400A bi-stable latch to a ZERO output. A ZERO on pin 1 of the 74123B multivibrator starts a train of pulses which are counted by the 74193B up-down counter. When the counter reaches a ZERO count, the borrow output goes to ZERO, resetting the latch output to a ONE, stopping the multivibrator.

The desired number of pulses, minus 1, is set into the 74193B data inputs through any unused equipment lines. All data inputs are held HIGH, to V_{CC}; therefore only a ground is necessary to provide the correct binary input. By tying another line to the input of the 74123B 1-shot and forcing a ground, the 1-shot output goes to a ZERO state for as long as the RC time constant of $R_{EXT} C_{EXT}$. This pulse loads the binary data input number into the 74193B and, at the same time, sets the 7400A latch output to a ZERO. A ZERO on pin 1 of the 74123B starts the train of pulses. When the 74123B counts down to ZERO, the borrow output will go LOW, resetting the latch to a ONE output, stopping the multivibrator. The circuit is now ready for the next command for clock pulses.

High-riding and low-riding clock pulses are available from pins 12 and 5, respectively, for DUTs which are triggered from negative or positive transitions of the clock.

The figure shows the circuit as it would be connected to an AAI 1000 tester. For this tester only three tests are necessary to generate up to 16 clock pulses.

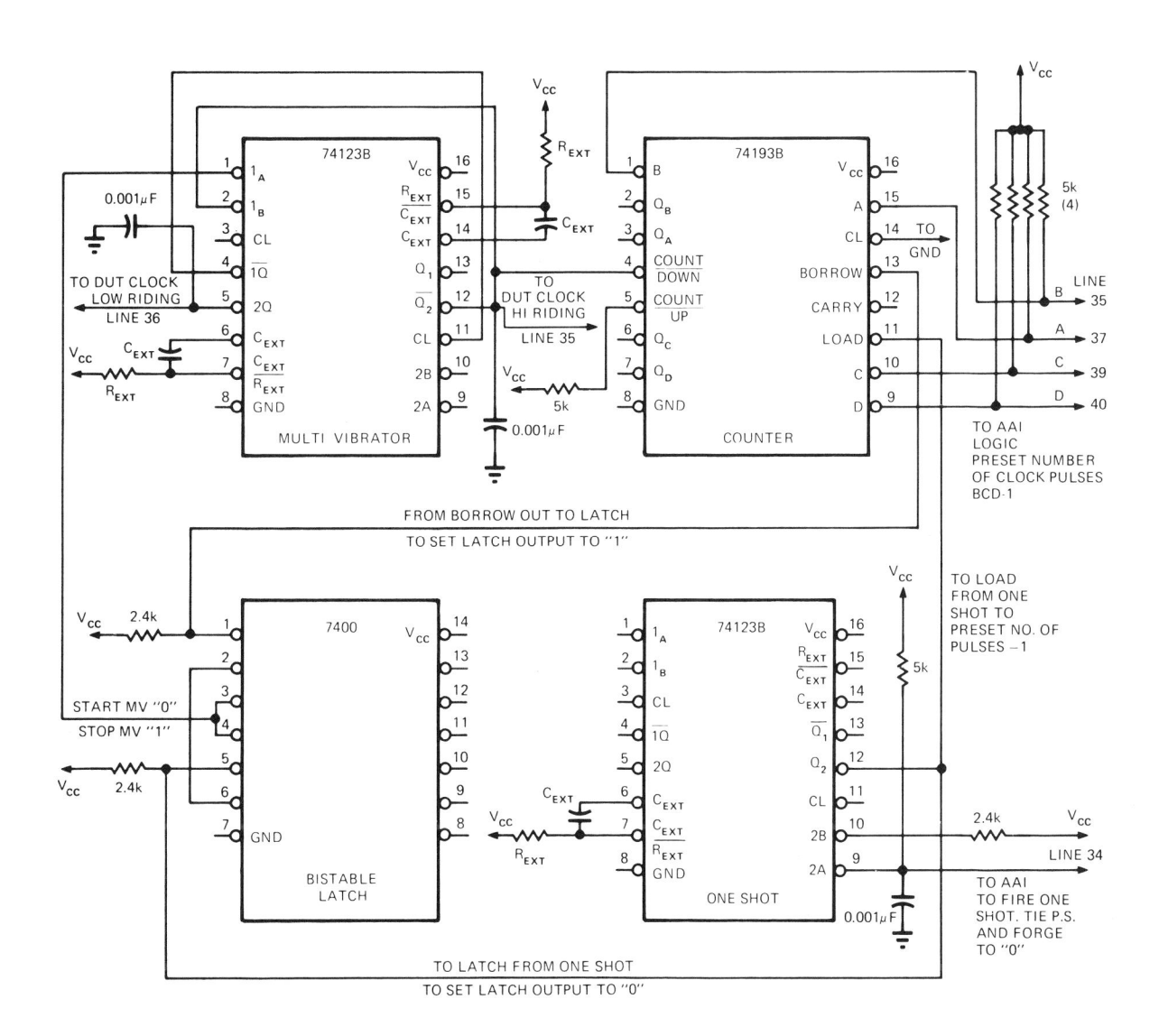

Programmable pulse generator can supply up to 16 pulses to preset states of devices under test by automatic test equipment.

Signal Sources

Low-power two-phase clock

by George A. Altemose
I.S.C.
Huntington, N.Y.

THE TWO PHASE CLOCK of **Fig. 1** requires only three milliwatts and is able to generate fast rectangular pulses into capacitive loads. This type of circuit is required for many MOS FET shift registers.

The circuit consists of two separate programmable-unijunction-transistor (DT 131) oscillators, each having a four-transistor driver stage. The two oscillators are cross synchronized by the 130-pF capacitor. Pulse-repetition frequency is a function of the PUT timing and bias networks. Output pulse widths are set by the 100-nanosecond delay lines. Fig. 2 shows system waveforms and timing.

Circuit characteristics are: prf 2.5 kHz, pulse width 100 ns, rise and fall time 20 ns each, power consumption 3 mW, V_{cc} 15 Vdc and capacitive load 240 pF.

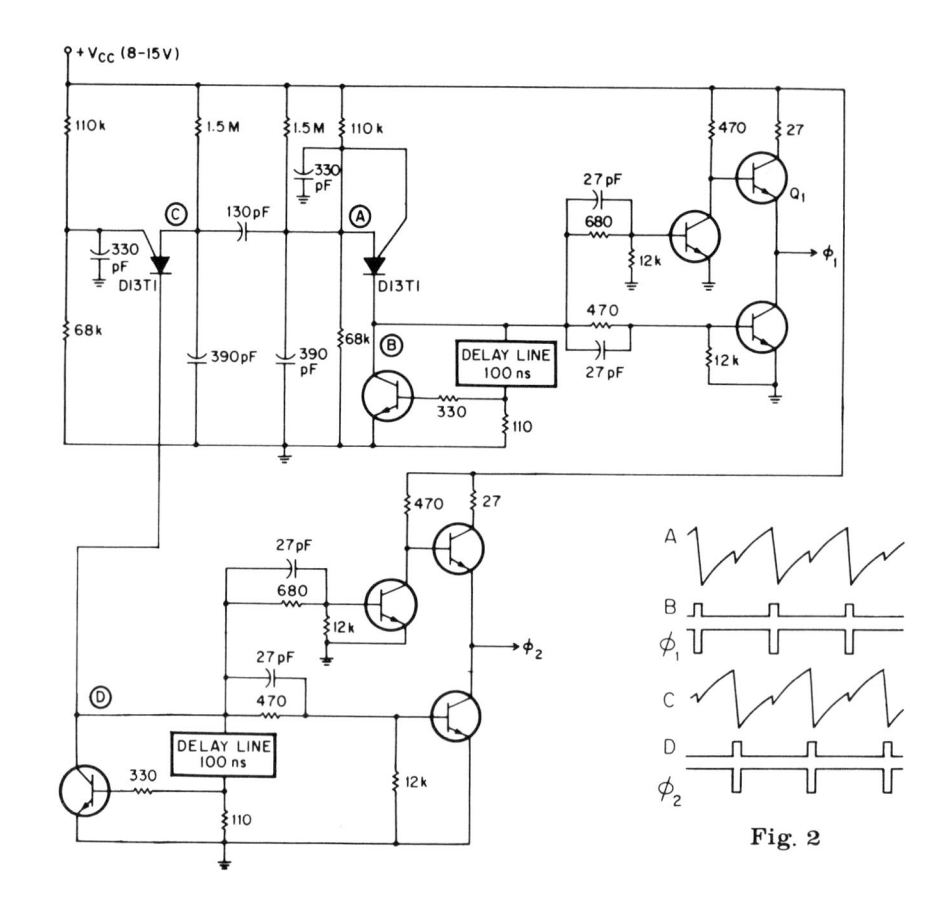

Fig. 1. This two-phase clock for MOS FET shift registers requires only 3mW. Unless otherwise noted, all transistors are 2N2369s.

Fig. 2. Typical waveforms and timing.

Voltage-Controlled Oscillators

By Robert Selleck
Magnavox Research Laboratory
Torrance, Calif.

THE USUAL LOW-FREQUENCY voltage-controlled oscillator is a relaxation oscillator delivering a rectangular or pulse waveform controlled by a dc voltage, with frequency dependent upon the magnitude of this dc control voltage. The circuit in Fig. 1, however, produces an excellent sine wave with very good linearity over the indicated 1000-cps range.

This circuit is basically a resistance-controlled three-section phase-shift oscillator. The frequency of the circuit, using untapered sections, is determined from:

$$f \simeq \tfrac{1}{2} \sqrt{6} \; \text{II} \; RC.$$

In this oscillator, field effect transistors Q_1 and Q_2 appear as linear voltage-variable resistors controlled by 0-7v.

The circuit in Fig. 2 uses conventional transistors in place of the FETs. Linear frequency response with respect to the contol voltage, does not exist in the circuit since Q_1 and Q_2 are operated in the knee region. However, lack of linearity when the control voltage is servo'ed may not be a disadvantage. Note that in each circuit an increasing voltage increases frequency.

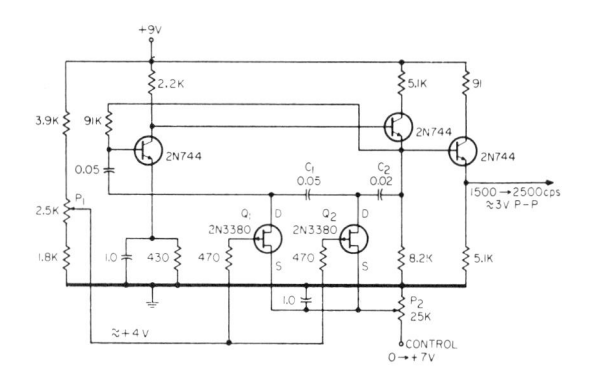

Fig. 1. Voltage-controlled oscillator using FETs.

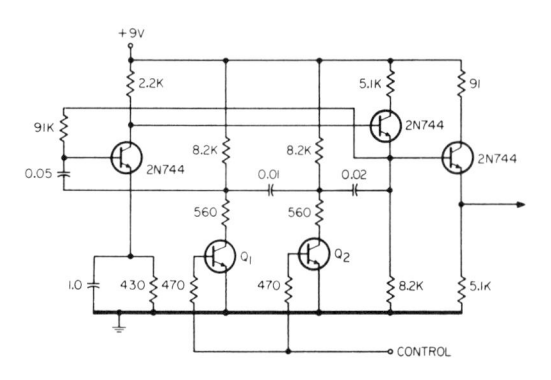

Fig. 2. Voltage-controlled oscillator using conventional transistors in place of FETs.

Triggered sweep features low dc offset

by David A. Meyer
Sperry Gyroscope
Great Neck, N.Y.

This triggered sweep combines the features available in the µA709 with those of a low-saturation-resistance FET to provide a 10-V linear sweep with very low dc offset.

In the circuit shown, Q_1 and Q_2 form a Schmitt-trigger input circuit. Transistors Q_3 and Q_4 provide the unblanking output and unclamp the FET Q_5 to initiate the sweep. Q_3 and Q_4 are complementary to provide protection for the base-emitter junctions. Transistor Q_6 allows the circuit to drive a low-impedance or capacitive load with little effect on the sweep timing.

Diode CR_2 and the 10-kΩ resistor from source to gate effectively isolate the FET from its driving circuit, thus providing the low dc offset voltage.

Sweep speed is determined by the $C(R + R')$ value. The value of $R + R'$ should be high with respect to the on resistance of the FET since small values of resistance produce a proportionally greater dc offset.

This circuit has been tried for sweep speeds from 1 Hz to 100 kHz. Dc offset voltages of 1 to 5 mV were obtained for these sweep speeds. Higher sweep speeds are limited by the slew rate of the µA709. The slow speeds are limited by the input resistance.

The dc coupling used throughout provides a circuit for use with hybrid microcircuit techniques.

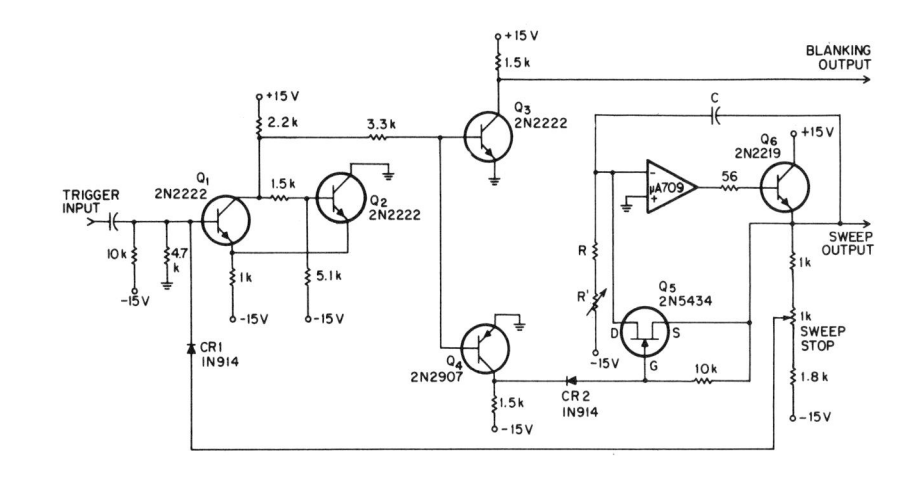

Triggered sweep circuit provides blanking as well as sweep output.

Keyed multivibrator produces symmetrical ac output

by Merle Converse
Southwest Research Inst.
San Antonio, Texas

Sometimes, the circuit designer needs to generate a keyed tone or pulse train using a compact low-cost circuit. Conventional keyed multivibrators are suitable, but they have the disadvantage that their output contains a dc level shift which causes severe distortion in an ac-coupled load. Fig. 1 shows a typical output waveform for a conventional circuit.

The improved circuit of Fig. 2, however, has an added transistor Q_3 which removes the level shift from the output. Also, this circuit starts instantly with a full-width first pulse.

In the circuit, transistors Q_1 and Q_2 are connected as a conventional astable multivibrator. This is keyed by switching the charging voltage for Q_1's base coupling capacitor. With values shown, the multivibrator oscillates at about 3 KHz when the gate input is +4 volts, and is turned off when the gate input is below 1 volt.

When the circuit is free running, the collector of Q_2 alternates between 0.1 volt and 4.8 volts. The junction of R_7 and R_8 alternates between about 2.4 volts and 4.8 volts, producing an open-circuit output of 2.4 volts peak-to-peak.

When the gate pulse returns to less than 1 volt, Q_1 is held off and Q_2 is held on by current through R_5. Normally, of course, the junction of R_7 and R_8 would fall to 2.4 volts. But with this modified circuit, Q_3 is also turned on. This places R_{11} in parallel with R_7 so that the junction of R_7 and R_8 is pulled up to about 3.6 volts.

With the improved circuit, the positive and negative excursions of the output pulses are symmetrical about the mean level of 3.6 volts. Therefore the output can be ac coupled without introducing low-frequency level-shift distortion.

Fig. 1. Level shift at the output of conventional keyed multivibrators causes distortion when the load is ac coupled.

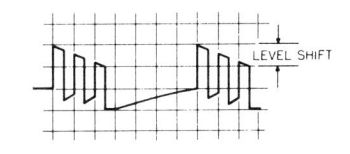

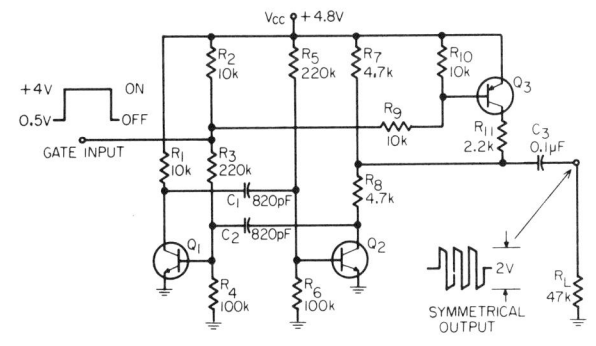

Fig. 2. This improved keyed multivibrator overcomes the level-shift problem. Added transistor Q_3 restores the output dc level to the mean value when the gate signal is low.

Stable low-distortion bridge oscillator

by Klaus J. Peter
H. H. Scott, Inc.
Maynard, Mass.

The Meacham bridge circuit can form the basis for a stable transformerless crystal oscillator. The oscillator described here produces a low-distortion sinewave output.

This type of oscillator requires a quartz crystal cut for operation in the series-resonant mode. Oscillator frequency can be adjusted upwards by adding trimming capacitance in series with the crystal.

Depending on the Q of the crystal and on the degree of temperature stabilization, accuracies of one part in 10^8 are possible with this type of oscillator.

In the circuit of **Fig. 1**, the Meacham bridge is nearly balanced at resonance and all arms appear resistive. As can be seen, the bridge is coupled to the differential inputs of a high-gain amplifier. The ratio of R_3 (crystal resistance) and R_2 determines the degree of positive feedback around the amplifier. The ratio of R_1 and R_4 determines the negative feedback applied to the amplifier's inverting input.

When power is first applied, R_4 consists of the channel resistance of the FET with zero gate bias. This has a value of between 200 and 300Ω. In this condition, the loop gain is high and oscillations begin to build up. When the negative peaks of the oscillations reach about 5.3V, the FET gate starts to go negative. This increases the drain-source resistance and alters the R_1/R_4 ratio so that negative feedback increases. At some point, an equilibrium is reached and the amplitude of the oscillation remains constant.

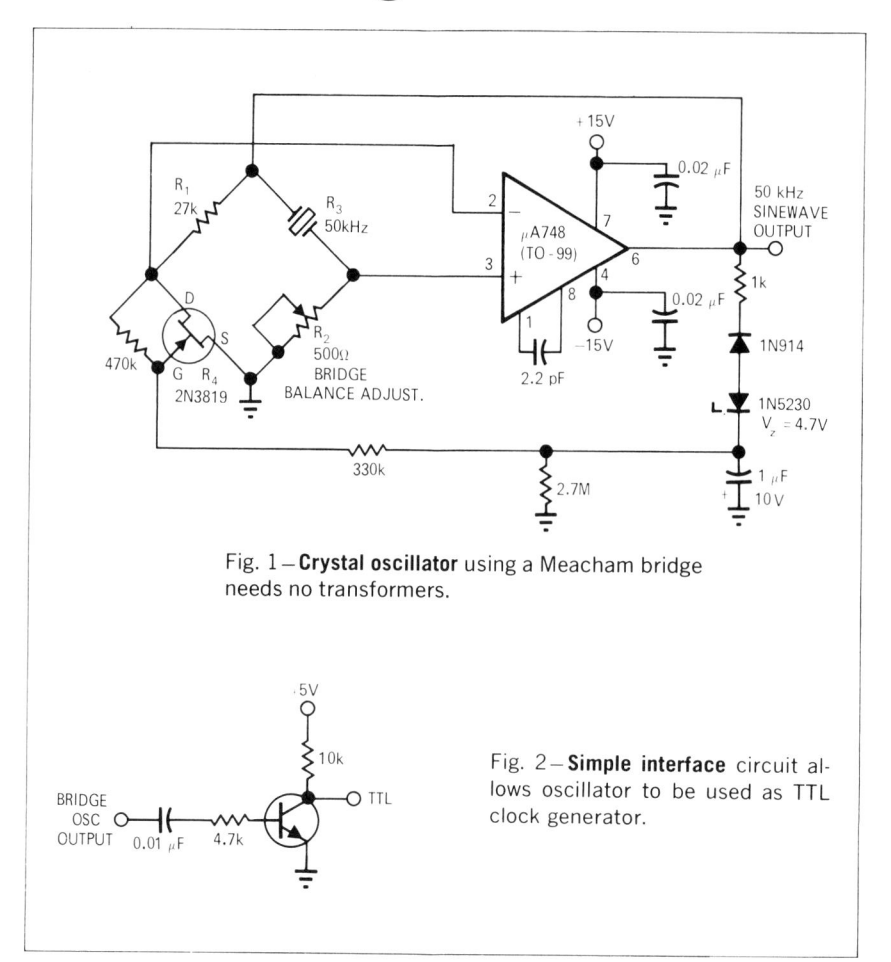

The component values shown in **Fig. 1** are for a 50-kHz crystal (+5° X-cut). The same circuit has been used (with minor modifications) for 100- and 200-kHz crystals. For frequencies above 200kHz, an op amp capable of higher slew rates must be used.

For the circuit shown, measured harmonic distortion was about 0.5%. Distortion consists mainly of second-harmonic products. To achieve low-distortion, the operating point should be chosen so that the ratio R_1/R_4 is greater than 100, thus reducing the nonlinear-resistance effects of the FET.

Output amplitude can be raised or lowered by changing the zener-diode value. If a split supply is not available, a resistive divider can be used across a 30V supply. To use the oscillator as a clock generator for TTL circuitry, a simple output stage, as shown in **Fig. 2**, can be added.

Fig. 1—**Crystal oscillator** using a Meacham bridge needs no transformers.

Fig. 2—**Simple interface** circuit allows oscillator to be used as TTL clock generator.

Gated 60 Hz clock avoids glitches

by Robert I. White
University of Wisconsin
Madison, Wis.

A poorly-designed gated asynchronous clock can create marginal pulses if the gate happens to be enabled or disabled during a clock pulse. The circuit described here, however, provides an inexpensive solution to the "glitch" problem. Using a single 7400-series IC plus a few discrete components, the circuit generates complementary gated clock pulses that are never less than 2 μsec in duration.

As can be seen in the schematic, 60-Hz timing signals derived from a 12V transformer are filtered by an RC network (to remove high frequency noise) before being discriminated by a programmable UJT. This part of the circuit is fairly conventional, and the approach has been described elsewhere (see *EEE*, May 1970, p. 105).

When the voltage on C_1 reaches about 8.5V, Q_1 fires and generates complementary pulses for the logic circuit. The negative-going pulse, at pins 1 and 2 of the IC, is considerably shorter than the pulse entering pin 13.

If the gate input (pin 5) is enabled during the short pulse, the flip-flop turns on, thus generating a logic-1 output at pin 8. After 2 μsec, the long pulse (entering pin 13) returns to zero, thus changing the flip-flop back to its original state (logic-0 at pin 8). Since the flip-flop can be triggered into the logic-1 state only when both the gate level and the short pulse are present, no output pulse of less than 2-μsec duration can occur.

Clock accuracy depends, of course, on line stability. Measurements have shown clock accuracies close to 0.01% for 1-sec time periods in Madison, Wis.

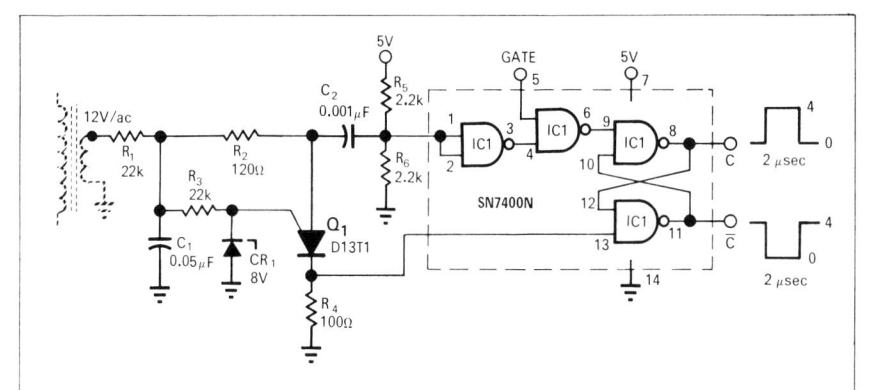

Clock generator provides output pulses that are always longer than 2 μsec, regardless of the relative timing of the gating signal and the ac line.

Wide-range pulse generator uses single IC

Mahendra J. Shah,
University of Wisconsin, Madison, WI

For designing digital circuits using DTL and TTL ICs, it is often necessary to have a pulse generator that has reasonably wide and independent frequency and pulse width ranges. The circuit shown (**Fig. 1**) is very simple, uses only one IC, is low in cost and has over eight decades of frequency and pulse width ranges—both independent of each other.

The IC, an SN74123N, is a dual, retriggerable one-shot multivibrator. These one-shots can be triggered on in less than 100 nsec after the end of a cycle. Taking advantage of this fact and by feeding the one-shots Q output back to its own B input an astable multivibrator is formed. Assuming that one-shot #1 is triggered on, the 1Q is HIGH and $1\overline{Q}$ is LOW. At the end of its timing cycle (controlled by $R_1 + R_2$ and C_f), 1Q goes LOW and $1\overline{Q}$ goes HIGH. This LOW-to-HIGH transition is slightly delayed (by less than 20 nsec) due to the added capacitive loading by C_d at $1\overline{Q}$. The delayed transition properly retriggers the one-shot to its ON state and the cycle repeats again.

The frequency of the astable multivibrator is given by:

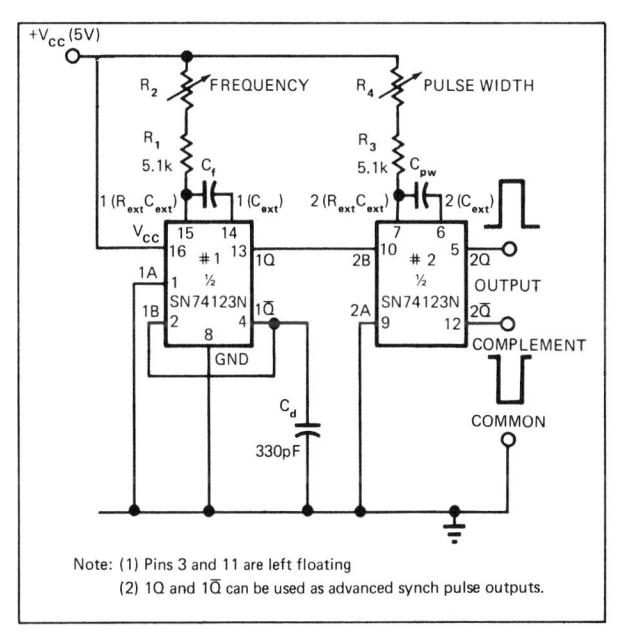

Note: (1) Pins 3 and 11 are left floating
(2) 1Q and $1\overline{Q}$ can be used as advanced synch pulse outputs.

Fig. 1—**Two one-shots** produce a pulse generator whose output can be independently adjusted in frequency and pulse width.

$$f = \frac{1}{(T_f + 60 \text{ nsec})}$$

where T_f is the on time of one-shot #1 (controlled by R_1, R_2 and C_f).

Signal Sources 135

Everytime 1Q makes a LOW-to-HIGH transition, one-shot #2 is triggered ON, which generates a pulse of width T_{pw}. Since 1Q makes LOW-to-HIGH transition of frequency f, the output pulses are generated having frequency f and pulse width T_{pw}. The pulse width is controlled independently by R_3, R_4 and C_{pw}. **Fig. 2** shows the graph of frequency verses timing capacitance (C_f). Note that for a given capacitance, continuous frequency range over a decade is obtained by changing ($R_1 + R_2$) over a range of 5.1 kΩ to 62 kΩ, except at the high frequency end. Only eight capacitors are required to cover a frequency range of over eight decades (0.054 Hz to 12.85 MHz), and only eight capacitors are requried to cover a pulse-width range of over eight decades (60 nsec to 18 sec). Pulse width for a given value of RC can be determined by obtaining the period (1/f) using the graph of **Fig. 2** and subtracting 60 nsec from it. Output pulse rise and fall times are less than 20 nsec.

A slightly advanced sync pulse is obtained at either 1Q or 1Q̄. Output 1Q has a negative-going transition that occurs approximately 60 to 80 nsec before the positive-going transiton of the output pulse. Output 1Q̄ has a positive-going transition that occurs approximately 10 to 20 nsec before the positive-going transition of the output pulse.

Manual trigger capability can be easily added by adding a non-bounce circuit triggering at 2A or 2B with the connection from 1Q to 2B removed. Square-wave output can be obtained by simply triggering a toggle flip-flop using the signal 1Q or 1Q̄.

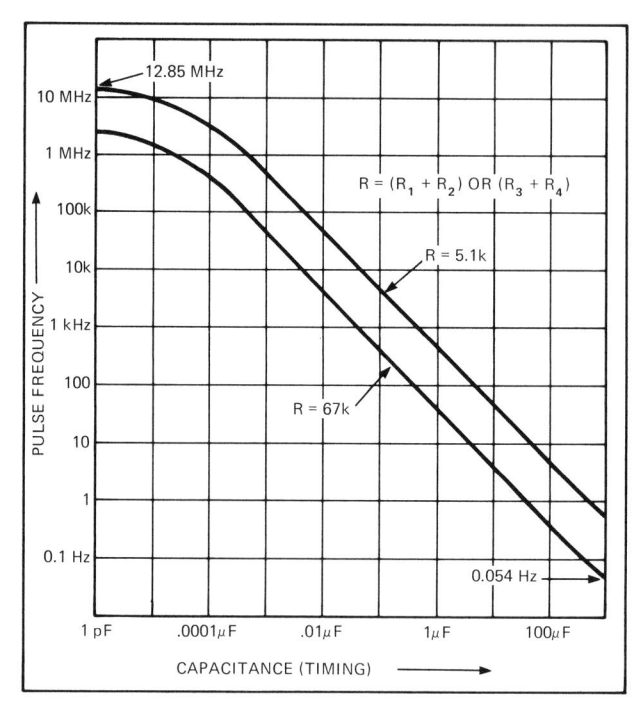

Fig. 2 — **Eight decades of frequency** and eight decades of pulse width can be obtained with just eight different capacitors.

Voltage control of frequency and pulse width can be obtained by connecting R_2 and R_4 to their respective control voltage lines rather than to +Vcc. Note that the control voltages should be limited to a range of 1.5V to 4.5V for V_{cc} = 5V. Also note that frequency will vary nearly linearly with control voltage, while the pulse width will vary nearly inversely with control voltage.

Emitter-Coupled Astable with Saturated Output

By David R. Hoppe
Collins Radio Co.
Cedar Rapids, Iowa

BECAUSE OF ITS OUTPUT CHARACTERISTICS, the emitter-coupled astable is not easily applied as a drive for logic circuitry that requires a low impedance-to-ground output (TTL for example). The modified version of the circuit shown here, however, accomplishes this. It also provides the advantages of being self-starting, has no recovery-time phenomena, and provides the good frequency stability found in emitter-coupled oscillators. In addition, high-speed saturated positive and/or negative polarity outputs or a current-mode logic output can be obtained.

Q_1 and Q_2 operate as an emitter-coupled oscillator. When Q_1 turns off, diode D_1 is reversed-biased and Q_3 is saturated. When Q_1 turns on, the base current of Q_3 is switched into the collector of Q_1. Diode D_1 clamps the collector of Q_1 to about -0.7 V so that Q_1 is not heavily saturated.

Diode D_1 may be removed and a pnp transistor Q_4 added as indicated to obtain either positive or negative or both polarity outputs. The base-emitter diode Q_4 now performs the function of diode D_1.

All circuitry in the dashed box may be replaced by Q_5 and its associated circuitry. The output at Q_5 swings from -0.75 to -1.55 V which is compatible with the present-day integrated-circuit current-mode logic.

Operating frequency may be varied from 50 Hz to 8.5 MHz by varying C_1. Symmetry may be adjusted by varying the ratio of R_{E1} and R_{E2}. Rise and fall times are 10 and 12 nsec respectively and vary little with operating frequency.

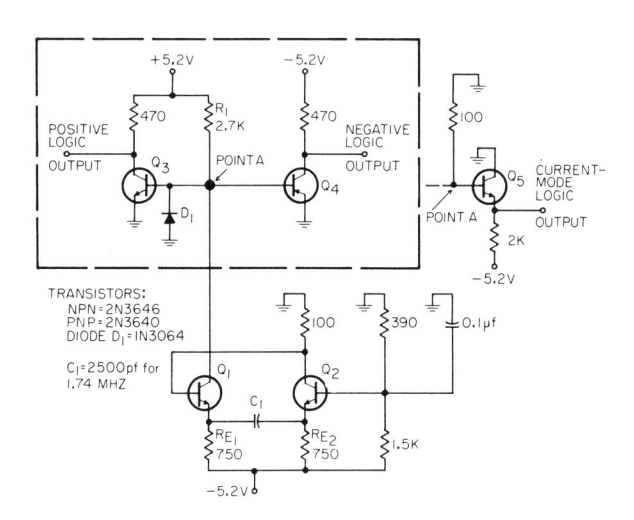

Emitter coupled astable; the circuitry in the dashed box may be replaced by Q_5 as shown.

Crystal controls rep rate of simple IC pulse generator

by Kuenseng Yu
Honeywell
Waltham, Mass.

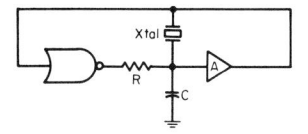

Fig. 1. Basic circuit of simple crystal-controlled pulse generator.

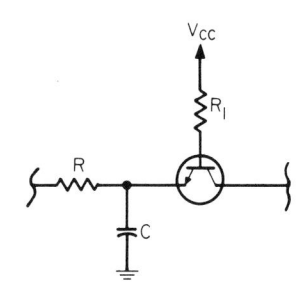

Fig. 2. Section of complete circuit shows how resistor R affects the input circuit of the IC gate.

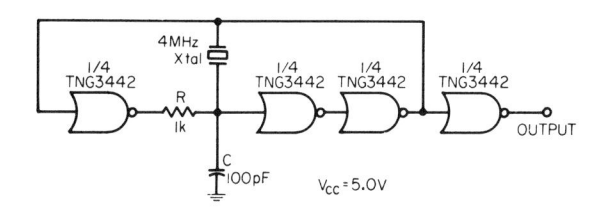

Fig. 3. Practical circuit for 4-MHz pulse generator, using quad 2-input gate. Spare input to the first gate can be used for stop/start control.

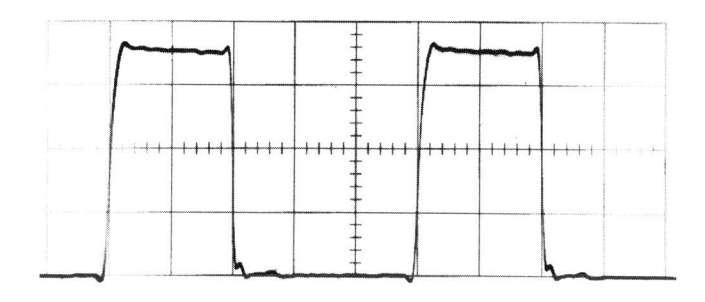

Fig. 4. Output waveform for the circuit shown in Fig. 3. Vertical scale is 1 V/cm and horizontal scale is 50 ns/cm.

To generate continuous pulse trains, circuit designers generally employ circuits like blocking oscillators, or sine oscillators followed by clipping or shaping circuits. Fig. 1 shows a much simpler approach that needs only one TTL IC, a quartz crystal, and an RC network. The crystal controls the repetition rate and the RC network determines the pulse width.

But the RC network can't be chosen arbitrarily to give any required pulse width. Because of the higher switching speeds of TTL ICs, the RC network must also act as a filter to prevent high-frequency instability. Also, the resistor must limit the drive current to the crystal.

The lower limits for the values of R and C depend on the speed of the IC. The low-pass filter combination suppresses the closed-loop natural frequency of the IC. The upper limit of resistance for R is determined by the input circuit and threshold voltage of the IC.

For practical circuits, we replace the lumped amplifier of Fig. 1 with additional gates as shown in Fig. 3. This reduces the physical size of the circuitry, because we can use a multi-gate IC instead of separate gates and amplifiers.

Figure 2 shows the RC network and the input circuit of the second gate. To ensure that this gate switches properly, the maximum value of resistor R must satisfy the following equation:

$$R = \left(\frac{V_T}{V_{cc} - V_{be} - V_S - V_T} \right) \times R_1 \qquad (1)$$

After finding an allowable value for R, we can select a suitable value for C. The value of C should be low enough to allow sufficient gain and feedback for oscillation. This sets an upper limit to the value. Provided the value is below the upper limit, C can be selected to give the required pulse width.

The practical circuit, shown in Fig. 3, uses a single Transitron TNG 3442, quad 2-input, gate. Crystal frequency is 4 MHz. In this circuit the fourth gate provides output isolation. Depending on the required output polarity, the fourth gate can either be connected as shown, or it can precede the third gate to reverse the pulse. Fig. 4 shows the output waveform for this circuit. The spare input of the first gate can be used as a stop/start switch.

With suitable component values, the basic circuit can accommodate a wide range of crystal frequencies from 10 kHz to around 10 MHz. Frequency stability depends primarily on the stability of the crystal. The circuit is insensitive to variations in crystal series impedance. Within the temperature range of the IC, the circuit shown in Fig. 3 has better than 0.02% stability.

Operational amplifier makes a simple delayed pulse generator

Dean T. Anderson,
Collins Radio, Cedar Rapids, Iowa.

It is often necessary to have a delayed pulse generated from a clock pulse, and use both pulses in gating circuitry. However, problems arise if the delayed pulse is generated as the clock pulse starts to fall before the clock is at its final state. This condition may generate an unwanted momentary output in the gate circuitry using these pulses.

A simple solution to this problem is to use an operational amplifier that will generate the desired delayed pulse from the clock pulse. Before the input pulse is applied to the circuit, the positive 1.1V bias on the operational amplifier, input C, keeps the output in its −15.5V state. As the clock or input pulse switches to +15.5V, both inputs B and C charge to 4.5 times larger than input C; the voltage on B will always be lower than C as long as the clock pulse stays positive. When the clock switches to −15.5V, input B and C will discharge to 0 and 1.1V respectively with C staying larger than B for 15μsec. After the 15μsec the voltage at input C will be less than at B for 1 msec and the output will switch to +15.5V for that time period. When the input voltage at B again drops below C, the output switches back to the −15.5V state.

With the circuit values shown, the output pulse is 1 msec wide and delayed from the clock pulse by 15 mμsec. Furthermore, any number of time constants can be used to generate different pulse widths and delays.

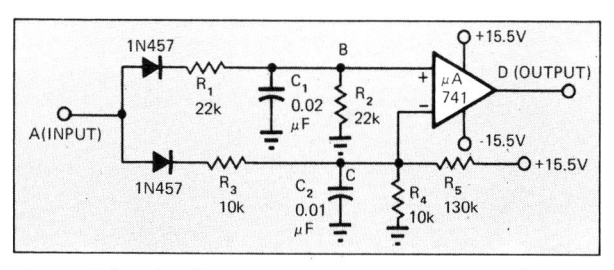

Fig. 1—Delayed pulse generator uses an op amp and is programmable for both delay time and pulse duration, by selection of the RC networks.

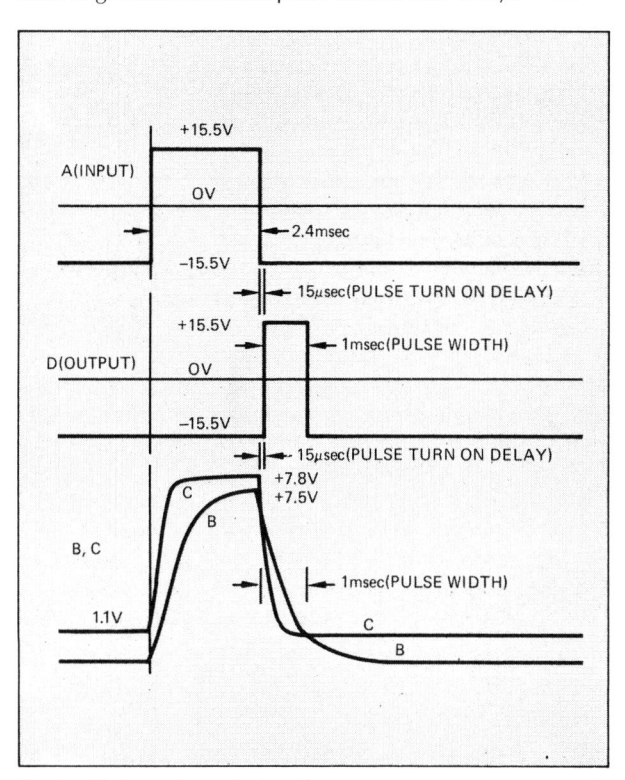

Fig. 2—Timing pulses obtained from component values given in Fig. 1 result in a 1 msec pulse output which is delayed 15μsec from the trailing edge of the input pulse.

Crystal Controlled Multivibrator

By Jay Freeman
Consultant
New York, N.Y.

THIS OSCILLATOR is useful as a system clock. It combines simplicity with crystal stability and uses no inductors or capacitors. The circuit can be built up from logic NOR gates with the only frequency controlling element the crystal itself.

Each transistor operates as a feedback amplifier. The Q_2 stage has unity gain. The gain of Q_1 is a function of the resonant impedance of the crystal. But since most crystals have a resonant impedance less than the 22-K feedback resistance, the gain of the Q_1 stage is considerably greater than unity.

The output at the collector of Q_1 is a rounded square wave with a peak-to-peak voltage dependent on the $B+$. The circuit oscillates at 1 mc from 2 v to 30 v. A wide range of frequencies from 3 kc to 10 mc can be accommodated without changing any elements in the circuit because of the sole use of resistance for both biasing and loop gain.

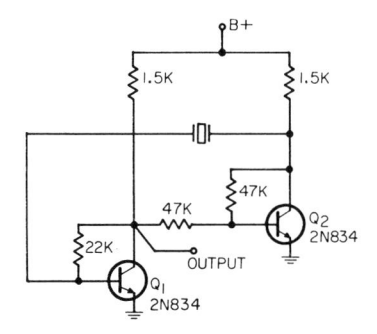

Crystal-controlled multivibrator.

The circuit operates over a temperature range of 0° to 60°C with less than 0.05 percent drift. The output can be loaded with a 15-K resistance without seriously affecting the waveshape or frequency.

Adjustable rectangular-wave oscillator interfaces with IC logic

by David E. Manners
Vision Laboratories, Inc.
Oriskany, N.Y.

With this inexpensive rectangular-wave oscillator, the frequency can be adjusted over a wide range and the ON and OFF times of the signal can be independently manipulated. The circuit requires only 5V dc for operation and can therefore operate from conventional logic power supplies. Because the current drain is only about 8 mA, the circuit can also be operated from a zener regulator fed by a higher voltage supply line.

Output levels of the oscillator are compatible with conventional IC logic. The circuit was originally developed to provide an input to a driving stage for low-frequency pulsing of an inductive load.

Another useful feature is that terminal 2 of the IC can accept logic signals which serve as an ON/OFF control. When held at logic 0, pin 2 forces the output to be a steady logic 1. If a steady logic-0 output is desired, another NAND gate on the same IC must be used. Removal of the logic 0 from pin 2 allows oscillation to occur. For normal operation, pin 2 is either open circuited or, preferably, connected to 5V dc.

Operation is fairly straightforward. When power is first applied, the SCR is nonconducting. Thus, pin 1 of the IC is presented with a logic-1 level. Pins 3 and 4 are therefore at logic 0. Pin 6, in turn, starts out in a logic-1 condition. This allows C_1 to begin charging from the voltage source via R_2 and R_3, in an attempt to reach a voltage level determined by the divider network formed by resistors R_1, R_2 and R_3. Thus, as the voltage level increases, the SCR is eventually triggered into conduction.

When the SCR starts to conduct, a logic 0 appears at pin 1 of the IC. This causes pins 3 and 4 to go to logic 1, and pin 6 becomes logic 0. The "near ground" on pin 6 causes C_1 to commence discharging through R_2 and also via R_1 and the gate-cathode path of the SCR. The capacitor voltage then decreases until the SCR is again cut off. This occurs because the current through the SCR is limited by R_4 and is always less than the required holding current. After the SCR stops conducting, the operating cycle repeats as described.

The table shows typical frequencies and duty cycles for selected values of C_1, R_2 and R_3. Note that there is a lower limit to practical values for R_3 (6 kΩ in this case) because otherwise the current-sinking capability of the IC NAND gate would be exceeded. Possible values of R_2 and R_3 are limited at the high end by the required gate voltage needed to turn on the SCR.

Total component costs for the circuit shown were about $8, with the SCR (at $5.75) contributing the bulk of the cost. Suitable SCRs can now be obtained for around $2, thus reducing the total parts cost to approximately $4.25.

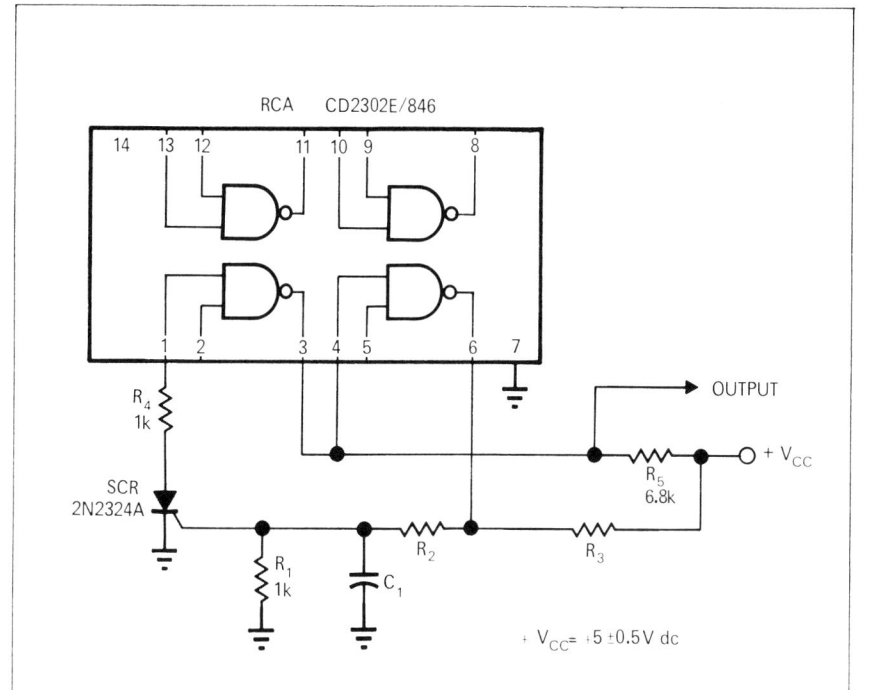

Both the frequency and the mark-space ratio of the output can be determined by suitable choice of component values in this oscillator.

OUTPUT FREQUENCY AND DUTY CYCLE FOR VARIOUS COMPONENT VALUES

C_1 (μF)	0.05	0.1	0.5	10.0	40.0	R_2 (Ω)	R_3 (Ω)	Approx. % ON
Freq. (Hz)	22.2 k	6.84 k	1.67 k	58.3	17.23	1 k	27.0 k	35
	23.6 k	8.62 k	2.08 k	75.7	22.00	1 k	13.5 k	45
	27.8 k	10.00 k	2.50 k	91.0	26.00	1 k	7.3 k	55
	7.8 k	3.92 k	0.96 k	31.2	9.26	2 k	27.0 k	45
	11.9 k	6.25 k	1.54 k	52.1	15.15	2 k	7.3 k	60

SECTION IV
Power Supply Circuits

Current Transformer Gives Fast Overload Protection

By S. J. Arnold
Welex, A Div. of Halliburton
Houston, Texas

In high-voltage transistorized power supplies, current limiting as a means of short-circuit protection is often inadequate. Large values of current and voltage will quickly exceed the control transistor's power rating. If a current transformer is used to sample overloads, a fast and effective short-circuit protection can be obtained that will reduce the power expended in the control transistor during overloads to a negligible level.

The current transformer T_1 samples the current supplied to the voltage regulator. Diode D_1 rectifies the sampled current and a voltage drop is produced across resistor R_1. R_1 is selected to trigger the silicon-controlled rectifier SRC_1, when a predetermined value of overload current is reached. Capacitor C_1 serves as a filter for the rectified signal.

When a current overload occurs, additional voltage is generated across R_1 which triggers SCR_1 on. With SCR_1 on, no base current is available to the control transistors Q_1 and Q_2, and these transistors are turned off. With Q_1 off, the transistor is required to sustain only the unregulated voltage E_{in}, thereby reducing its expended power to a negligible amount.

The circuit can be reset by removing the input voltage or by providing a manual reset switch SW_1 to eliminate the holding current of the SCR.

The current transformer used in the circuit shown was wound on an EI-187 core with an 8-turn primary and a 120-turn secondary.

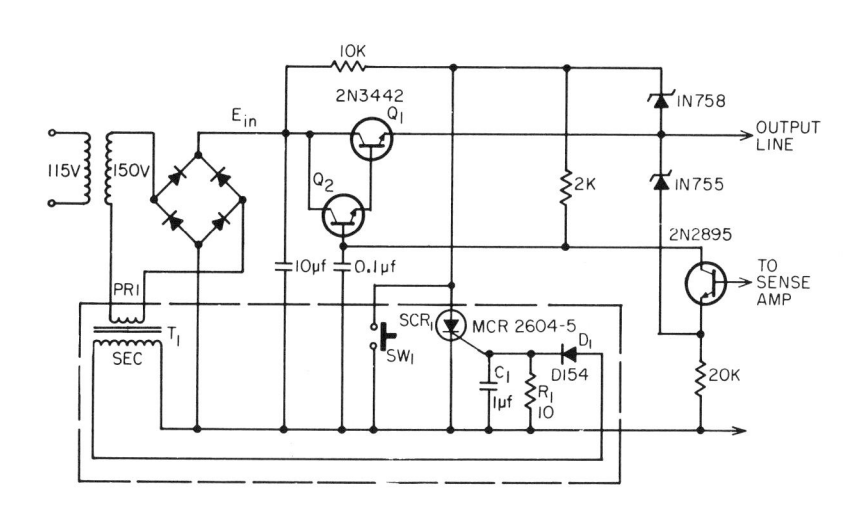

Fast reacting overload protection circuit uses current transformer to sense current to voltage regulator.

One transistor improves IC voltage regulator

Carlo Venditti, The Charles Stark Draper Laboratory, Cambridge, Mass.

A parameter to consider in using IC voltage regulators is the minimum input-to-output voltage differential. Typical values range from 2 to 3V for regulators like the National LM105 and Fairchild µA723. This sets the minimum amplitude that can be regulated at a value of 2 to 3V above the desired regulated output. You can decrease this differential to only 0.5V above the output by using a voltage divider and a transistor. Fig. 1 shows how a µA723 regulator and a single NPN transistor can be used to provide this reduced differential.

The values of R_1 and R_2 are selected, with current I flowing into node a, so that the voltage at node a is 3V (minimum) less than the desired E_0. The composite regulator will give a regulated E_0 for $V = E_0 + 0.5$. The input-output voltage differential for the µA723 is held greater than 3V, assuring its proper operation.

This circuit technique allows dc voltage sources that did not meet the previous minimum input-output voltage differential to be regulated directly. Further, the input working voltage can be decreased, for a given E_0, minimizing power dissipation and increasing overall efficiency.

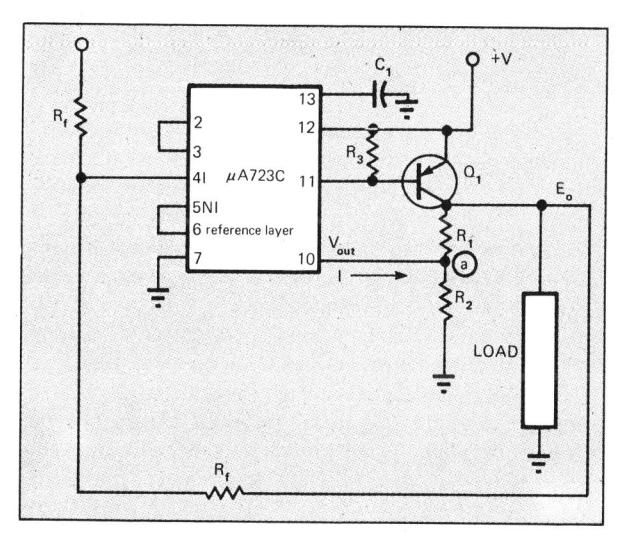

Fig. 1 — IC voltage regulators can be operated with as little as 0.5V input-to-output differential with the addition of a single NPN transistor. 2 to 3V differential is normally required for proper functioning of this type of regulator.

Power Supply Circuits

Simple ±15V regulated supply provides tracking

Hank Olson
Stanford Research Inst., Menlo Park, CA

A simple ±15V regulated supply which provides tracking can be built using only four semiconductor packages. The output current capability as shown is 100 mA, but this can easily be increased to about 200 mA using a larger rectifier-filter section.

The µA7815 is a positive, fixed output regulator with built-in current limiting at about 230 mA. It would seem a simple task to use two of these power ICs to obtain ±15V, but this requires two isolated transformer windings. The circuit shown uses only one center tapped transformer winding and an integrated bridge (as two full-wave rectifiers). By using the National LM304 and an outboard current-carrying PNP, and the +15V from the µA7815 as an external reference, a simple −15V companion regulator is obtained. The LM304 does not even require compensation, as do most IC regulators. Depending on the application and construction methods, the 0.22 µF and two 4.7 µF capacitors may not be needed.

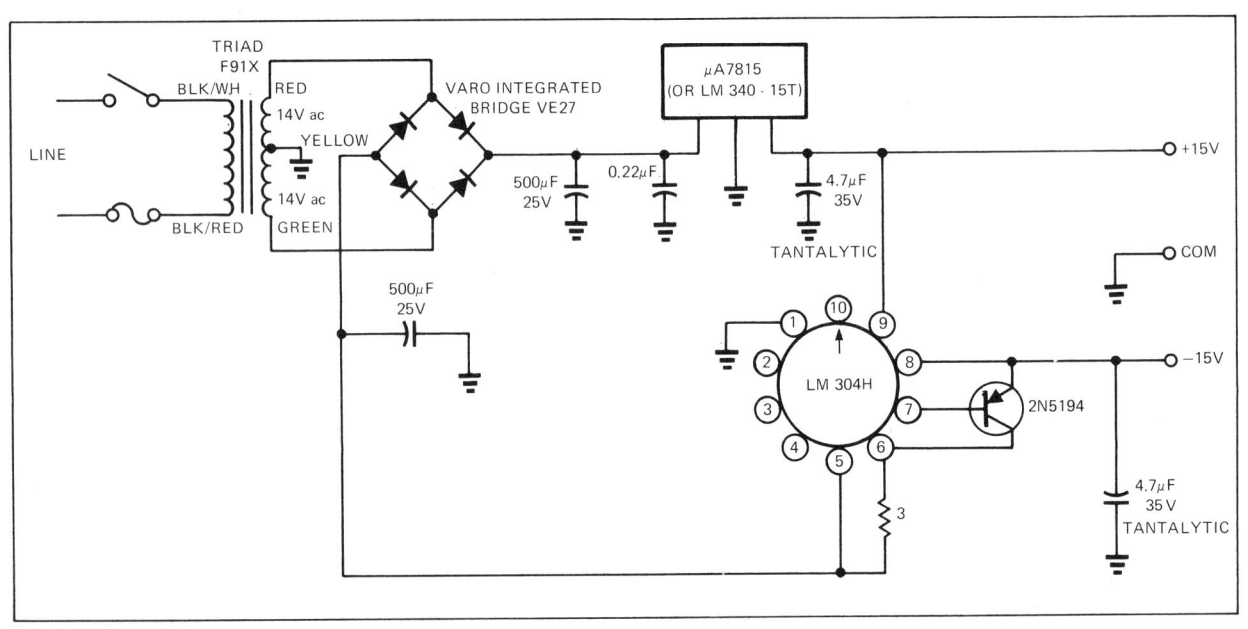

Tracking 100 mA regulator provides ± 15V and can be boosted to 200 mA output by using a larger rectifier section.

AC Power Interlock

By Charles J. Ulrick
Collins Radio Co.
Cedar Rapids, Iowa

IF YOU START YOUR DAY turning on a lot of lab equipment, or if you would like to control many ac powered units from a central point, this circuit will make things easy.

Any unit plugged in receptacle 1, drawing anywhere from 5 w to the amount allowed by the $D_2 - D_3$ rating, will produce a 60 cps square wave at the base of Q_1 when the unit is turned on. Q_1 and D_4 will energize RY_1 on the negative cycles of the ac line, and C_2 will hold RY_1 in on positive cycles. Closure of RY_1 applies power to receptacle 2. Thus, the power switches of units in receptacle 1 control the power to units in receptacle 2 with no re-wiring. D_1, C_1 and R_1 produce positive bias across D_2, which holds Q_1 off when units in receptacle 1 are all off. The x-x and y-y points indicate additional receptacles, if needed.

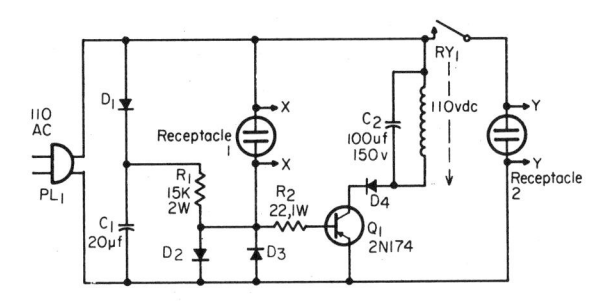

AC power interlock.

In the circuit shown, the diodes have the following ratings:

$$D_1, D_4 : 0.5 \text{ amp}, 200 \text{ v}$$
$$D_2, D_3 : I = \frac{P_{max} \text{ (Recep 1)}}{130}, 50 \text{ v}$$

Under-voltage sensing circuit

by Richard J. Buonocore
Dynell Electronics
Melville, N.Y.

Using a hex inverter IC, it is possible to build a circuit that monitors levels of several different input voltages. The basic circuit technique is quite flexible and can easily be modified for different input voltages (either positive or negative) and expanded to accept additional inputs. Circuit operation is as follows:

First consider what happens with a negative input voltage. Whenever one of these input levels falls below the breakdown voltage of the associated series zener diode, a high impedance is presented to the input of the respective inverter. This means that a logic 0 appears at the inverter's output (i.e. at the WIRED-OR connection). Because the lamp-driving output inverter has a logic 0 at its input, it no longer sinks current from the lamp and hence the lamp is extinguished, thus providing a "no-go" signal. (Note that the circuit is "fail safe" in the event of a lamp failure.)

If a positive input voltage (for example, V_4 on the schematic) falls below a predetermined value set by the associated resistive voltage divider, a logic 0 will again be presented to the output converter, causing a "no-go" signal as in the first example.

Thus, by suitable modifications to the input circuits, it can be arranged that the lamp will be extinguished whenever one (or more) of the inputs is below a specified critical level. With the components specified, and without special trimming, an accuracy of ±30% or better can be expected. Refinements could be incorporated to improve this figure.

The SN7406N is an open-collector hex inverter capable of sinking 40 mA. This device can handle standoff voltages up to 30V.

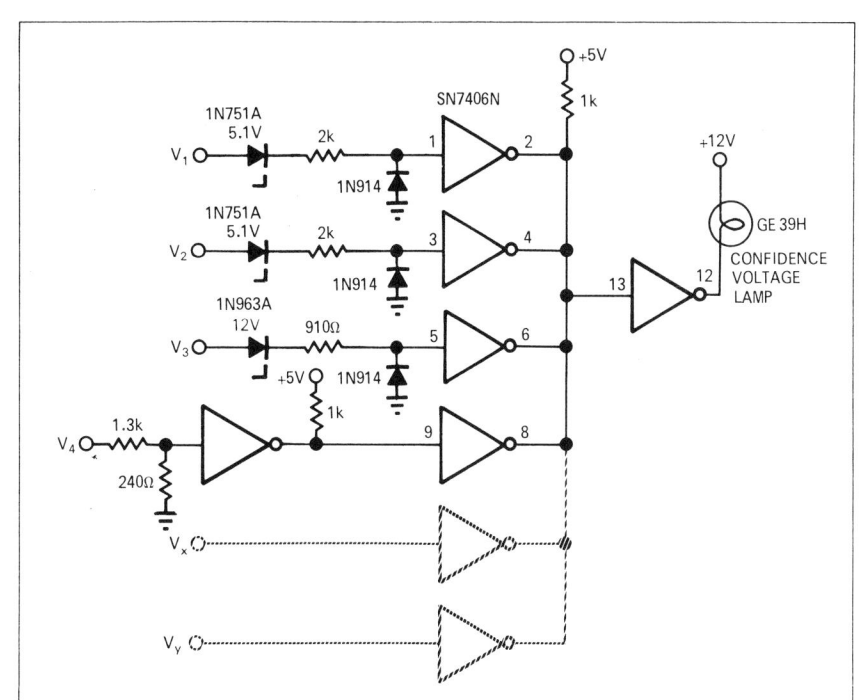

Simple indicator circuit provides a continuous confidence indication provided none of the input voltages falls below a predetermined level. Circuit can be easily expanded to accept additional inputs.

OPERATING LEVELS

INPUT	NOMINAL VOLTAGE	CONFIDENCE–VOLTAGE LAMP: TURNS OFF AT:	TURNS ON AT:
V_1	−5V	−4.8V	−4.9V
V_2	−5V	−4.3V	−4.4V
V_3	−12V	−11.3V	−11.5V
V_4	+8V	+6.6V	+6.7V
V_x	OPTIONAL ADDED INPUT		
V_y	OPTIONAL ADDED INPUT		

Spare IC gate serves as regulator

by Bob Horn
Forney Engineering
Dallas, Texas

AN UNUSED gate in a multiple-gate IC can serve as a zener regulator. In one case, it was necessary to derive 12 Vdc (for a digital IC and a reed relay) from an ac line that varied from 75 to 125 V. To cut costs, the 12-V supply was developed from the ac line by a half-wave rectifier and a line-dropping resistor.

But the wide variation in line voltage and variable demands of the load required some kind of regulator and a zener would have been too costly.

In the circuit shown, an unused gate in an Amelco 303CJ (a quad 2-input NAND buffer capable of sinking 60 mA), was used as an active regulator. The input switching level in the 303CJ is quite constant from one IC to another and also very temperature stable. Other NAND or NOR gates could be used as well.

In the circuit, R_3 and R_4 bias the gate in its active region while R_5 reduces power dissipation in the output transistor. If V_{cc} tries to rise, the gate input rises. This makes the output go lower, thereby drawing more current through R_2, which causes V_{cc} to decrease.

The circuit supplies 6 to 10 mA and V_{cc} varies from 11.4 to 11.5 V as the line varies from 75 to 125 V. The circuit works well from 0 to 75°C.

The circuit can be used, too, where the supply voltage is constant, but load current changes. It is then necessary to change values of R_1, R_2, R_5 and C_1. Supply cost, excluding the gate, is about $2 in single quantities, $1 in 100-up.

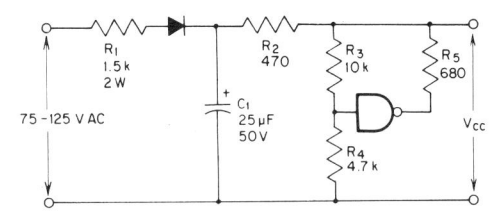

This simple circuit uses a spare IC gate from an Amelco 303CJ to regulate a +12-V supply derived from a widely varying ac-line voltage.

High voltage dc-to-dc converter

By Walter Coyle
Locus, Inc.
State College, Pa.

THIS CIRCUIT CONVERTS a rectified line voltage (about 160 Vdc) to ±15 Vdc. A conventional converter uses transistors with collector ratings of 400 V but this converter by a clever design technique, uses 135-V transistors.

A cascade of three dc-to-dc-converters is constructed on a single tape-wound core. There are three input windings connected in series and one output winding. Since all input windings are on the same core and are magnetically coupled, all three converters share the input voltage equally. The single output winding eliminates load hogging. This unit delivered +15 V at 50 mA and −15 V at 50 mA. Package size is 2.5 in.³ including line rectifiers.

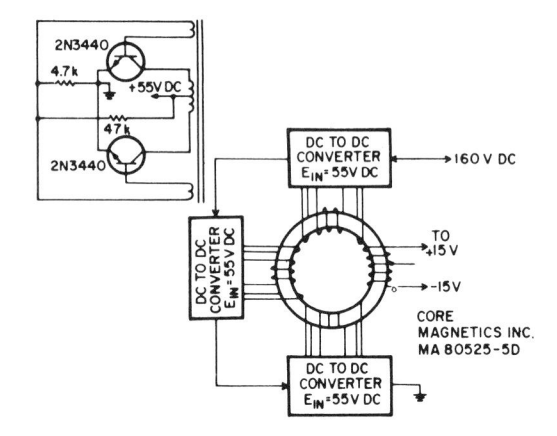

A novel cascade scheme eliminates high-voltage transistors in this dc-to-dc converter.

Resistor reduces switching regulator losses

Walter S. Duspiva
IBM Corp., Kingston, NY

Transistor switching losses (power dissipated during transition) become significant in switching regulators operated at relatively high frequencies. By placing a resistor in series with the inductor in a switching-regulator LC network, a voltage point may be derived for providing an almost constant OFF-drive.

In one typical form of switching regulator, transistor Q_1 is provided with ON-drive pulses to its base in order to deliver current pulses to an output filter consisting of inductor L, capacitor C, free-wheeling diode D_1 and load R_2.

A single resistor, R_1 is usually connected from base to emitter to provide OFF-drive (negative-base drive) for Q_1. With this configuration, however, V_{BE} determines the amount of OFF-drive. As Q_1 begins to turn off, V_{BE} drops, consequently reducing the OFF-drive increasing fall time, and thus causing considerable switching losses.

By adding a small resistor, R_3, in series with R_1,

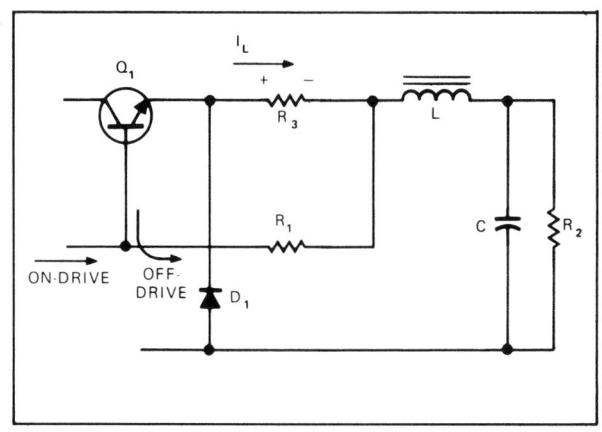

Addition of resistor R_3 decreases switching losses by causing the voltage drop across R_1 to remain almost constant.

the OFF-drive becomes less a function of V_{BE}. Inductor L and diode D_1 operate to fix inductor current I_L through R_3 at an almost constant value for a fixed load. Accordingly, the voltage drop across R_1 is returned to an almost constant negative voltage.

For some applications, OFF-drive fall time can be reduced by about two-thirds, thereby decreasing switching losses by about 50%.

Diodes avoid accidental damage from power supply sensing

Alfred W. Zinn, Kearfott Div.,
Singer-General Precision, Inc., Pleasantville, New York

When remote power supply sense leads are accidentally disconnected from the output voltage terminals, the output voltage in some supplies may rise uncontrollably to its maximum value. Components which are still connected to the output lines, such as LOAD A in **Fig. 1**, can be damaged, or the overvoltage protector will be tripped. In any case, circuit operation will be impaired.

In order to avoid this situation, two diodes can be connected at the power supply output terminals as shown in the figure. When the sense leads are connected, they effectively short the diodes and do not affect the sensing operation. However, when a sense lead is opened, the diode takes its place. The only difference is that the output voltage will be increased by an amount equal to the forward voltage drop of the diode.

If it is found that the voltage drop along wire path A-B-C is larger than 0.5V, zener diodes can be used in place of the 1N914's. Simply choose a zener voltage larger than the wire voltage drop.

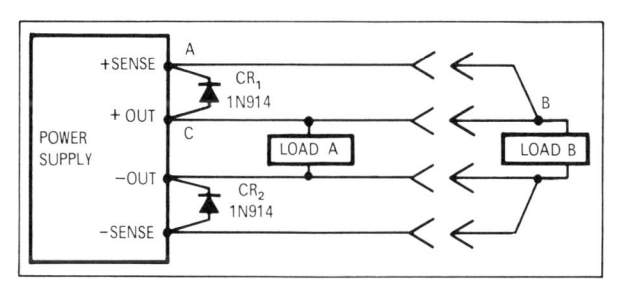

Fig. 1—Remote sensing feature of some power supplies can cause damage to unsensed loads. If the sense leads are accidentally disconnected, the voltage may rise to maximum. Diodes D_1 and D_2 will limit this rise and protect the unsensed loads.

IC timer makes transformerless power converter

Maxwell Strange
Goddard Space Flight Center, Greenbelt, MD

The excellent drive capability of the 555 IC timer is used to advantage in this simple transformerless power converter. It is so compact and inexpensive that it can often save considerable power-distribution wiring by generating many system power requirements locally.

The circuit consists of a 555 timer in the self-triggered mode as a square-wave generator,

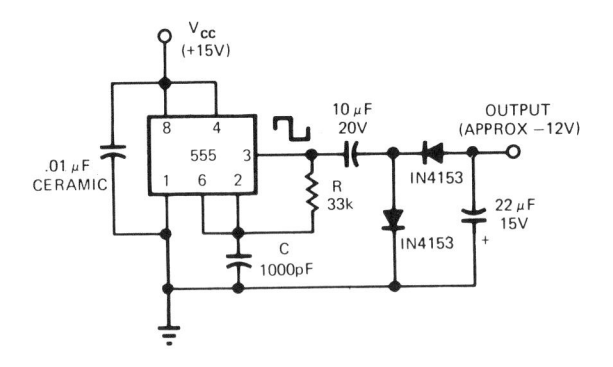

Fig. 1—**IC timer in self-triggering mode** generates squarewave to produce a negative output voltage from a positive supply.

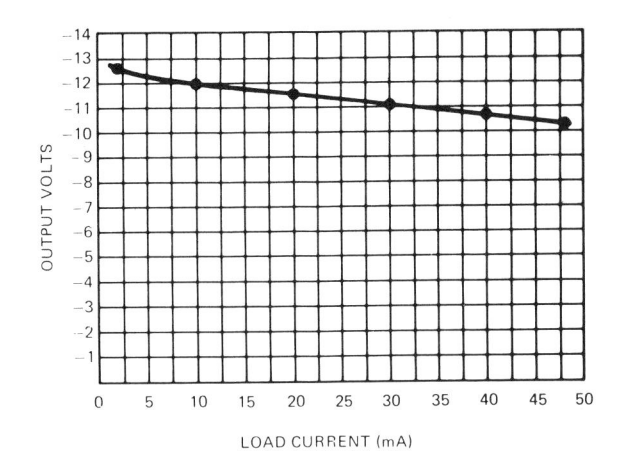

Fig. 3—**Load regulation for circuit in Fig. 1** is good, due to the drive capability of the 555 timer.

followed by a voltage-doubling rectifier. Frequency, determined by R and C from the manufacturer's data sheet, is set at approximately 20 kHz to provide good filtering with relatively small capacitors.

One configuration, **Fig. 1**, derives a negative output from a positive power source. The configuration of **Fig. 2** produces a positive output referenced to the input to provide the same polarity as the source but with approximately doubled voltage. For higher voltages, a voltage-multiplying rectifier can be used.

Fig. 3 shows that the load regulation is surprisingly good. For better regulation, a 3-terminal IC regulator is easily added. The maximum practical load current is about 80 mA at normal ambient temperatures.

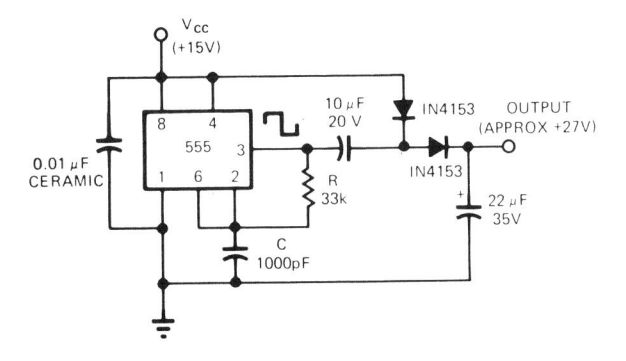

Fig. 2—**Positive output voltage approaches 2V$_{CC}$** under light loading. Ripple is only 50 mV at 80 mA.

Opto isolators provide foldover indication

Jim Burns
Tektronix, Inc., Beaverton, OR

Current limiting and current foldover have become essential parts of supply regulators. Many times it is desirable to know when limiting takes place via a fault lamp or some alarm, but placing such a fault indicator in the path of the limiting current can be difficult. Usually one has little choice but to power

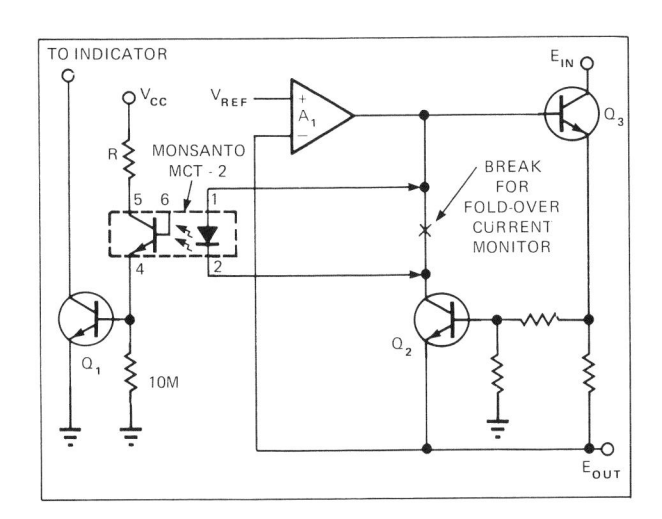

Opto-isolator sensing of power supplies allows the use of neon, LED or incandescent indicator alarms without drawing excessive power from the monitored device.

148 Circuit Design Idea Handbook

the indicator from the supply that it is monitoring. Power losses then become the determining factor in selecting this indicator. The use of neons becomes mandatory when working with higher voltages.

One solution to the above problem is to operate the indicator from a ground referenced device that is interfaced to the regulator via an optical coupler. This leaves the designer with complete freedom to choose a monitoring device. Power to operate the indicator is no longer a problem and can be taken from another source.

Q_1 in the drawing has very high gain and must be shunted with the 10 MΩ emitter-base resistor. This is done to sink leakage currents of the phototransistor. This circuit configuration will turn Q_1 full on with less than 40 μA flowing through the LED portion of the optical coupler. The LED can carry 150 mA (absolute max.) yielding a current range greater than 3000:1!

Versatile circuit behaves like SCR

A.J. Baracz,
Picatinny Arsenal, Dover, NJ

The circuit shown is both a versatile electronic switch and a logic element which behaves similar to an SCR but has the advantage of several activation and deactivation modes and exhibits immunity to transients and common-mode voltages. The extreme stability of the circuit arises from the fact that the two complementary transistors are simultaneously driven into saturation or, alternately, into cutoff. Thus, the gain of the circuit in both stable states is zero, except for the brief transition period when a rapid cumulative action takes place.

The circuit can, in many cases, replace an SCR and can be used as a logic element by suitable routing of pulses. Operation always starts from the nonconducting state when the supply voltage is applied. A monostable action can also be produced by adding a simple time delay to supply the secondary triggering pulse.

The circuit can be activated in the following ways:
(1.) by application of a +Ve pulse to the base of Q_1
(2.) by shorting momentarily the collector of Q_2 to the ground.
(3.) by application of a −Ve pulse between the +Ve end of the supply and the base of Q_1
(4.) by shorting momentarily the collector of Q_1 to the +Ve end of the power supply.

The circuit can be deactivated:
(1.) by applying a −Ve pulse to the base of Q_2

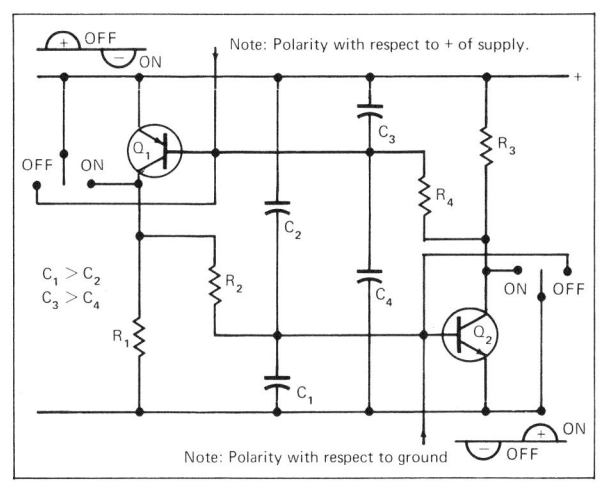

Combined electronic switch and logic element can be activated and deactivated by a variety of signals. When the switches are used, the switch action is momentary.

(2.) by shorting momentarily the base of Q_2 to ground
(3.) by applying a +Ve pulse from the +Ve end of the power supply to the base of Q_1
(4.) by shorting momentarily the base of Q_1 to the +Ve end of the power supply

The circuit capacitors C_1 through C_4 provide a potential divider for the high-frequency component of the transient common-mode voltages. Transients exceeding ±50% of the supply voltage do not affect the circuit.

Logic-supply crowbar

Lee Strahan
Exact Electronics, Inc.
Hillsboro, Ore.

Here is a simple circuit that can provide "crowbar" overvoltage protection for a 5V, 1A power supply. It can be adjusted to trigger at 10% overvoltage (5.5V). The circuit is fast-acting, temperature-stable, and requires no external power source.

Tunnel-diode CR_1 is the level-sensing element. Resistors R_2 and R_3 bias the diode near its peak-point current. Trimming resistor R_3 adjusts the bias current so that at precisely 5.5V, the tunnel diode switches slightly past its valley point. The voltage across the diode then increases. This voltage, combined with the voltage drop across R_1, is sufficient to bias Q_1 into saturation.

When Q_1 turns on, it supplies gate current to the SCR which fires and continues to conduct until power-supply current is removed. The power supply that is being protected must include a fuse and current-limiting circuitry. Resistor R_5 prevents leakage current from activating the SCR.

Operating speed of the crowbar circuit depends almost entirely on the SCR. An overvoltage transient with a duration of under 0.5 μsec can trigger the crowbar, and the SCR can discharge a typical filter or line capacitance in less than 1 μsec.

Trip-point stability of the circuit depends on the peak-point current stability of the tunnel diode. Actual measurements have indicated that the trip-point voltage shifts by less than 2% over a temperature range from room ambient to an upper test limit of 150°F.

If desired, it is possible to remove CR_1 and R_1 and replace these components with a single 150-Ω resistor. This modification yields a less expensive circuit which can react at the same speeds but which does not have the trip-point stability of the circuit shown.

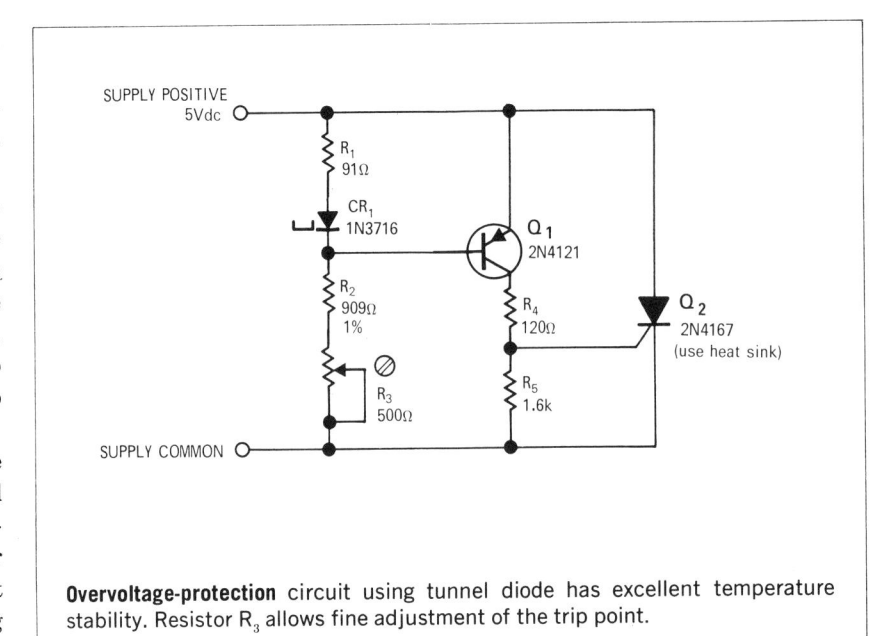

Overvoltage-protection circuit using tunnel diode has excellent temperature stability. Resistor R_3 allows fine adjustment of the trip point.

Adjustable-overvoltage circuit breaker

by Virgil R. Laul
Disc Instruments
Santa Ana, Calif.

THIS CIRCUIT serves as an overvoltage protector and, at the same time, generates trigger pulses for an alarm or other functions.

The pass transistor, Q_1, is normally on; it is turned off when SCR Q_2 fires. Zener D_1 keeps the gate of the programmable unijunction Q_3 at a constant voltage. Q_3's anode voltage is kept below the zener voltage. Where the supply voltage rises and Q_3's anode voltage exceeds the gate voltage, Q_3 fires, turning on Q_2, which cuts off Q_1 and the load.

By selecting a suitable R_2/R_3 ratio, one can set Q_3's anode voltage and thus, the voltage at which the circuit will trip. When Q_3 fires, it generates complementary pulses A and B which can be used to trigger other circuitry.

After the circuit trips it stays off until the supply voltage is shut off, at which time Q_2 commutates and resets the circuit.

In many cases, it may be desirable to actuate a relay or lamp without controlling a load. In these applications, one can eliminate Q_1 and the load, substitute a relay or lamp for R_1 and use more than one circuit to trip at different levels.

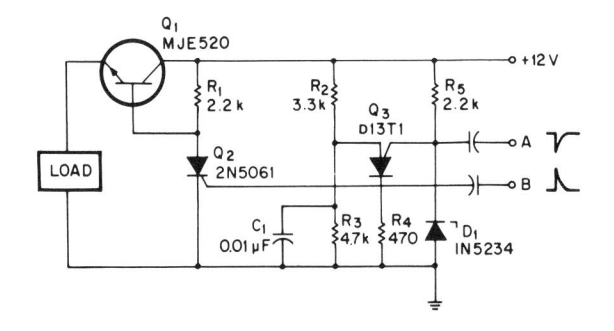

The R2/R3 ratio determines the trip point of this overvoltage protector.

150 Circuit Design Idea Handbook

Window detector uses one IC regulator

Neal Pritchard
ACDC Electronics Inc., Oceanside, CA

Window detectors typically use two voltage comparators whose outputs are combined to form a single logic output. The unique detector presented here uses only one 723 IC voltage regulator to perform the same function.

The detector circuit compares the output voltage of two separate voltage dividers with a fixed reference voltage. The resultant absolute error signal is amplified and converted to a logic signal which is TTL compatible.

The 723 has an internal reference voltage output which is only capable of sourcing current, not sinking it. Since current sinking is a necessity for proper operation, additional components are required. The reference voltage is applied thru D_3 to the base of Q_1. The emitter of Q_1 is now equal to the reference voltage and is temperature compensated by D_3. If a lower voltage is needed, a resistor may be inserted in series with D_3.

The voltage divider for the lower limit of the detector is R_1 and R_2. This voltage is compared to the reference voltage at point A. During normal operation D_1 is back biased resulting in no error signal. When V_{IN} lowers to a point where D_1 is forward biased, current flows thru R_9, R_5, D_1, and R_2. This produces a negative signal on pin 4 of the IC regulator which turns on Q_2 and Q_3, resulting in a logic ZERO at V_{OUT}. The voltage divider for the upper limit is R_3 and R_4. Similar to the lower limit operation, D_2 is back biased until V_{IN} reaches the upper limit, at which time current flows thru R_3, D_2, R_6, and Q_1. The positive signal developed across R_6 is amplified by the IC regulator and converted to a logic ZERO at V_{OUT}.

A resistor R_F may be connected between the collector of Q_2 and pin 5 of the regulator. This positive feedback will produce hysteresis in the upper and lower limit settings, thus eliminating noise problems at crossover.

An additional feature of the IC regulator is the internal overload transistor which can be used for blanking. A logic ONE at the blanking terminal will cause V_{OUT} to be logic ONE regardless of the state of V_{IN}.

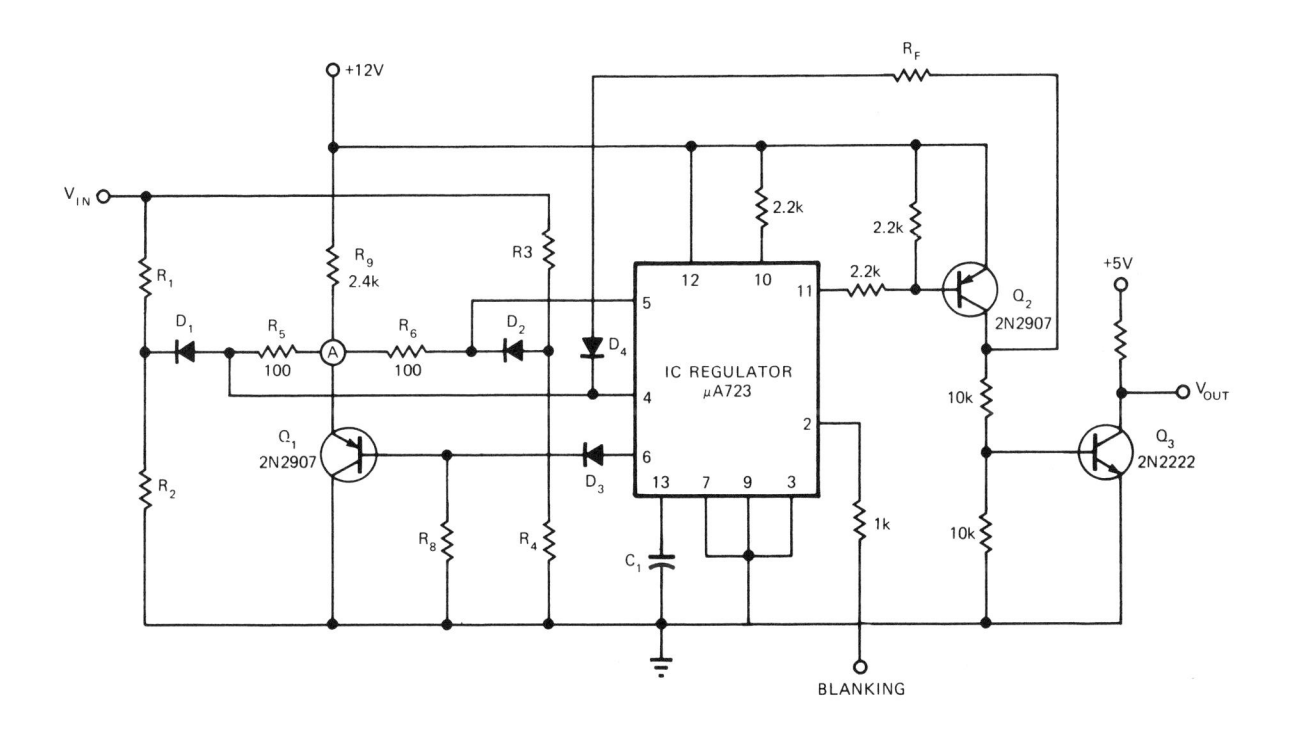

IC voltage regulators can also monitor input voltage. When V_{IN} is within predetermined limits, V_{OUT} will be a logic ONE. If V_{IN} is above or below limits, V_{OUT} will be ZERO.

SECTION V
Switching Techniques

High-voltage triacs reverse capacitor motor

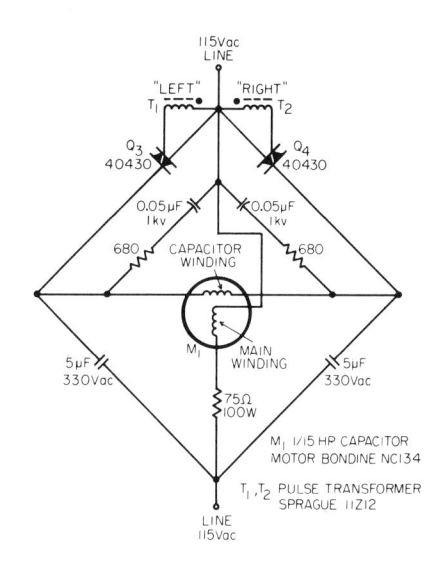

Fig. 1. Simple triac bridge controls direction of rotation of capacitor-type motor.

Fig. 2. Control circuit for the motor controller. Depending on the input logic, pulses from the unijunction oscillator are transformer-coupled to one of the two triacs in Fig. 1.

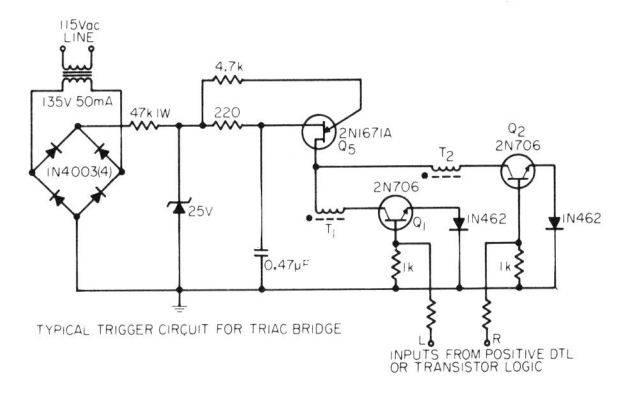

Charles H. Harris
Alden Research
Westboro, Mass.

Two triacs and two capacitors are the essential components for a simple bridge circuit that can control rotation-direction of a fractional-horsepower, capacitor-start, motor. With a suitable trigger circuit, also described here, motor direction can be controlled by IC logic voltages.

Figure 1 shows the bridge circuit, and Fig. 2 shows a suitable trigger circuit together with its power supply. The bridge and trigger circuit are coupled together by transformers T_1 and T_2. No external power supply is needed, other than the 115-Vac supply for the motor.

Low-level logic, from an IC flip-flop, forward-biases either Q_1 or Q_2 at points L or R respectively; resulting in either "left" or "right" motor drive.

Unijunction oscillator Q_6 provides pulsed emitter current which is conducted by either Q_1 or Q_2 depending on the input logic. One of the two pulse transformers then couples the pulsating signal to the corresponding triac Q_3 or Q_4. Triac conduction determines the phase of the capacitor winding relative to the main winding. This, in turn, determines the direction of rotation.

Pulsed output from Q_6 is synchronous with the ac line, because the supply voltage for the unijunction is full-wave clipped ac derived from the combination of transformers, rectifiers and zener. Of course, the pulse train ceases each time the line voltage crosses the zero axis. There is a short delay before the unijunction oscillator can restart with a new line-voltage cycle. This is because the timing capacitor must recharge to the firing voltage of Q_5. Delay is sufficient to allow the voltage to build up across the triac, thus ensuring reliable triggering.

The 75-ohm resistor, shown in series with the main winding, allows motor direction to be reversed while the motor is running. This resistor is not required if the motor need only be reversed from standstill rather than while running. Also, the resistor is not needed if the motor is always heavily loaded. Use of a series resistor causes a sight reduction of starting torque.

A disadvantage of the simple two-triac controller is that it doesn't allow phase control of the motor — to vary the starting torque. This is because the nonconducting triac sees an extremely high voltage across the near-resonant motor-and-capacitor circuit. Though the 400-volt triacs used here are adequately rated for full-phase running, they would not be able to withstand the dV/dt conditions encountered with delayed triggering. If phase control is needed, a third triac in the main winding could be phase-controlled from a separate trigger source.

The two-triac controller is foolproof. Simultaneous triac triggering, or a shorted triac, will stall the motor — which then draws about two-thirds of its normal running current. With the specified motor, current in the capacitor winding ranges from 500 mA rms at idling speed, down to 300 mA with both triacs shorted. Current in the main winding is 1.6-A idling and 500-mA stalled.

RC networks across each triac minimize the possibility of unwanted dV/dt triggering. The values depend on the inductance and back emf of the particular motor used.[1] With the motor specified here, the no-load back emf is about 800 V, resulting in dV/dt of 2 V/μs.

The same basic circuit can be used to control motors other than the one specified. Larger motors would probably increase the dissipation of the conducting triac, necessitating a heat sink. Very small motors could be controlled by a simpler circuit with a triac connected from one winding or the other, directly to the line.

Reference
1. "G.E. SCR Manual, 4th Edition," General Electric Co., 1967, p. 136.

Simple zero-crossing solid-state switch

by A. J. Marek
LTV Aerospace Corp.
Dallas, Texas

MINIMUM RFI is generated when ac power is switched on and off to a load at the zero-voltage crossover point. This switching can be accomplished by a simple circuit consisting of four diodes, an SCR, a transistor and three resistors connected as shown in Fig. 1A.

The positive output of the bridge (Fig. 1B) drives Q_1 on during each half cycle of the supply voltage except when the voltage is at or near zero. The SCR can be triggered on only when Q_1 is off. Closing of switch S_1 provides continuous voltage to the collector of Q_1 but a trigger pulse to the SCR gate is provided only at the zero-crossing point as shown in Fig. 1C. When switch S_1 is open, the SCR will commutate off at the zero-voltage point. The values of resistance for R_1 and R_2 are selected to provide the required pulse width for the SCR being used.

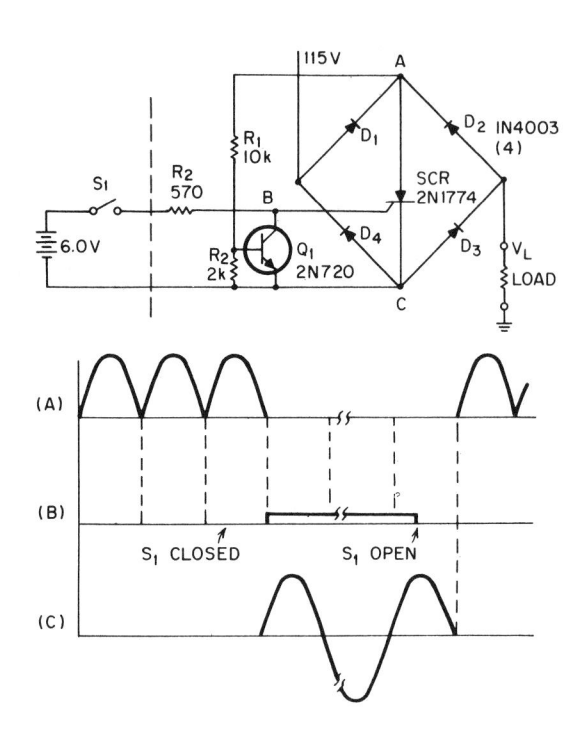

Fig. 1. Schematic and waveforms of zero-crossing solid-state switch.

Digital interlocking switch is inexpensive to build

Bjorn Brandstedt,
McDonnell-Douglas Corp., St. Louis, MO

This circuit is an electronic equivalent of a mechanical interlocking switch assembly. Its advantages, relative to the mechanical interlocking switch assembly are:

1. Cheapness
2. Less bulk
3. Elimination of switch bounce
4. TTL compatibility (in this case)

When one of the push-buttons is depressed, it automatically restores to normal the RS flip-flop associated with the push-button that was previously actuated. When a channel is selected its, and only its, output is HIGH. Mechanical switches required are DPST types, one for each channel.

The cross-coupled two-input NAND gates act as RS flip-flops and provide two functions, one to reset all other flip-flops and one to be used as an output. When the circuit is at first turned on, none or any one of the outputs may be HIGH. Also, if two or more push-buttons are simultaneously activated, those channels activated will all be HIGH, but only the last one released will remain on. The circuit is expandable to any number of channels.

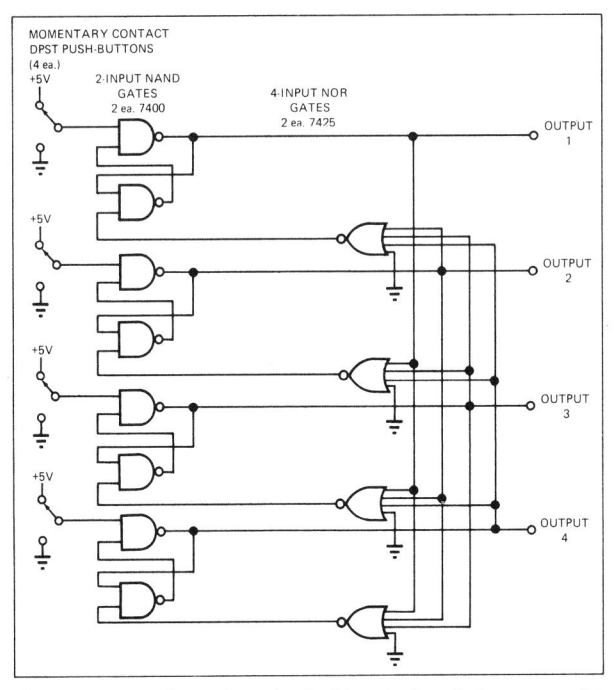

Electronic equivalent of mechanical interlock switch uses one RS flip-flop for each switch channel. The circuit can easily be expanded to a larger number of channels.

Monolithic timer makes convenient touch switch

J. Courtenay Heater,
Claremont Men's College, Rolling Hills, CA

A versatile touch switch for security or convenience purposes can be constructed from the Signetics SE/NE 555 Monolithic Timer and just a few additional components.

Some of the virtually unlimited applications of the touch switch include: switchless keyboards, thief accunicators, activators for the physically handicapped, bounceless electronic switches (with no moving parts) and novelty controls.

The timer itself features either free-running or one-shot capabilities which can be controlled through the use of TRIGGER and RESET inputs. The characteristics of the output pulses are fully adjustable over a large duty cycle, with timing periods adjustable from μsec to hours. The output is capable of sinking 100 mA for either electromechanical activation or other interfacing applications. Supply voltages are non-critical as the device is specified for operation between 4.5 and 16V dc. For 5V operation, the device is directly TTL compatible and draws only 3 mA making it suitable for battery operation. At 16V, the timer draws on the order of 8 mA.

The trigger input on the device is the key feature in touchswitch applications. Requiring only 500 nA to fire at 1/3 the V_{cc} supply voltage (referenced to circuit ground), the device is easily triggered by the voltage differential found between a floating (non-grounded) human body and the circuit itself. This is 20V or more, depending on static build-up. The touch plate can be any conducting material with virtually no size limitations.

Once triggered, the device cannot be retriggered and it will time out. However, if the duration of human contact exceeds the RC time constant of the timer, random spikes occur in the output after the time out. This can be avoided by making fairly large time constants, so that the device will not time out before contact is removed.

A touch plate at the trigger input converts the SE/NE 555 Timer into a versatile touch switch.

The duration of the output pulse is controlled by both the RC time constant and the control voltage input (pin 5). By varying the voltage at pin 5 the timing can be changed by about one decade. If the entire RC network is omitted and pin 7 is connected directly to pin 6, the circuit will latch ON when "touched".

Perhaps the best feature of all, though, is the low price (less than $1 in single units)

Triac gating circuit

Entry by Charles A. Farel and David M. Fickle
IBM Corp., Boulder, Colo.

Triacs have become increasingly popular for controlling ac power to resistive and inductive loads such as motors, solenoids and heating elements. Compared with competing devices such as relays, triacs offer important advantages in the area of cost, reliability, packaging and electromagnetic interface.

This simplified triac gating circuit switches ac power from the supply to the load. The circuit's chief innovation lies in the combination of the basic switching network with suitable dV/dt commutation suppression.

The circuit applies power to the load when the low-power switch (or relay) S1 is closed. Current through resistor R1 turns on the triac in the conventional manner.

When the switch is opened, the circuit automatically turns off the power. Resistor R1 and capacitor C1 suppress the dV/dt commutation pulse to ensure safe and effective turnoff. This suppression is essential when switching power to inductive loads.

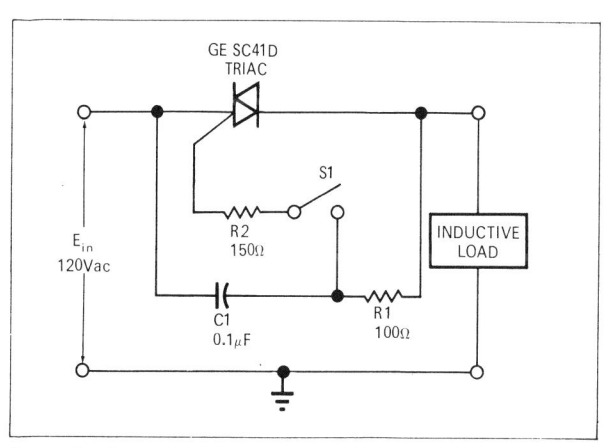

Triac circuit switches ac power to inductive loads. Components R1 and C1 provide suitable dV/dt suppression.

The circuit shown can be extended easily to polyphase applications.

Set-reset latch uses optical couplers

Robert N. Dotson,
Motorola Inc., Phoenix, AZ

This optically coupled set-reset latch provides almost complete isolation between each input and the output as well as between the inputs. Momentary application of about 2V at 14 mA to the SET terminals will allow up to 150 mA to flow between E and E' with a 1.6V drop. This current will continue to flow, even after the SET LED is turned off, until approximately 2V at 15 mA is applied to the RESET terminals, or until the current from E to E' is reduced to less than 1 mA. Note that when the circuit is off, a maximum of about +30V or −9V is all that can be applied from E to E'.

Basically, the MPS6518 and the SET coupler's Darlington make up a discrete approximation of an SCR. The two resistors are used to reduce the gain of the transistors so that the circuit will not automatically latch up. RESET is accomplished by shunting the base of the SET coupler with the RESET coupler, or by reducing the current from E to E' until the betas of the SCR transistors drop low enough to unlatch the circuit.

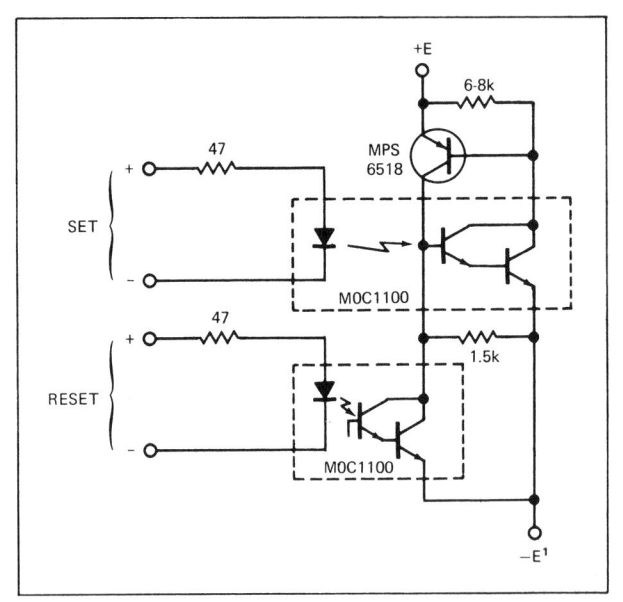

Almost complete isolation between the inputs and output of the set-reset latch is provided by the optical couplers.

Noise-Rejecting SCR Trigger Circuit

By Sumner B. Marshall
Sprague Electric Co.
N. Adams, Mass.

IT IS OFTEN difficult to obtain reliable SCR triggering when inductive switching transients are present. The circuit described here, however, can discriminate between data pulses and random transients.

An integrator is combined with a voltage comparator to detect the difference in voltage-time areas in the two types of pulses (random transients have a smaller $V \times t$ area than data pulses, which have a fixed $V \times t$ area).

An approximate integration of the input pulses is performed by $C_1 R_1$. The complementary pair Q_1, Q_2 functions as a voltage comparator, delivering an output whenever V_c is greater than V_{ref}.

Without an input, the voltage of C_1 will be zero, and thus Q_1 and Q_2 will be cut off. When an input $V(t)$ is applied, the emitter voltage of Q_2 (V_c) will rise. If the (voltage) × (time) area is sufficiently high, Q_1 and Q_2 will turn on when V_c is slightly greater than V_{ref}. Since both Q_1 and Q_2 are saturated and R_B is shunted by R_3, C_1 will discharge rapidly, producing a 10-μsec pulse across R_3 and firing the SCR.

If the applied pulse is too narrow to fire Q_1 and Q_2, C_1 will discharge through D_1, preventing an accumulation of noise pulses from firing the circuit. However, note that if the pulse at the input is too wide, multiple outputs may occur.

In the circuit shown, data pulses are 8 V high and 0.5 msec wide. The reference voltage is chosen to cause firing before the exponential slope becomes too flat.

Thus, with $V_c = 5$ V at 0.5 msec,

$$5 = 8(1 - e^{T/\tau})$$

Or τ should be about one time constant.

$$\tau = RC = T = 0.5 \times 10^{-3} \text{ sec}$$
$$C = 0.05 \text{ μF}$$

Note that R_B must be much greater than R_3 to obtain a large change in V_{ref} when the transistors switch on.

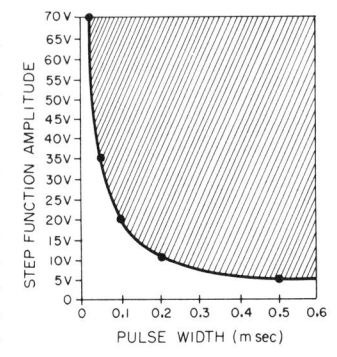

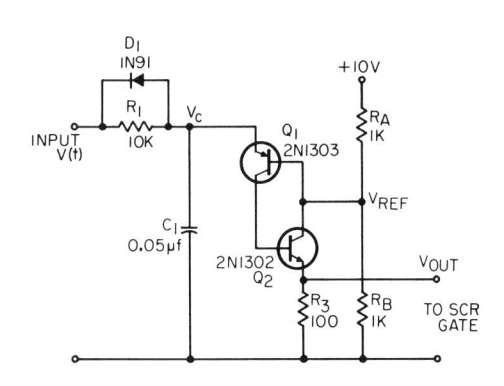

Locus of possible firing points

Data-pulse discriminating trigger circuit.

158 Circuit Design Idea Handbook

Putting the "thumb" on thumbwheel switch multiplexing

Eric Breeze
Fairchild Semiconductor, Mt. View, Calif.

Thumbwheel switches are becoming increasingly popular for remote programming of counters, displays, industrial control systems, etc. To reduce the number of interconnections between the switches and the destination, it is desirable to use multiplexing techniques. Ten decades of BCD thumbwheel switches unmultiplexed would require 50 interconnections, while a multiplexed system would require only 10 interconnections.

The usual method of multiplexing is to use BCD (or any 4-bit code) thumbwheel switches, each with a diode in series with the four outputs. These are connected to four parallel bus lines to the system output. The wiper arm of each switch is then selected to get decade location.

This system uses standard-low-cost, single-pole decade switches with all ten outputs of the switches parallel bussed and the wiper arm connections brought out separately. The ten parallel outputs are fed into a simple encoder comprised of four NAND gates to generate the desired 4-bit output code (diagram shows BCD). Only the "true" connection of the ten outputs are required, as ambiguous outputs cannot be generated due to the mechanical design of the switches.

The wiper arm of each of the switches is addressed from an active low open-collector decoder. Collector pull-up resistors are placed on the encoder NAND gate inputs. The use of an open-collector decoder prevents undefined logic levels whenever two or more switches are in the same position.

In operation the 3-bit input address determines which decade switch is addressed. Switch position then determines which encoder NAND gates are activated. The common of each of the switches is addressed from an active low open-collector decoder.

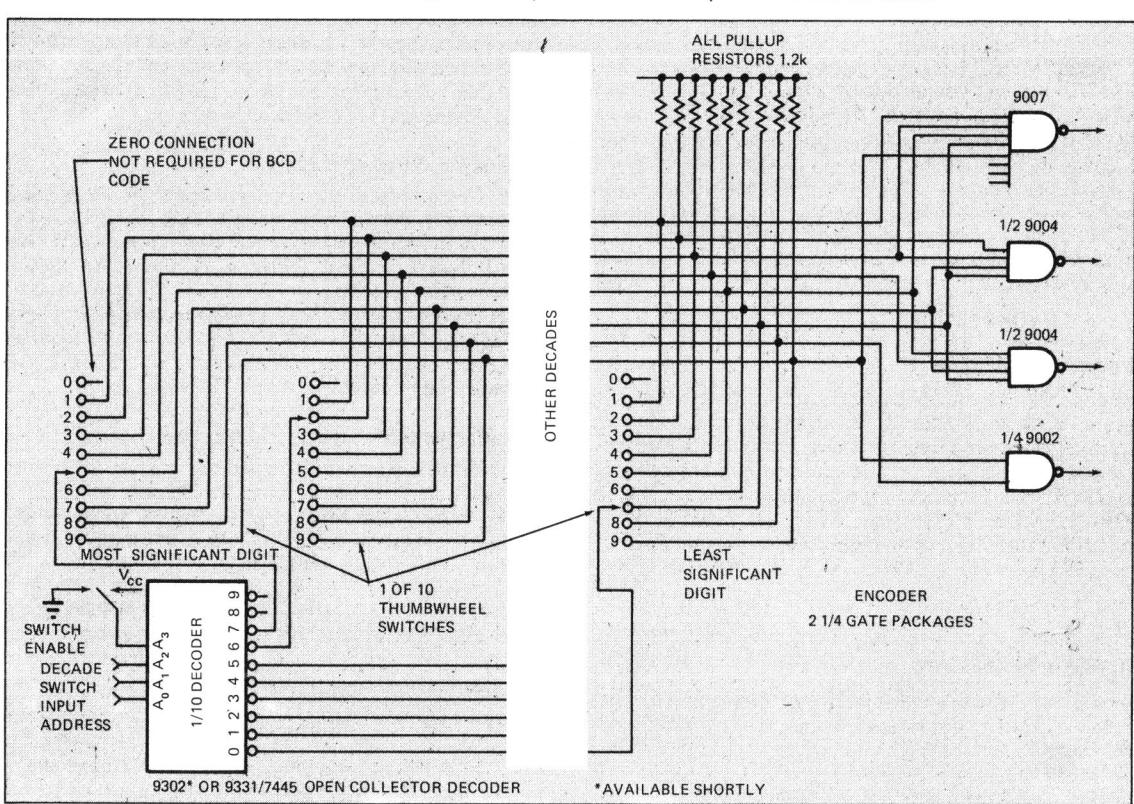

Thumbwheel switches multiplexed in the method described here provide system economies by allowing use of low-cost "1-of-10" switches, rather than encoded switches. Wiring interconnections are also reduced, from 50 to 10 in this case, offering reduction of assembly costs. Reliability is increased appreciably since there is only one moving contact per switch.

Battery-saving remote-command detector

Peter J. Hof
Battelle-Pacific Northwest Labs., Richland, Wash.

This simple remote command detector requires only 2μW of standby power. It features electrical isolation, excellent noise immunity, CMOS compatibility, and can be hermetically sealed. No physical connection to the outside world is required.

The circuit provides excellent results for remotely controlled equipment requiring a low-command duty cycle. A single 6V battery provides power for both the command detector and the equipment it controls.

Phototransistor Q_1 detects light from an LED or incandescent light source. Light enters the hermetically sealed unit via a light pipe or line-of-sight. Resistor R_1 sets the sensitivity. R_2 and C_1 integrate the input signal to provide noise immunity. R_2 can be as great as 330kΩ and C_1 can be as small as 0.001μf. The PUT (programmable UJT) Q_2 provides a 3V pulse to set a power-up latch which in turn powers the beam width detector. Resistors R_4 and R_5 in the gate circuit of the PUT form a 20 MΩ impedance across the battery until light enters the detector, resulting in a leakage current of approximately 300 nA. Resistor R_3 determines the pulse amplitude.

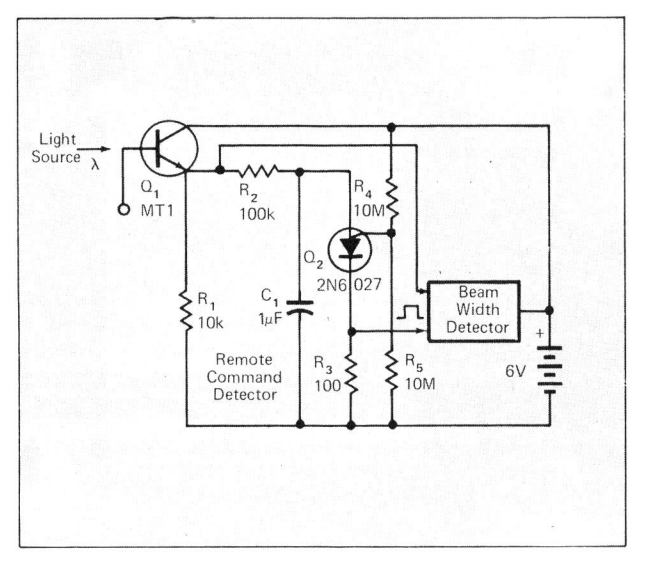

Light activated command detector for remote control devices uses an inexpensive PUT to achieve near zero power consumption in standby mode.

600V switching circuit provides good speed and isolation

T.J. Dewees and **J. Raamot**
Western Electric Co., Princeton, NJ

The circuit shown switches either ±300V source to a load within one μsec. A common design of such a circuit is to insert a 600V transistor between the load and each voltage source and to insure that only one of the transistors is in the conducting state at any one time. Faster switching times are obtained by replacing each 600V transistor with two 300V transistors connected in tandem.

Such a design has a problem: because transistor turn-on times are faster than turn-off times, the switching has to occur at logically separate instances to prevent both sets of transistors from conducting simultaneously. At the first instance, the ON-transistor capacitance maintains the output at the previous voltage level, thereby retarding the turn-off time. Later, when the other transistors turn on, this voltage is discharged into the other power supply, causing large current transistors.

This problem is solved with the circuit shown. In the tandem transistors, as the transistor connected to the 300V supply turns off, the one connected to the load is forced to conduct and discharge all voltages to ground. Thus, all capacitances discharge to ground, while the 600V isolation between power supplies is maintained. Furthermore, all switching can occur simultaneously because the current is limited in the transistor connected to the 300V source.

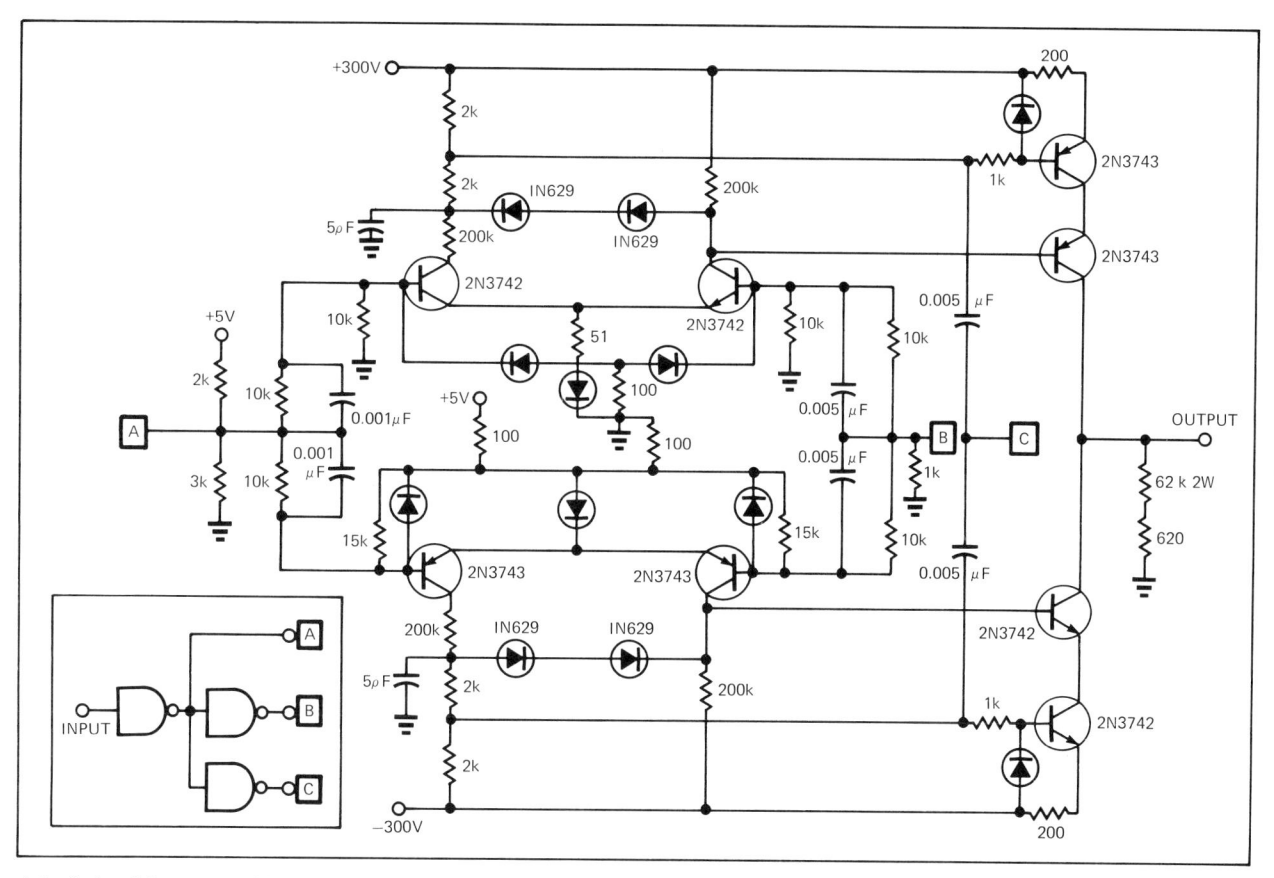

A logic-level input provides isolated switching between the two power supplies. Power supplies of up to ±300V can be safely switched in this manner.

Remote-Controlled Exclusive OR Gate

By Russell F. Sherwin
Melabs
Palo Alto, Calif.

THIS OR gate will direct one of several low-impedance audio sources to a single high-impedance output, while inhibiting all other sources. One application was in a paging system which had several push-to-talk carbon microphones feeding a common amplifier. Control, bias for the carbon microphones and audio are all served by a two-wire line to each location.

When a given input is closed to a low-impedance source, the corresponding UJT is triggered. The resulting emitter current causes the emitter voltage to stabilize at a point below the trigger level of the other stages. Audio from the source can now feed through the forward-biased emitter junction to the output.

Best operation will be obtained with the supply voltage greater than 15 V and with R_1 adjusted to supply a minimum of 2 mA emitter current. The audio sources should have a dc resistance less than 1 K.

None of the components used in this circuit is critical. Almost any UJT will work here.

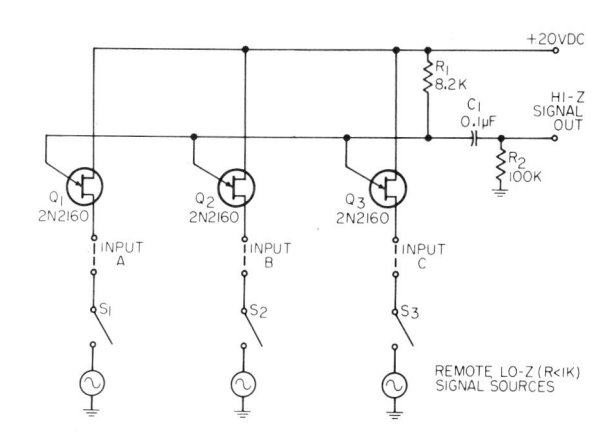

Exclusive OR gate for use with carbon microphones, or any audio source providing a low-impedance dc-return path.

Switching Techniques

An improved rotary-switch interlocking circuit

by A. T. Nasuta
Westingouse Electric
Baltimore, Md.

SOME SORT of interlock circuit is often needed for use with rotary switches. For example, if a switch selects voltage inputs to a multi-range measuring instrument, it would be desirable to prevent switch rotation until the measuring instrument has been set to the correct range — otherwise the instrument could be damaged by excessive voltage.

There are various ways of interlocking switches, but many of them have serious disadvantages. For example, one widely-used method is to interlock a relay-coil voltage through its own contacts, and through the switch as shown in Fig. 1. When S_1 is depressed, K_1 is energized, allowing the interlocked signals to pass through the remaining contacts. When S_1 is rotated K_1 de-energizes momentarily, thus interrupting the interlocked function until K_1 is again energized.

But this method can be unreliable because of relay switching time. Most relays take about 10 ms to de-energize (and longer if arc suppressing components such as diode CR_1 are used), and the travel time of most rotary switches ranges from 1 to 15 ms.

Another possible interlock method is to use a special mechanically interlocked rotary switch. The problem with these switches is they're often expensive and clumsy to use

Fig. 2 shows an improved interlock technique that could overcome the disadvantages cited above. The circuit uses a silicon controlled switch (SCS) Q_1. When S_2 is depressed, voltage divider R_1 and R_2 supplies 1.2 Vdc to turn on Q_1. This provides a ground return for K_1, energizing it and thus allowing signal current to flow through the relay contacts to the monitoring instrument.

When the switch is rotated to another position, point 2 goes from +28 V to ground for 1 ms. This interrupts the holding current of Q_1 thus turning it off. With no ground return, the relay is de-energized and therefore disconnects the monitoring instrument.

The differentiating network of C_1 and R_4 converts the negative-going pulse at point 2 into negative and positive spikes at point 3. Fig. 3 shows the waveforms at this point. Diode CR_2 conducts only the positive spike to point 4 and turns Q_1 off.

In the circuit of Fig. 2, R_3 suppresses the rate-effect of the SCS. If the anode gate of the SCS (point 4) is left open circuited, the SCS has a tendency to trigger on when +28 V is applied to the anode of Q_1. The filter capacitor C_2 bypasses spurious switching transients at the cathode gate (point 1). This prevents the SCS from triggering due to pick-up transients at point 1.

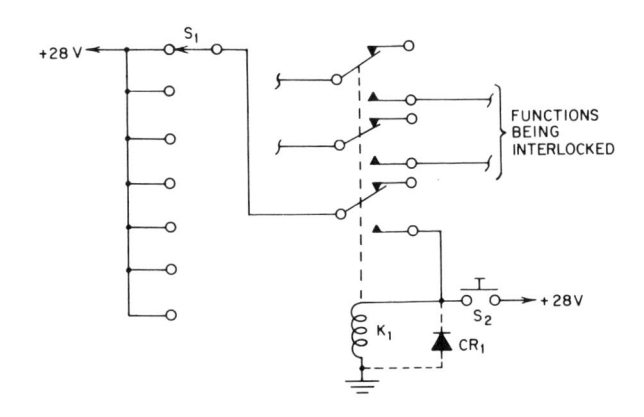

Fig. 1. Commonly-used circuit for sensing the rotation of rotary switch.

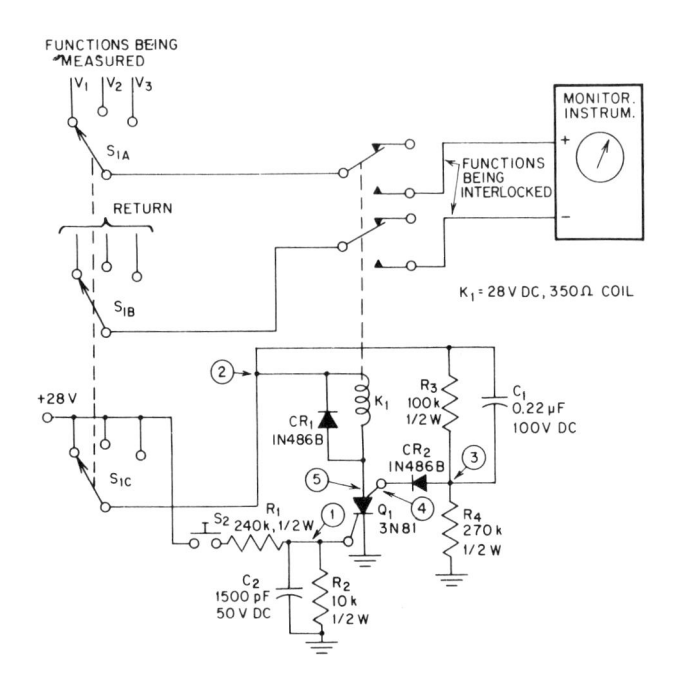

Fig. 2. A more reliable rotary-switch interlocking circuit using a silicon-controlled switch.

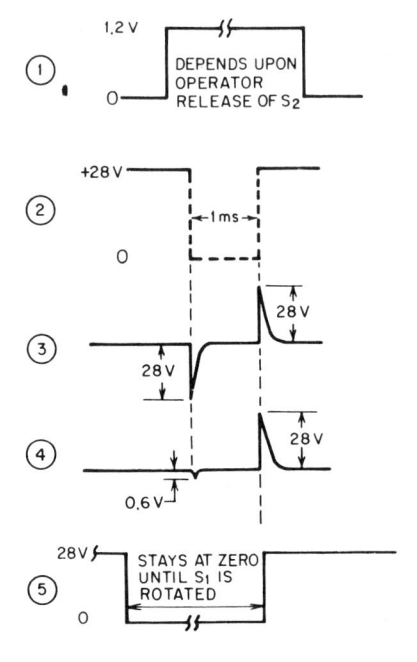

Fig. 3. Waveforms at the various points of the circuit shown in Fig. 2.

Make-and-break bounceless switching

By Leo F. Walsh
and Thomas W. Hill
State University of New York
Syracuse, N.Y.

This circuit eliminates switch bounce problems during closing as well as opening. When the switch closes (Fig. 1), the Q output from the flip-flop goes to logic-1 state and remains in that state until a period of time after switch release. Turnoff delay is determined by the RC time constant.

When the switch is closed, it immediately causes Q = 1 and $\overline{Q}$ = 0. The $\overline{Q}$ level tries to close the NAND gate but the switch level, although it bounces, holds the NAND gate open and its output remains at logic 0.

Releasing the switch operates the NAND gate causing its output to go to logic 1. This charges capacitor C through resistor R until the reset level is reached. At that time, the flip-flop resets, producing Q = 0 and $\overline{Q}$ = 1. The sequence of operations can be seen by referring to the timing diagram of Fig. 2.

Values for components R and C should be chosen according to the bounce duration of the switch used. For a typical spst switch with 1A contact rating, values of R = 10,000 and C = 0.47 μF have proved satisfactory.

Fig. 1 – Switch bounce eliminator circuit.

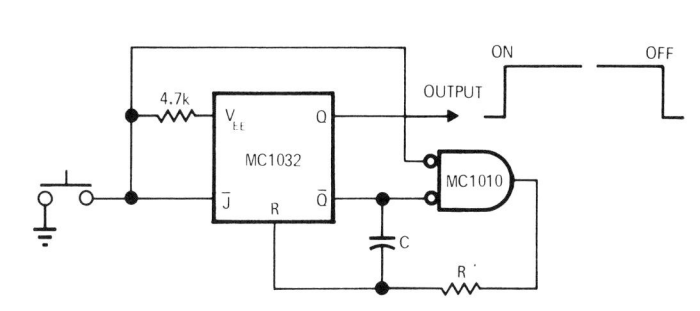

Fig. 2 – Output waveform is free of switch bounce.

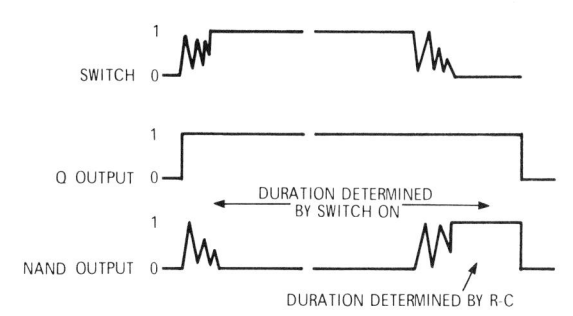

Low-Cost Bistable Relay Circuits

By G. Richwell
Reflectone Electronics
Stamford, Conn.

BISTABLE CIRCUITS using ordinary relays are cheap, and are quite adequate for many low-speed applications in binary registers etc. They may also be used as a substitute for latching relays. Here are two circuits, using capacitor transfer (Fig. 1) and relay transfer (Fig. 2).

In the circuit shown in Fig. 1, capacitor C is initially charged to the supply voltage. When the "set" contacts are activated, capacitor C will discharge through relay coil R_k and the relay will be energized if the time constant R_kC is large. Before the capacitor is completely discharged, the "hold" contact will close and the relay will continue to be energized by the current flowing through resistor R. The "set" contacts may now be deactivated, whereupon the capacitor will discharge rapidly through the grounded relay contact. When the "set" contacts are again activated, the capacitor will provide an effective short across the relay coil, and the relay will be deenergized. After the "set" contacts are returned to the deactivated position, the capacitor will be recharged to the supply voltage and the circuit is ready to repeat the cycle.

The circuit of Fig. 2 is similar to that of Fig. 1 with the exception that the capacitor has been replaced by a relay to perform the transfer-storage function. This transfer relay is initially energized through the "set" contacts and the contacts of the bistable relay, making the supply voltage available to energize the bistable relay when the "set" contacts are activated. After activation, the current through the coil of the transfer relay is maintained through a limiting resistor and through the transfer relay contacts. Upon deactivation of the "set" contacts, the coil of the transfer relay will be shorted to ground through the contacts of the bistable relay, thus deenergizing the transfer relay. The transfer relay contacts will now be grounded and the coil of the bistable relay will thus be deenergized when the "set" contacts are again activated. When the "set" contacts are deactivated, the transfer relay is energized and the circuit is ready to repeat the cycle.

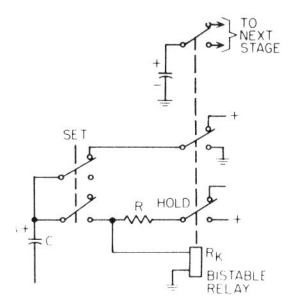

Fig. 1. Bistable relay circuit with capacitor transfer. When used for binary register applications, the "set" contacts shown are part of the bistable relay for the preceeding stage.

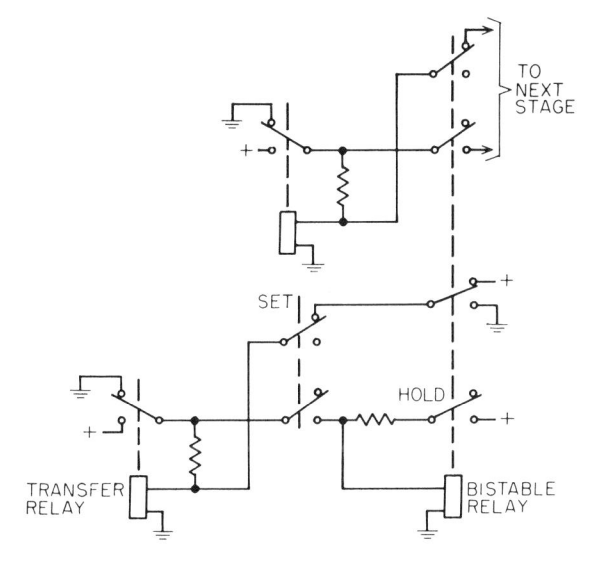

Fig. 2. Circuit similar to that shown in Fig. 1 except that a second relay is used instead of the capacitor for transfer storage.

Switching Techniques 163

Efficient and simple zero-crossing switch

by A. S. Roberts
and O. W. Craig
Univac
Bristol, Tenn.

A zero-crossing switch eliminates the high surge currents and transients that normally occur when switching line voltage to a load. For example, the surge current in an incandescent lamp that is switched on at peak line voltage is many times the peak current for a lamp switched on at a zero crossing. Another advantage of the circuit described here is that it needs no dc power supply (see **Fig. 1**).

When the line switch is closed, power will not be applied to the load until the next time the voltage goes through zero.

Identical circuits control each half cycle of the ac line, so only one of the two possible sequences of events will be described here. Assume that the line switch is closed at a time when the switch side of the line is positive and the voltage is greater than 1.4V. Transistor Q_1 immediately conducts, thus prohibiting C_1 from charging to the voltage required to trigger SCR_1. During the remainder of the half cycle, the high resistance of R_1 limits the load current. Diodes CR_2, CR_3 and CR_4 limit the voltage across the transistor circuit to about 3V, while R_2, CR_5 and R_4 provide proper bias for the transistor. Resistor R_3 limits Q_1's collector current and controls the charge rate for the capacitor.

When the line voltage crosses zero, current flows through resistors R_5 and R_6, thus beginning to charge C_2. This capacitor charges to the SCR trigger voltage (0.4 to 0.8V) and triggers SCR_2 within 0.02 msec (assuming a 115V, 60-Hz line). This occurs before the line voltage reaches an amplitude of 1.4V, which would be sufficient to turn on Q_2 and hold the SCR gate voltage below the other half of the circuit.

Because SCR turn-on occurs approximately 0.013 msec after zero crossing (less than a third of an electrical degree) and each SCR conducts for almost a full half cycle, essentially no power is wasted in the circuit. Because of this low dissipation, the circuit lends itself to fabrication as an IC. Note also that the circuit is never subjected to full line voltage except during the time interval between line-switch closure and the first zero crossing.

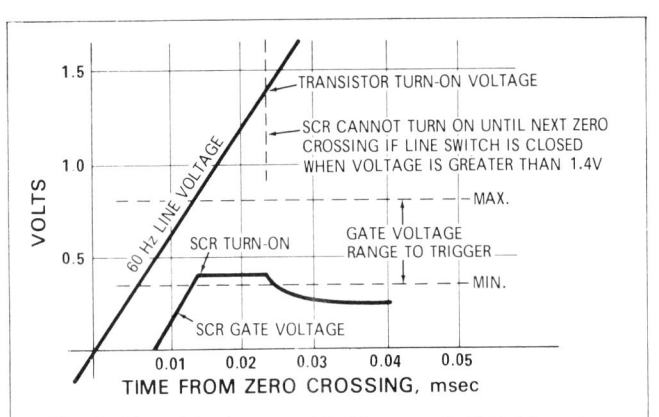

Fig. 2—**Transistor turn-on** at 1.4V prevents SCR triggering until after the next zero-crossing point. Timing diagram shows sequence of events after zero crossing.

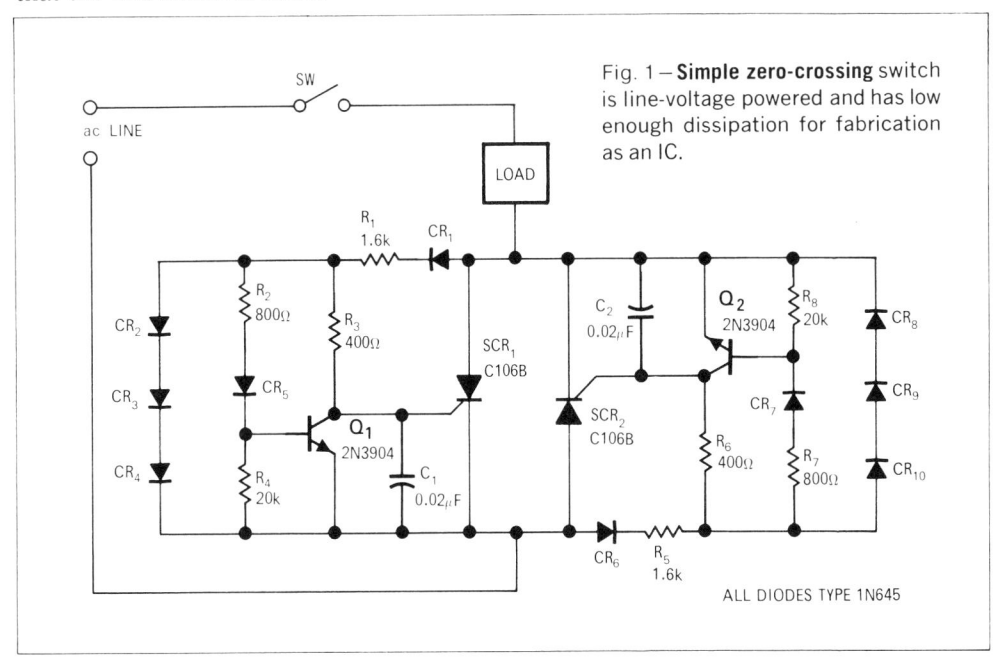

Fig. 1—**Simple zero-crossing** switch is line-voltage powered and has low enough dissipation for fabrication as an IC.

Battery saver has automatic turn-off

Dave Weigand
Gulf and Western, Swarthmore, PA

An automatic turn-off battery saver is often required for alarm, remote control, toy or unattended electronics. The versatile circuit shown here can be designed for a wide range of voltage, power and delay.

The SCR is triggered on by external control logic or contact closure. This connects ground to the load and to the shutdown unijunction timer, Q_2. After the C_1R_1 delay, unijunction Q_2 fires, discharging C_1 timing capacitor into R_3. The resulting pulse across R_3 is coupled by C_2 into the base of commutating transistor Q_3. Q_3 commutates the SCR by momentarily shunting the SCR current below holding current. With the SCR commutated OFF, only leakage currents through the reverse SCR junction will drain the battery.

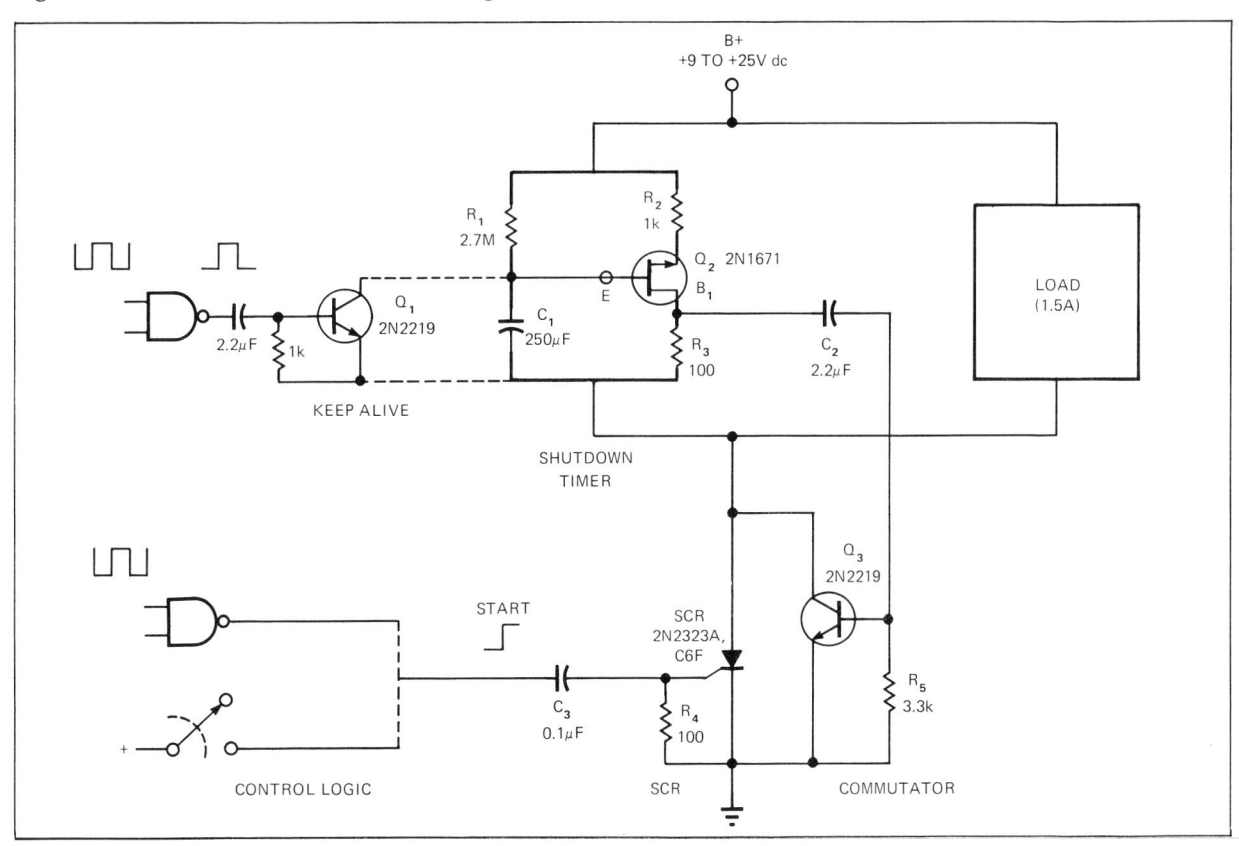

Low-power timer drives stepping relay

By Larry R. Eaton
Desert Research Institute
Reno, Nevada

THE EXTREMELY low forward-blocking current of some SCRs (for example, GE's C106B, with $I_{fx} \simeq 0.1\ \mu A$) allows the design of inexpensive timing circuits which can drive inductive loads (such as stepping relays), yet which require a minimum of continuous power.

The circuit shown has a current drain of only 25 μA when cycling at 1 pulse/min. This allows battery operation in portable applications. For dc operation, a 90-V battery is suitable. Alternatively, the circuit will work directly from rectified ac (half-wave or full-wave), using a 110-V line.

Timing depends on the supply voltage as well as on the RC time-constant. With the component values shown (R = 10 MΩ, C = 25 μF), the circuit cycles at 5-minute intervals using a 90-V battery, or at 10-minute intervals using half-wave rectified ac. For optimum efficiency and timing accuracy, a low-leakage capacitor should be used.

Circuit operation is extremely simple. Capacitor C charges from the supply line via resistors R_1 and R_2. When the capacitor has charged to around 65 V, the neon fires and triggers the SCR. The charge on the capacitor is dumped through the relay coil, activating the relay. The discharged capacitor then starts to recharge, commencing a new cycle. Resistor R_1 allows adjustment of the cycling rate, while resistor R_2 determines the minimum cycle period.

In addition to its efficiency, the circuit has other advantages. the SCR is self-commuting — no special turnoff circuitry is needed. The circuit reliably drives inductive loads

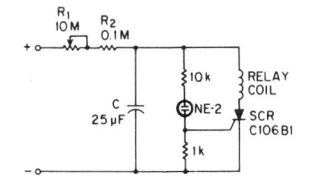

This timing and driver circuit for stepping relays offers high efficiency, provided a low-leakage capacitor and SCR are used.

— the trigger signal is present until the SCR is fully conducting. Recycle time can be very short — the capacitor is discharged through the SCR and not through the neon or through a resistor.

Bistable switch with zero standby drain

by Don B. Heckman
AMF Electrical Products
Development Div.
Alexandria, Va.

In this bistable switching circuit, the transistor switch Q_3 turns on upon receipt of an ON command at input A, and turns off upon receipt of an OFF command at input B. Current drain is zero in the OFF state. The circuit is especially useful in battery-operated equipment where supply drain must be kept as low as possible.

When point A is driven positive, this triggers SCS (silicon controlled switch) Q_1, at time t_o, thus turning on transistor switch Q_3. Components R_2 and C_1 provide the necessary coupling, while R_1 is needed to discharge C_1. Diode CR_1 protects the gate-cathode junction of Q_1 when point A is driven negative. While Q_1 is on, point C is near ground and point D is near V_{cc}. Holding current for Q_1 must be provided through R_4, therefore Q_1 should be chosen for low-holding-current characteristics. With Q_3 on, C_3 is charged to almost V_{cc}, with $+V_{cc}$ at the anode of Q_2.

When point B is driven positive, the other SCS, Q_2, is gated on, at time t_f, via R_6 and C_2. Components R_5 and CR_2 have similar functions to R_1 and CR_1 as previously described. The anode of Q_2 drops to near zero potential at time t_f, thus placing $-V_{cc}$ at the anode of Q_1, and turning off Q_1. Resistor R_8 is chosen large enough to allow less than the required holding current for Q_2. After C_3 has been charged through R_4, at time t_d, Q_3 and Q_2 turn off. The complete circuit is then in an OFF state and only transistor leakage current (typically 2 to 3 μA) flows.

The circuit was designed, with the component values shown, to supply up to 7 mA of switched current from a 12V dc supply. Operation is satisfactory over the range -10 to $+60°C$ with 12V trigger signals.

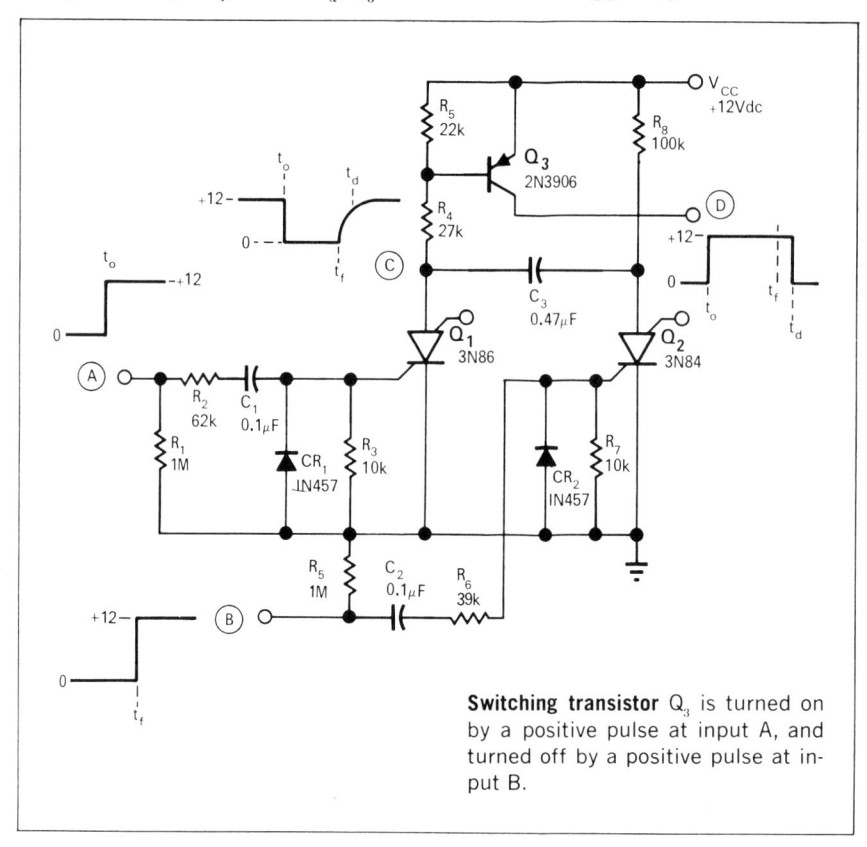

Switching transistor Q_3 is turned on by a positive pulse at input A, and turned off by a positive pulse at input B.

Solid-State Relay

By Jack E. Frecker
University of Arizona
Tucson, Ariz.

THIS CHOPPER-DRIVER circuit switches at rates from dc to 10 kHz. Its switched output is completely isolated from the input control signal. The circuit has many of the qualities of a conventional relay, but is much more reliable because it is totally solid state.

The left-hand part of the circuit shown, is a common-base oscillator. Frequency of oscillation is about 10 MHz. Input logic signals bias the base of Q_1, turning the oscillator on and off. Output from the oscillator is transformer coupled, to provide dc isolation. After rectification and filtering, the oscillator output controls the bipolar switch Q_2, Q_3. The transistor switch shown has an "on" resistance of about 20 ohms, plus a few millivolts offset. "Off" leakage current is less than 1 μA.

The transformer is wound on an Arnold A4-134P toroidal core. Number of turns for each winding is shown in the schematic.

Input logic levels are -3 and -11 V, which are standard for the EECO T-series and similar logic modules.

The basic driver circuit can control other circuits instead of the simple bipolar switch shown here. For example, it can drive a dual-emitter chopper such as the 3N74.

Also a dc load can be controlled via a Darlington-connected switch. For an ac load, the rectified output from the driver can control an SCR circuit[1], which in turn will control the load.

In all the above applications, the basic driver circuit remains the same. Thus it can be encapsulated to form a versatile module. Only the switching transistors are left outside the module (shown dotted) as these may need to be changed for different applications, or replaced if they get damaged by overload.

Reference

1. "150-W Voltage Controller," *SCR Manual*, 2nd Edition, General Electric Co., 1961, p. 116.

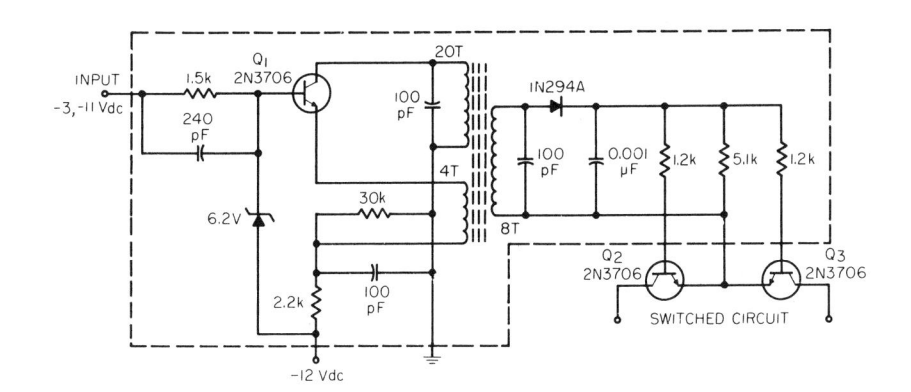

Solid-state switch driver replaces conventional relays.

SECTION VI
Test Circuits

One CMOS package makes universal logic probe

Filip Bryant
Hudson, New Hampshire

Why design, or buy, a logic probe for each type of logic you must check? The CMOS logic ICs in RCA's CD4000A series (or their equivalent) will operate from supplies of 3 to 15V. The defined "ZERO" and "ONE" levels for CMOS correspond closely to RTL, DTL, TTL and HTL when operating at the V_{cc} levels common to those logic families, but a level translator will be required for ECL interfaces, or for any logic levels over 15V. One very attractive feature if you're using CMOS for probing around logic circuits is that its input impedance — typically $10^{12}\Omega$ — isn't likely to disturb the circuit under investigation.

The CD4009AE hex buffer costs a little over $4 in small quantities, and as shown in **Fig. 1** only 2 resistors and 2 LEDs are required to convert it into a logic probe. The basic circuit consists of inverters B, F & E in a Schmitt trigger configuration. Resistor R_1 prevents latch-up of the circuit, and its value is somewhat arbitrary. If R_1 is less than 50 kΩ, switching speeds can be increased, but hysteresis will be introduced into the circuit. Inverter D merely provides an output complementary to the Schmitt trigger, and inverters A and C provide extra current sinking capability to drive the LEDs.

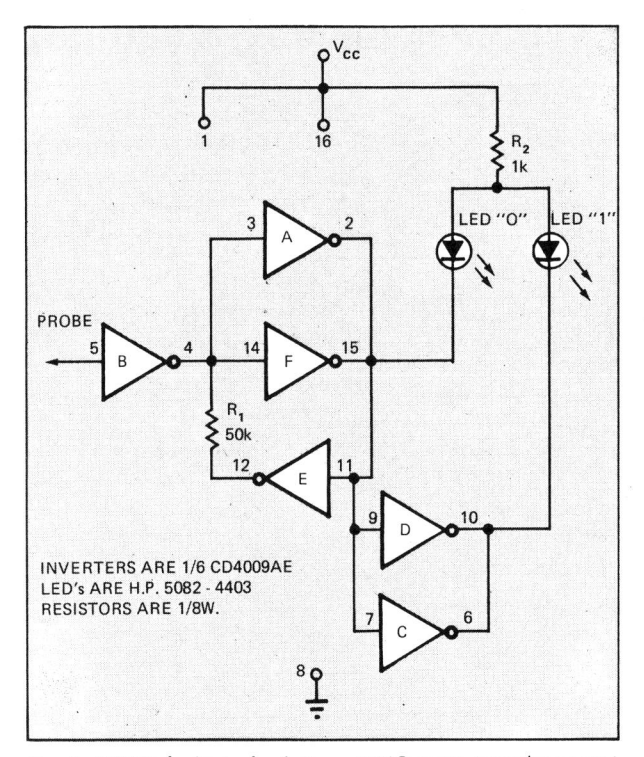

Fig. 1 — **CMOS logic probe** features $10^{12}\Omega$ input impedance and covers 3 to 15V range. While LEDs are visible at all voltages, a 1 kΩ pot in place of R_2 will allow user to increase brightness at lower voltages.

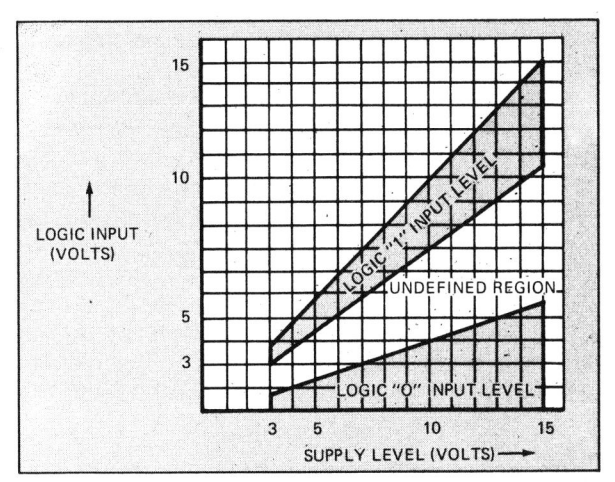

Fig. 2 — **Logic levels of CMOS gates** vary directly with supply voltage. The undefined region shown here for a single gate is greatly reduced in the circuit in **Fig. 1** because of the regenerative feedback through inverter E in the Schmitt trigger.

Technique locates shorted gates on common bus

Paul M. Kintner
Cutler-Hammer, Milwaukee, WI

One of the most difficult diagnostic problems in printed-circuit testing is the location of a ground short on a common bus. Both the outputs and the inputs of gates can be bussed. The first arrangement is commonly used for wired-OR logic; the second for register selection. In either case, a short to ground at any one of the gates keeps the bus from going HIGH and with no direct indication which gate is at fault.

The following procedure has been found useful in quickly locating a ground short (see illustration): Force the bus to HIGH from a 5V power supply set for a current limit of 100 mA. Most internal gate shorts will draw a current of between 60 and 100 mA and will heat within a minute or so to the point where they can be located by touch. Note that the driving gates must all be disabled for a valid test.

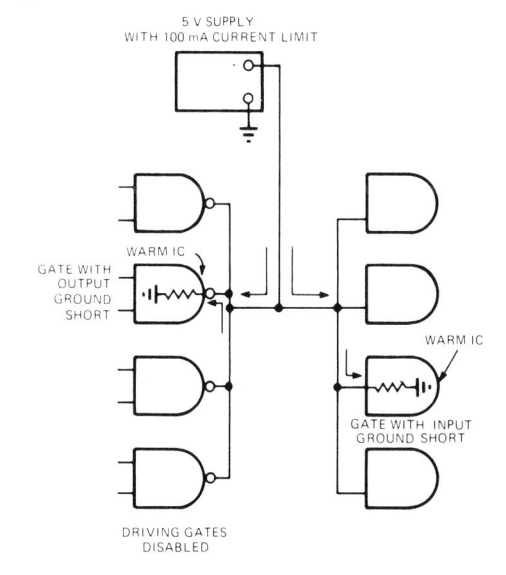

Test Circuits 171

Simple DC voltmeter uses single op amp

by Richard S. Burwen
Analog Devices, Inc.
Norwood, Mass.

A low-input-current op amp, such as the AD503K, can be used to build a simple dc voltmeter. This voltmeter circuit is useful as a general-purpose laboratory meter.

With the attenuator network shown in the schematic, the voltmeter has full-scale input ranges of 10V, 1V or 100 mV. Higher and intermediate voltage ranges can be added if required. Voltage is displayed on a 0-to-100 μA meter. Also, an output terminal simultaneously provides a full-scale output of ±10V for driving a chart recorder.

Basically the voltmeter circuit consists of the op amp connected for a closed-loop gain of 100. Limiting within the amplifier protects meter M_1 against overloads. Resistors R_4 and R_5, together with capacitors C_1 and C_2, form a low-pass filter that prevents the amplifier from overloading on large ac input signals and allows the circuit to read the dc component alone. The series resistance of this filter, in conjunction with diodes CR_1 and CR_3, protects the amplifier from input overloads up to 1000V.

Potentiometer R_7 zeros the amplifier over a span of approximately 50 mV, with the range limited by resistor R_9. Variable resistor R_{10} adjusts the full-scale reading to compensate for the tolerance of the micrommeter.

Circuit stability is such that, in a typical laboratory environment, potentiometers will not have to be reset after the initial calibration.

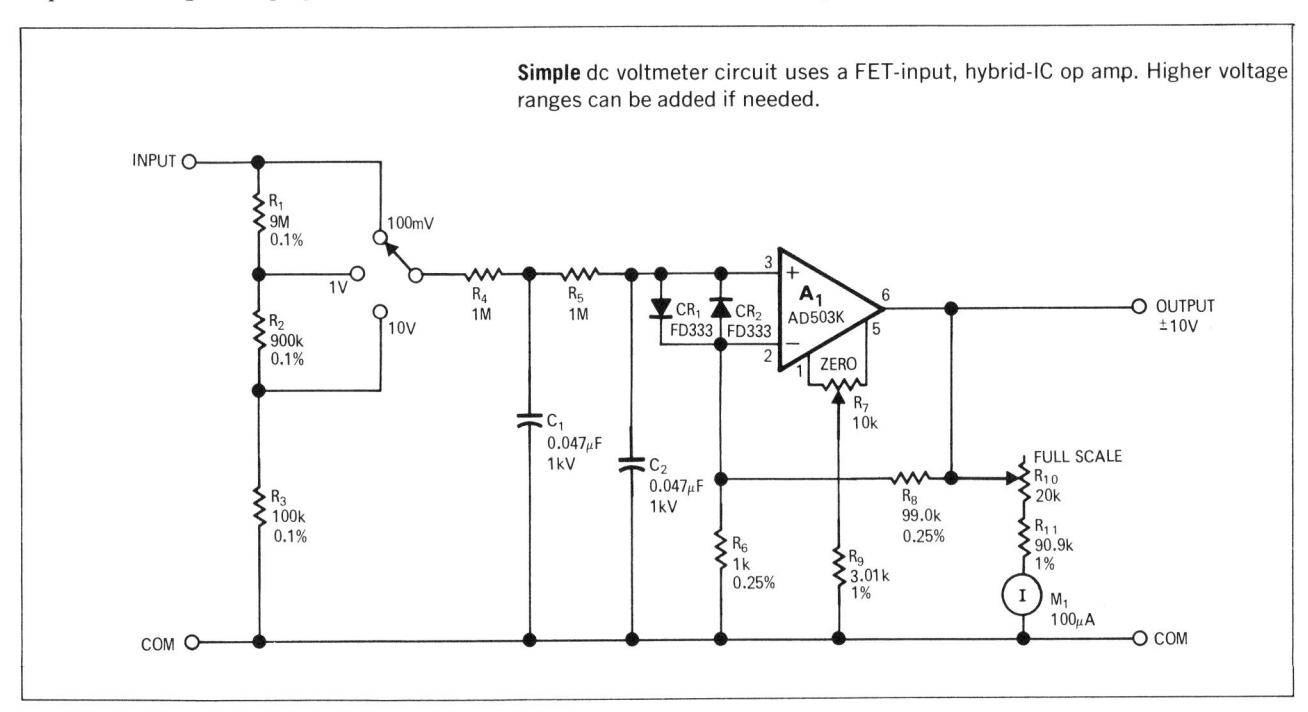

Simple dc voltmeter circuit uses a FET-input, hybrid-IC op amp. Higher voltage ranges can be added if needed.

Simple jig determines transistor gain bandwidth product

Michael J. Salvati
Sony Corp. of America, Long Island City, NY

This simple jig greatly facilitates determining a transistor's gain-bandwidth product (f_T), beta cut-off frequency (f_B), alpha cut-off frequency (f_α) and ac beta by reducing the number of discrete connecting leads, components and test fixtures required for the conventional "hay-wired" test setup. It saves time and aggravation by circumventing the errors usually accompanying a hastily-assembled test setup.

The measurement procedure for employing the jig directly measures ac beta and beta cut-off frequency. The gain-bandwidth product is then determined by the formula $f_T = f_B \times$ ac beta. For all practical applications, the alpha cut-off frequency is numerically equal to the gain-bandwidth product.

The principle by which the jig measures ac beta is quite simple. The series impedance of the transistor's base circuit is approximately 2500Ω, and the collector circuit impedance is 25Ω at the test frequency, a 100:1 ratio. Since voltage gain in the common-emitter configuration is equal to B (Z_{out}/Z_{in}), the circuit voltage gain is X 0.01 per unit input voltage per beta unit. This produces an output voltage numerically the same as the transistor's ac beta when the input voltage to the jig is 100 mV. Thus, a transistor whose ac beta is 70 produces 70 mV output voltage, and a transistor whose ac beta is 220 produces 220 mV output.

The procedure for using this jig is as follows:

1. Connect the signal generator cable to the oscilloscope with a 50Ω feed-through termination. Adjust the generator for 100 mV p-p trace amplitude at 20 kHz.

2. Remove the generator cable from the feed-through termination and connect the cable to the jig INPUT connector. Connect the jig OUTPUT connector to the feed-through termination on the oscilloscope.

3. Adjust the power supply for the desired collector voltage, then adjust the 100 kΩ pot for the desired collector current. The oscilloscope trace amplitude now indicates the small-signal ac beta.

Fig. 2—**The test jig** can be packaged into a simple, compact unit.

4. Adjust the oscilloscope's step and vernier attenuators for a trace height of seven divisions.

5. Increase the signal generator frequency until the oscilloscope trace amplitude decreases to five divisions (the −3 dB point). The signal-generator frequency dial now indicates the transistor's beta cut-off frequency.

6. Multiply the ac beta value found in Step 3 by the frequency found in Step 5 to obtain the gain-bandwidth product.

With the parts values shown, the jig is most accurate for transistors operated at 10-50 mA collector current, having betas in 50-200 range and beta cut-off frequencies above 20 kHz. However, the jig can be optimized for nearly any other class of transistor by a suitable choice of component values. For high-frequency use (as shown), construct the jig so that the parts shown in heavy lines are physically close.

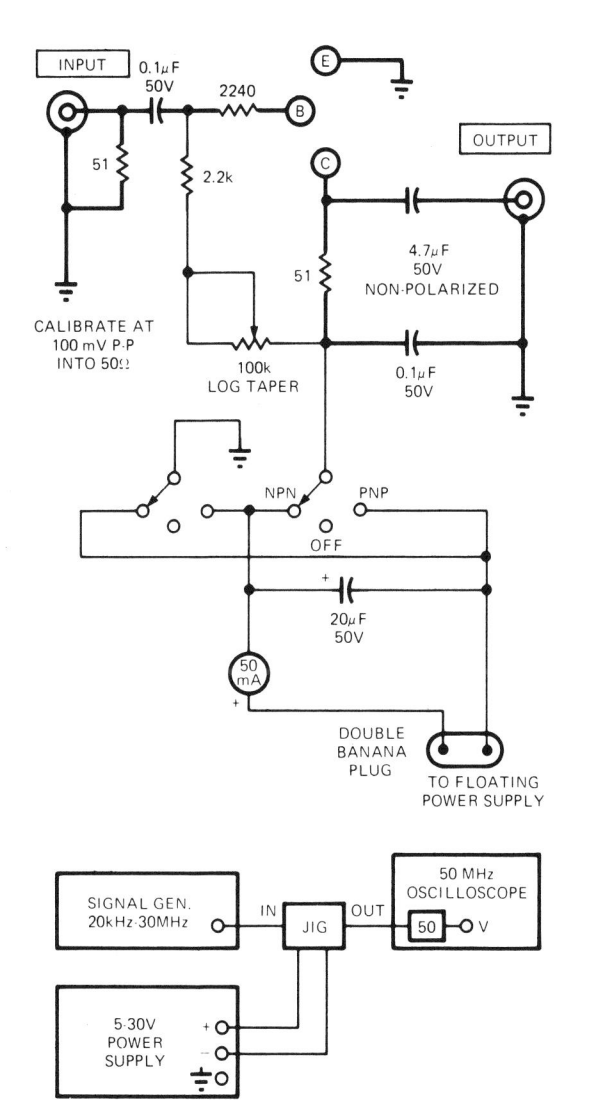

Fig. 1—**Most important transistor parameters** are easily measured with this test jig circuit.

Test Circuits 173

High-gain ac/dc oscilloscope amplifier

by **Glen Coers**
Texas Instruments Incorporated, Dallas, Tex.

You can extend the vertical sensitivity range of your scope at minimum cost by adding this simple adapter. It can also be used to extend the voltage sensitivity of your VOM.

The circuit uses a SN72741 op amp, which has internal frequency compensation and requires no external compensation. The ac input is through a 1.0μF capacitor C1 that is connected to the noninverting input of the op amp. The voltage at the output will be in phase with the input. A switch is connected across the capacitor to provide ac (switch open) or dc (switch closed) operation.

The 1 MΩ resistor R1 establishes input bias current for the op amp. Because of its high resistance value an offset voltage is developed, but this can be nulled with the 5 kΩ pot, R2 (offset adj). The output of the op amp is connected to a 100Ω resistor, R3, so that if coax or shielded cable are used, the output of the device will not have to drive their highly capacitive load directly. (Direct drive sometimes causes instabilities.)

The frequency and output voltage limitations are listed in **Fig. 1**, along with the schematic. Input impedance is approximately 500 kΩ.

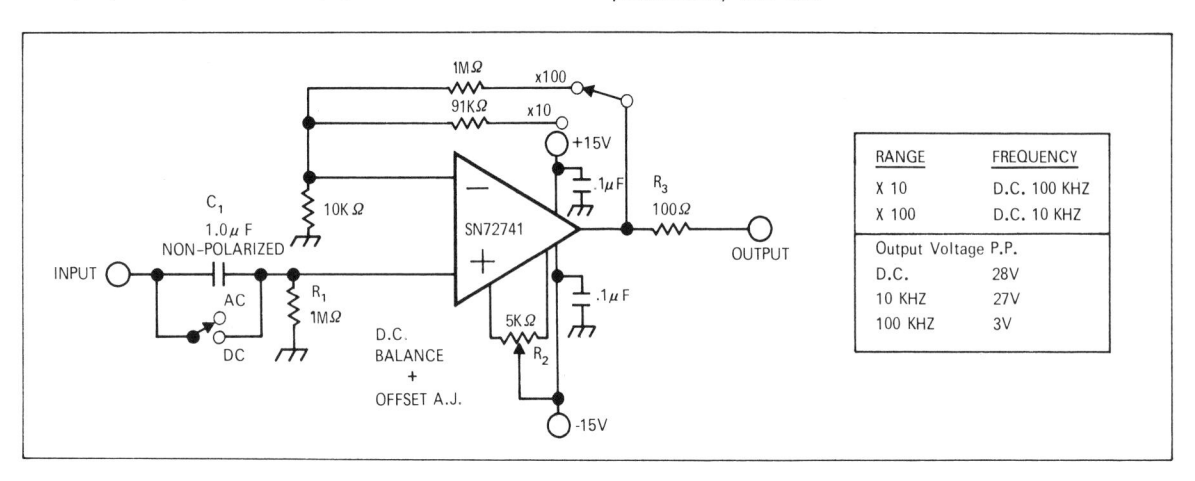

Fig. 1 — **This simple op amp circuit** permits an oscilloscope to monitor signals that would normally be below its input sensitivity.

Logic probe tests three-state logic circuits

Leslie K. Torok, University of Toronto
Toronto, Ont. Canada

This logic probe can be neatly packaged into a P-6025 or other suitable probe-tip and only takes about an hour to assemble. The probe is powered by the circuit under test, and in addition to indicating "1", "0" or "Hi-Z" logic states it will indicate "overvoltage" from the supply if V_{cc} exceeds 5.7V.

Circuit operation is extremely simple, as shown in **Fig. 1**. If the device under test is in the "high impedance" state, there will be no base drive to either Q_1 or Q_2 and neither LED will light. If the logic state is "zero" (below 0.3V) Q_1 is biased off and Q_2 on. Current will be drawn through CR_1 and LED "0" will light. When Vin exceeds 2.4V (for a logic "one") Q_2 will be off and Q_1 will turn on, providing a current path for CR_3 and lighting LED "1".

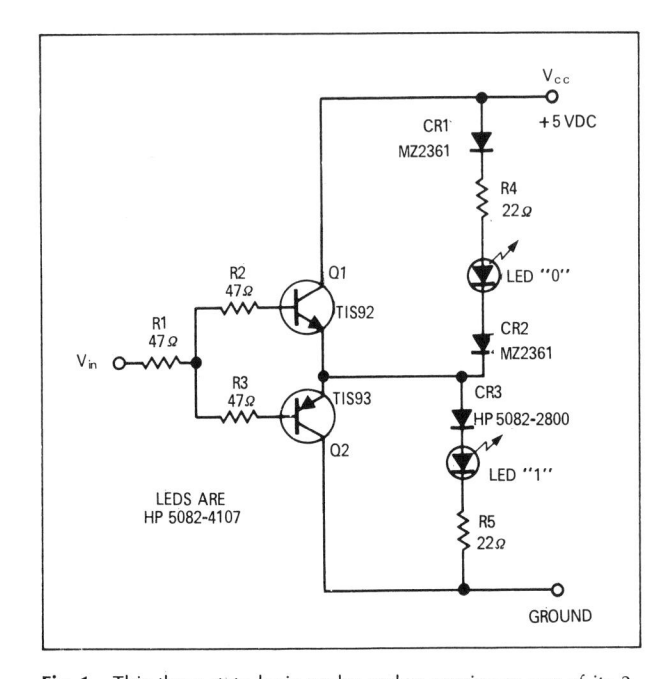

Fig. 1 — This three-state logic probe makes maximum use of its 2 LED bits by providing four distinct messages concerning the circuit under test. If there is a logic "1" at the test point, Q_1 will light LED "1". If there is a logic "0", then Q_2 will light LED "0". If the test point is in the third, or "Hi-Z," state neither transistor will have adequate base drive, and both LEDs will be unlighted. If the supply voltage exceeds +5.7V both LEDs will light before the probe is touched to the test point.

555 timer makes simple capacitance meter

Richard Horton
Teradyne, Inc., Boston, MA

The circuit in the diagram shows how the 555 timer can be used to make a linear-scale capacitance meter. Two characteristics of the 555 make it well-suited for this application. The first is its wide frequency range; the second, its high output current (both sourcing and sinking).

The circuit operates as follows: the timer is connected as an astable square-wave multivibrator, with frequency determined by R_A, R_B and C_1. When the output of the 555 is high, the unknown capacitor, C_X, charges almost to V_{CC} with the current passing through the meter circuit. When the output of the 555 goes low, C_X discharges to approximately 0V through D_1. Since charge on a capacitor is equal to capacitance times voltage, the net charge, Q, per cycle equals $C_X V$. The effective current through the meter circuit (I_{eff}) then equals the charge per cycle times frequency, or I_{eff} equals Qf. Thus $I_{eff}=C_X Vf$, and the meter scale is linear. For $V_{CC}=10V$, the accompanying table shows full-scale capacitance range vs. frequency for a 100-μA meter.

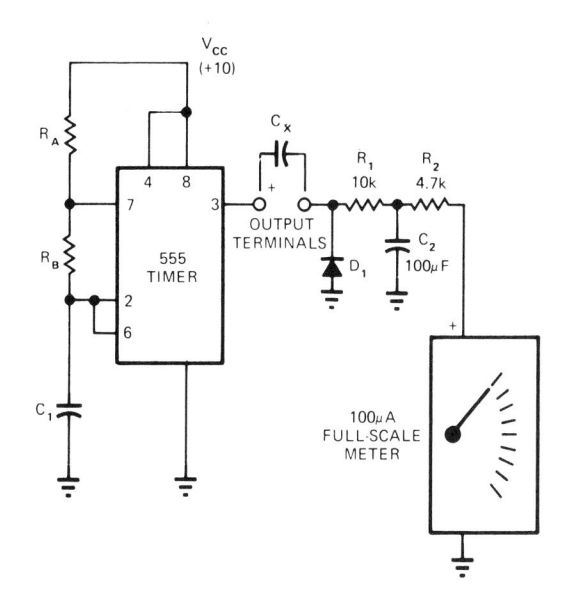

Pulses from the 555 timer IC charge the capacitor under test, which is discharged by D_1 during the 0V-input portion of the cycle. R_1, R_2 and C_2 serve to damp the meter movement.

TIMER FREQUENCY REQUIREMENTS

C_X (FULL-SCALE)	FREQUENCY
100 pF	100 kHz
1 nF	10 kHz
10 nF	1 kHz
0.1 μF	100 Hz
1 μF	10 Hz
10 μF	1 Hz

R_1, R_2 and C_2 damp the meter movement enough so that at f=10 Hz and above the meter appears to read a constant, even though the charge is a series of pulses. At f=1 Hz, the pointer bounces slightly, but still allows a rough interpolation. The meter damping could be increased if desired, but this would lengthen settling time.

Since the time needed to charge the unknown capacitor through the meter circuit is much longer than the time needed to discharge through D_1, an asymmetrical duty cycle is desired. With $R_A \approx R_B$ the charging time is twice the discharge time.

Since the unknown capacitor is charged to 10V during part of the cycle, it is important that polarized capacitors be oriented properly. Also, because the meter reading is directly proportional to the supply voltage, a regulated supply should be used for best accuracy. If this is done, it is simple to make accuracy better than you can read on a meter.

In-Circuit Capacitor Tester

By Emeric L. Major
Cohu Electronics, Inc., Kin Tel Div.
San Diego, Calif.

WITH THIS SIMPLE CIRCUIT, capacitors, surrounded by other components in the circuit, can be checked dynamically for open or short conditions.

If the test capacitor, C_x, is active in the circuit, the indicator light is off. If the capacitor is open or shorted, the light turns on.

The figure shows the tester circuit. Pulses from the multivibrator (MV) drive two transistors, Q_1 and Q_2. An integrating differential amplifier, formed by Q_3 and Q_4 detects any difference between the signals on the collectors of Q_1 and Q_2. Shunting either collector resistor with purely resistive loads (between infinity and about 1 K) does not produce a difference signal between the two stages, which are being driven between cut-off and saturation. A difference signal will be produced, however, by loads having a capacitive component.

A fixed capacitor shunting the collector of Q_1 produces a small off-set signal which lights the indicator if the test capacitor C_x is open or shorted, either of which condition

In-circuit capacitor tester.

leaves Q_2 with a purely resistive load. With a highly capacitive load shunting Q_2 (C_x operating), the differential or off-set signal changes polarity, keeping the indicator light off.

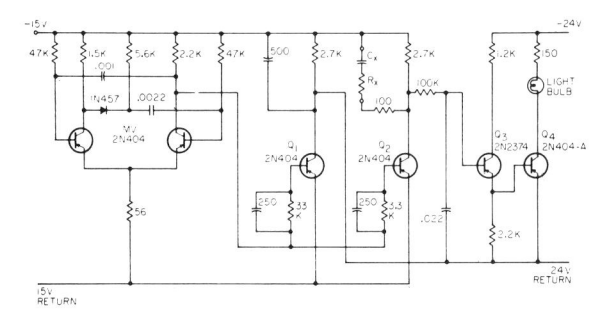

The capacitive circuit under test is basically equivalent to a simple RC circuit, as indicated. Tester circuit parameters such as the resistance in series with circuit under test, the capacitor shunting Q_1, the capacitor-resistor values on the base of Q_3 and the multivibrator frequency, may be modified as necessary to accommodate capacitive circuits differing from the one shown; the procedure is not critical.

IC power supply provides test spikes or level shifts

Larry Latham, Texas Instruments, Inc.,
Attleboro, Mass

There are times when power supply level shifts or voltage spikes are needed to check the performance of equipment subjected to these varying conditions. The circuit shown in **Fig. 1** accurately provides these test signals. Spike or step width depends directly on the input pulse width, and the reference voltage is programmable by means of R_2.

The component values chosen for the circuit illustrated provide a reference voltage range of approximately 2.0V (4.7V to 6.7V). However, a much greater variation in reference voltage is possible (0 to 18V) if different component values are used.

Amplitude of the spike or step is varied by R_5 (for a negative spike or step). With the components shown, maximum variation is approximately 5V with respect to reference. However, this variation can also be increased by changing values.

The circuit shown in **Fig. 1** can be used in any application requiring less than 100 mA of current. In applications requiring less than 20 mA, the circuit configuration shown in **Fig. 2** can be used, allowing faster spikes. The pulse input amplitude required for a negative spike or step output is zero to +3.0V. For a positive spike or step output a negative-going pulse from +3.0V to zero is required.

The circuit is primarily a voltage follower with a switched voltage divider on the input. R_7 and R_3 set the voltage variation range of R_2. A_1 and Q_2 are connected in the unity gain voltage follower configuration with resistor R_8 acting as an output limiter. Q_1 acts as a switch, connecting resistors R_5 and R_6 in parallel with the divider resistors R_2 and R_3.

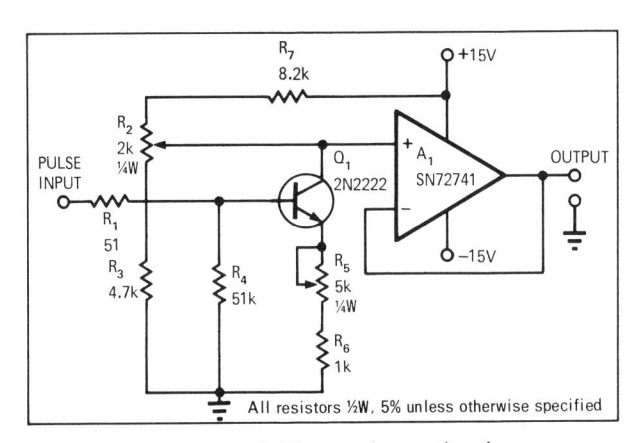

Fig. 1 – Power supply level shifts or voltage spikes, for test purposes, can be programmed into IC circuits requiring less than 100 mA with this circuit. Using the values listed, the output reference voltage level is 4.7 to 6.7V.

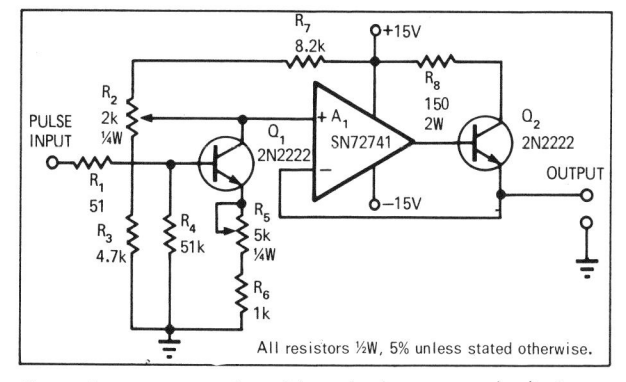

Fig. 2 – Low power version of the pulsed power supply eliminates Q_2, and can supply up to 20 mA. Faster test spikes can be generated with this circuit. Either circuit can provide spikes or level shifts of up to 5V.

Bias supply circuit provides constant current

Glen Coers, Texas Instruments, Inc.
Dallas, Texas

When testing transistors on an RX meter, G.R. bridge, or "S" parameter set up, two power supplies are usually required and collector current for each device tested must be adjusted individually. With the aid of the constant-current supply in **Fig. 1**, devices with a wide beta range can be biased automatically to within one percent of the desired collector current.

Operation is very simple: just select the proper range with S_2 and adjust that potentiometer for the desired current. A reference voltage is established by the 1N914 diodes and Q_1, the 2N3819 FET. The error is sensed across the potentiometer and R_7. The 47Ω resistor, R_7, is for current limiting when the potentiometer is at minimum resistance. The 2N4058's and 2N3702's form a differential amplifier and, due to the 150Ω resistors R_5 and R_6 in series with the collectors, the amplifier does not saturate when there is no device under test. Each collector resistor carries about 4 mA, and the small amount of base current required to drive the U.U.T. is subtracted from that 4 mA. Therefore, the amplifier operates in it's most active region. The 9V battery can be replaced by an ungrounded power supply.

V_{CC} range is 3 to 25V
I_C range is 12 μA to 20 mA.

It is also possible to measure beta by placing a current meter in series with the base of the U.U.T., and to measure V_{be} (active) by placing a voltmeter across the emitter-base terminals.

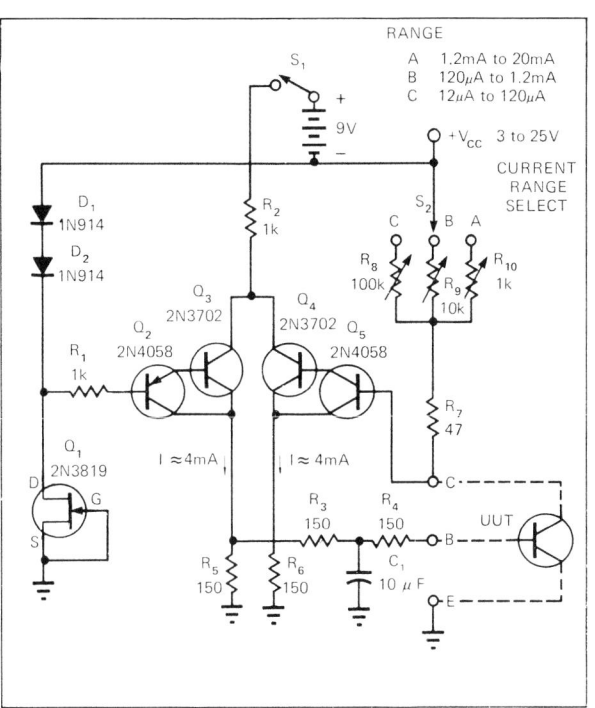

Fig. 1—**Constant current supply** consists of a differential amplifier, Q_2 through Q_5, and a voltage reference, Q_1, D_1 and D_2. Base drive to the transistor under test, through R_3 and R_4 maintains the desired collector current even when testing devices with a wide beta variation.

2 CMOS gates convert counter into capacitance meter

Roger Melen Stanford University,
Stanford California

In many applications it is very convenient to have a direct digital readout of capacitance. The circuit shown in **Fig. 1** will convert any counter into a digital readout capacitance meter with good accuracy for values measured between 1000 pF and 10 μF.

Two CMOS gates (A and B) are connected to generate a 1 Hz squarewave which drives the SN74121 monostable multivibrator. The monostable, when triggered, generates a pulse, the width of which is determined by the capacitor under test. The pulse output of the monostable turns on a 100 kHz oscillator (made up of gates C and D). The output, when a capacitor is being measured, is a 100 kHz burst every second. The direct measurement of capacitance is possible because the duration of the burst, hence the number of pulses in the burst, is proportional to the capacitance under test.

An ordinary dc voltmeter connected to the output terminals can also be used to measure capacitance, since the average value of the output waveform is directly proportional to the capacitance under test.

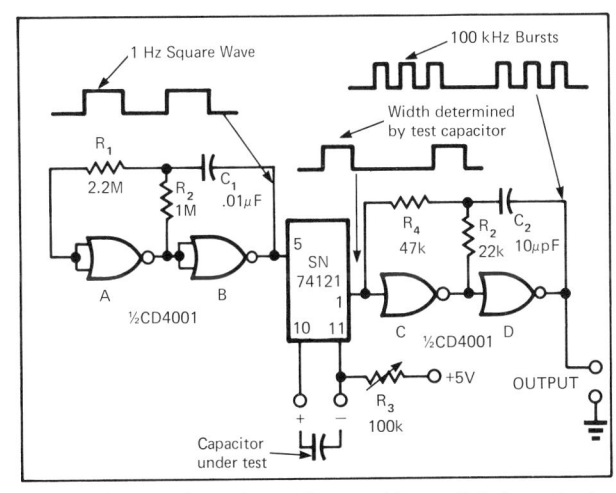

Fig. 1—**Direct readout of capacitance** with any digital counter is a very quick operation with this CMOS circuit. The system is calibrated with a known capacitance by adjusting R_3 so that one output pulse equals 100pF.

Temperature-coefficient measuring circuit

by **Richard C. Gerdes**
Optical Electronics, Inc. Tucson, Ariz.

Temperature testing of amplifiers and other analog circuits is usually a tedious process. The output voltage must be measured at discrete temperatures and then the temperature coefficient must be calculated by subtraction and division. Here is a circuit that performs the required measurements and calculations automatically and continuously.

Silicon diode CR_1 senses the temperature of the device under test. FET-input amplifier A_1 converts the forward voltage drop of the temperature-probe diode into a high-level analog voltage. The diode specified has a forward drop of approximately 2 mV/°C and the temperature coefficient is linear within 3% over the range from −55 to +125°C. With the component values shown, the output from A_1 changes by 325 mV/°C and is −10V at +55°C and +10V at −5°C. If the 10-MΩ scaling resistor is changed to 3.6 MΩ, the output from A_1 will be −10V at +110°C and +10V at −60°C.

The output from the temperature-sensing circuit is applied to a sample-and-hold circuit, while the analog voltage from the device under test is applied to a second sample/hold. Actuation of S_1 causes the voltage and temperature information to be stored in the sample/hold circuits at the commencement of the test. The specified sample/hold circuits have no memory decay, hence there is no upper time limit for the temperature-coefficient measurements.

Data stored in the sample/hold circuits at the beginning of the test bias amplifiers A_2 and A_3. During the test, the initial bias voltages are automatically subtracted from the input voltages to the amplifier. Thus the amplifiers respond solely to the changes in temperature and output voltage of the device under test.

The outputs from amplifier A_2 and A_3 can, of course, be either positive or negative depending on the direction of the temperature change and on the performance of the device under test. Unity-gain absolute-value amplifiers A_4 and A_5 ensure that the inputs to the analog divider are always positive.

The analog divider calculates the temperature coefficient. Its output is the ratio of voltage change to temperature change. Scaling resistor R allows the output to be calibrated so that it is a convenient analog of the desired engineering units (μV/°C, mV/°C, mV/°F, etc.).

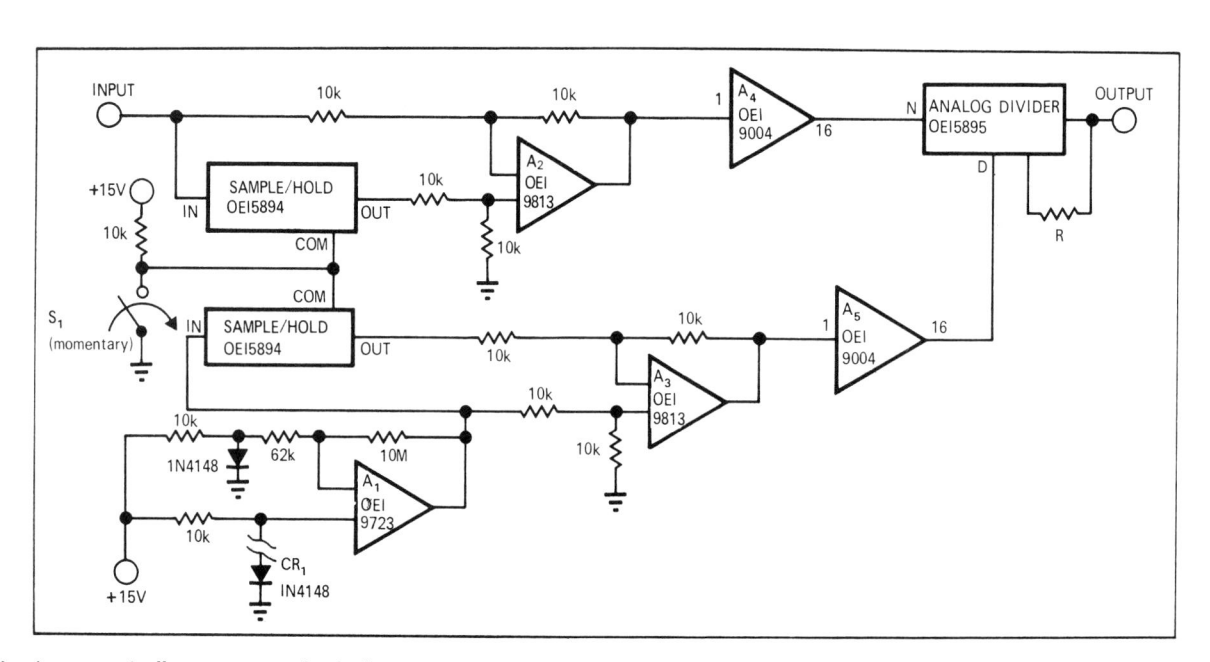

Circuit automatically measures and calculates temperature coefficients of analog circuits or devices. Diode CR_1 is used as a temperature probe. Resistor R allows the output scale factor to be adjusted for the desired measurement units.

FET probe drives 50Ω load

M.J. Salvati
Sony Corp. of America,
Long Island City, NY

This impedance converter probe was originally designed to drive the 50Ω input of a spectrum analyzer but has since been used with ac voltmeters, video amplifiers, frequency counters and other devices which cannot use an ordinary scope probe to minimize loading.

As shown, the probe has an input impedance of 10 MΩ shunted by 8 pF. Eliminating the protective diodes reduces this to about 4 pF. The frequency response of the probe extends from dc to 20 MHz (−1 dB), although higher frequency operation is possible through optimized construction and use of a UHF-type transistor.

Zero dc offset at the output is achieved by selecting a combination of a 2N5246 and source resistor that yields a gate-source bias equal to the V_{be} of the 2N3704 (approximately 0.6V).

At medium frequencies, the probe can be used unterminated for near-unity gain, although for optimum high-frequency response, the cable must be terminated into 50Ω. If a 75Ω system is desired, use a 62Ω output resistor and Belden 8218 cable. In either case, the voltage gain when properly terminated is precisely 0.5X.

By using miniature parts and careful construction, this circuit can be neatly and conveniently packaged in an aluminum cigar tube. The input end of the probe is fitted with a 6-32 threaded stud, so Tektronix probe tip accessories can be used.

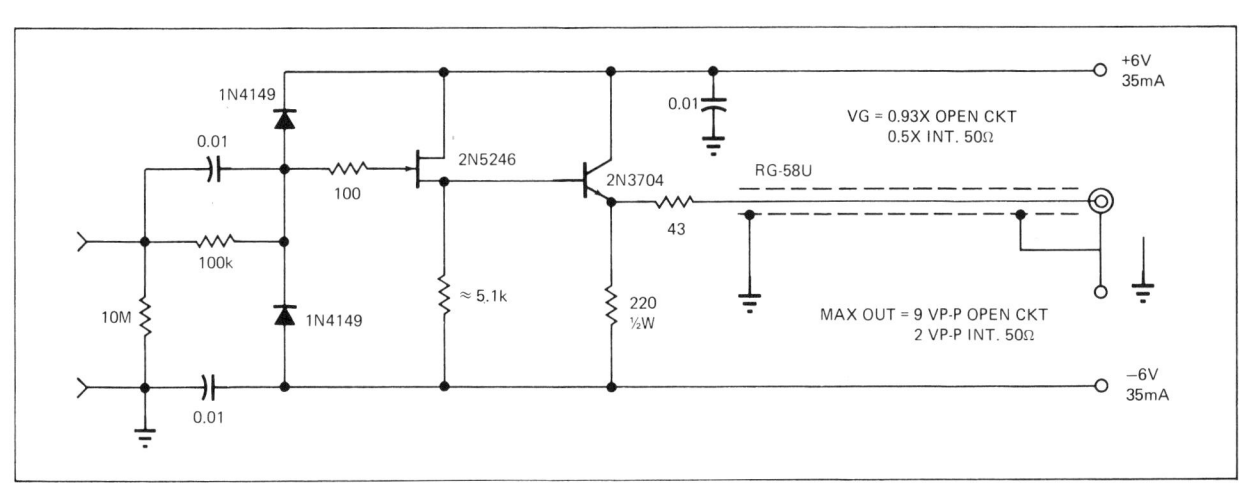

Impedance converter exhibits 10 MΩ input and 50 Ω output with a voltage gain of precisely 0.5. It was designed to drive test equipment.

Test box indicates logic levels for entire IC

Richard G. Sullivan,
Control Data Corp., Rochester, Mich.

This circuit performs the function of a logic-level indicator, and works as follows:

There are 16 identical circuits, each having one connection to a pin of the IC under test. The circuits independently determine which pin is +5V and connects this to the voltage bus, and which pin is ground and connects this to the ground bus. The two diodes can be considered a decision gate network directing the +5V pin and the ground pin to the proper bus. The diodes are Germanium to allow a greater voltage on the voltage bus, about +4.5V in this case. It doesn't matter where +5V and ground are, the circuits automatically connect the proper voltage to the proper bus.

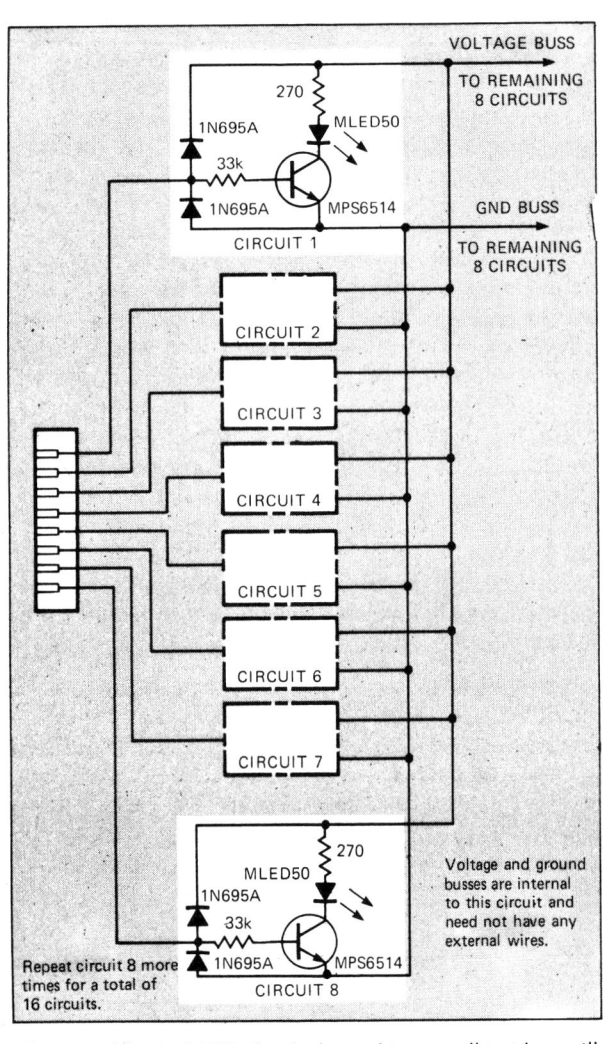

Fig. 1—16 identical LED circuits housed in a small test box will indicate logic state of each pin in a DIP package. Diodes automatically switch input to the proper ground or +5V bus.

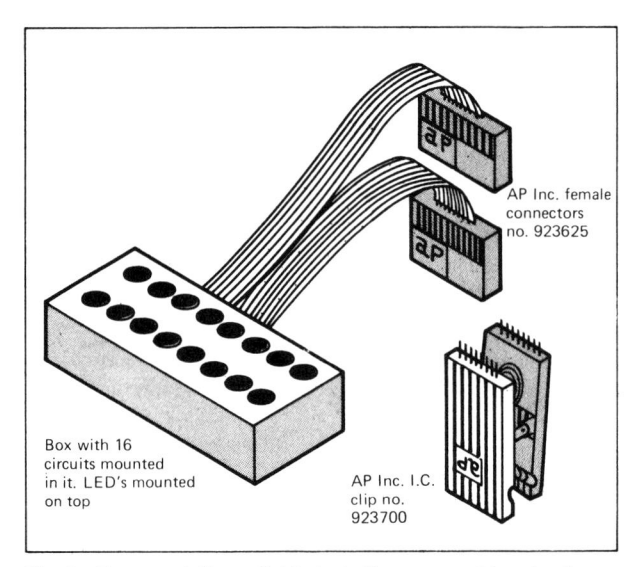

Fig. 2—Commercially available test clips are used for simultaneous probing of all pins in the DIP. No internal supply is required since power is drawn from the supply circuit for the IC under test.

Any pin at a logic "ONE" level will provide enough current through the 33Ω resister to turn on the MPS6514 transistor, and hence the LED will glow red. If the pin under check is at a logic "ZERO" the MPS6514 will not turn on and the LED will be off. This circuit draws power from the IC and uses 10 mA per LED for a typical brightness of 200 foot lamberts.

To make the circuit useful as a trouble shooter, put the 16 circuits in a small box, with the LEDS on top. Two harnesses come from the box terminating in AP Inc. female connectors number 923625 or equivalent. These are 8 socket connectors that mate with their Logic Clip number 923700.

Low-speed logic probe

by John W. Hamill
Brookhaven National Lab
Upton, N.Y.

Using two IC monostable multivibrators and four light-emitting diodes, one can build a handy and inexpensive logic probe. The circuit is useful for low-repetition-rate applications where single narrow pulses are encountered. Such situations occur in computer-interface and logic-control systems. The one-shot multivibrator stretches pulses as narrow as 50 nsec so that they provide a clear LED indication.

Switch S_1 allow the user to determine the steady state of the circuit under test. If the steady state is low, lamp CR_1 will light when the switch is in the LOW position. Conversely, if the steady state is high, CR_2 will light with the switch in the HIGH position.

The other two lamps indicate the presence of a pulse superimposed on the selected steady-state signal. For example, if the steady-state is low and the switch is in the appropriate position, then the leading edge of a positive-going pulse will trigger monostable IC_1 thus lighting CR_3. Alternatively, a negative-going pulse will light CR_4.

External connections to pins 9, 10 and 11 of the monostable ICs determine the durations of the output pulses and hence the length of time that the pulse-indicating LEDs will remain illuminated. With the connections shown (using the 2-kΩ internal timing resistor and a 200-μF external capacitor), the ON time is 200 msec. If the pulse repetition frequency is high, the pulse light will remain on continuously.

This circuit can be compactly packaged either in a pen-shaped container or in a small box. The lamps should be labelled as indicated on the schematic. A battery can be included to provide the necessary 5V power source, or the voltage can be derived from the logic under test.

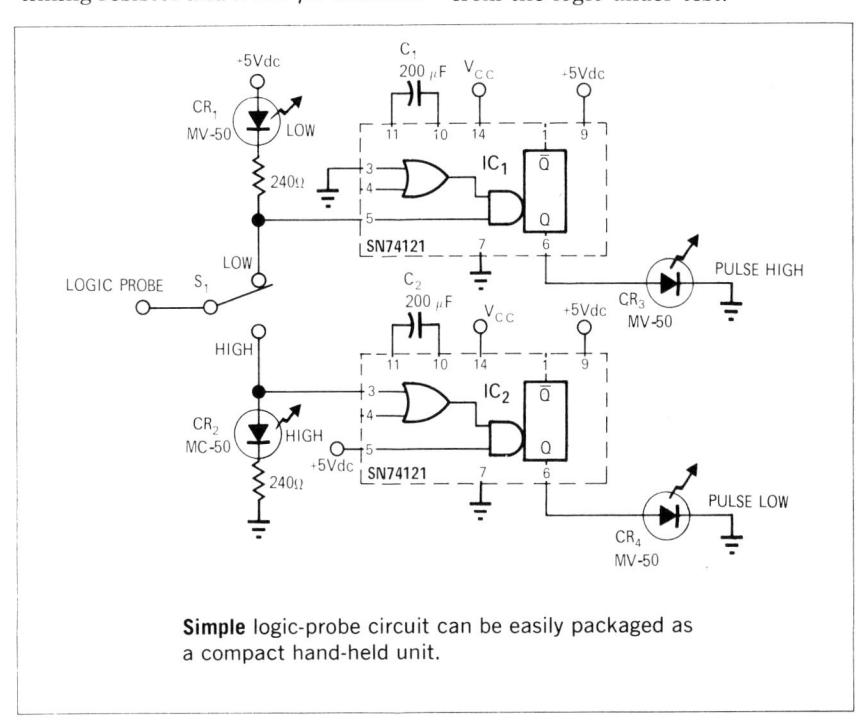

Simple logic-probe circuit can be easily packaged as a compact hand-held unit.

Short-Circuit Alarm

By Joseph J. Russo
Burroughs Corp.
Paoli, Pa.

DURING THE WIRING of today's complicated backplane for computers, where point to point wiring is used, short circuits between voltage buses or a voltage bus shorted to ground because of wiring errors or solder splashes are commonplace. If these shorts are not detected and removed as soon as they occur and are left in the computer to be found after the wiring is completed, many hours of debugging time could result. The circuit described here sounds an alarm as soon as a short occurs between any two of five different voltage buses or any one of the voltage buses to ground.

The circuit uses a model SC-628 Sonalert device for the alarm, which oscillates at 2.5 kHz when the circuit is activated due to a short. These units can be obtained in a variety of frequencies.

Proper circuit operation is made possible by the use of the SG22 stabistors. The consistent EI curves for these devices are essential for proper stabilization.

While no shorts occur, Q_1 conducts and Q_2 is off. As soon as a short occurs between any two of the six input lines, Q_1 will shut off and Q_2 will conduct. This causes the alarm to sound until the short is removed.

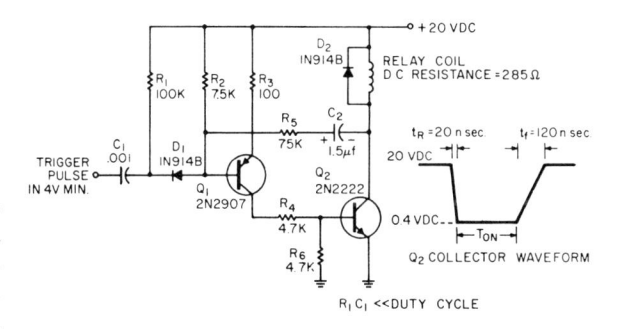

The supply can be any voltage between -6 and -20 v. The higher the voltage, the louder will be the alarm.

With 20 volts for $-V_{cc}$, the current when an alarm is obtained is only 8 ma, which makes this alarm circuit suitable for battery operation. The standby current is about 1 ma.

The alarm circuit cannot be used unless the voltage buses are floating. This does not present a problem because the power supplies do not get installed until after the rack wiring is complete.

Short-circuit alarm.

SECTION VII
Miscellaneous

Novel clock circuit provides multiplexed display

George Smith
Litronix, Inc., Cupertino, Calif.

There are many clock circuits using TTL counters to count the 60-Hz line. The "can-count" and power consumption can be substantially reduced by a different MSI approach that yields a multiplexed display suitable for LED readouts.

Counting is accomplished by circulating the data through a binary adder, and adding one to the digit at the required times. Two 9328 packages give 8 BCD digits of storage, and a 7483 4-bit adder is used to add one when the carry flip-flop N is set.

A 9301 counter keeps track of the digit position, and via the 9301 decoder, drives the digit-select transistors. When the counter reaches state 9, the carry FF is enabled, and then set on the next clock pulse. The incremented count is reloaded into the shift register and the carry FF is reset, unless the adder output has reached 10. In this case gate G_1 switches the data select line to Do, and a zero is loaded. The carry FF reset is also inhibited, so the Cin is carried forward to the next digit. For every other digit, $Q_1 = 1$ and the gate G_2 forces a carry on 6 instead of waiting for 10. In this way, the 10, 6, 10, 6, 10, 6, 10, 6 sequence is generated. When the count reaches 10 o'clock, flip-flop M is set on every cycle. Gate G_3 then detects when the time goes to 13 o'clock, and asynchronously clears the shift register. The carry FF remains set so a 1 is loaded into the hours digit, and the transition from 12:59:59 to 1:00:00 is accomplished. The seven segment decoder driver looks at the shift register output and drives the LED segment lines. The leading hours digit is blanked, using the RBI input on the 9317.

Using push-button switches, the carry FF enable can be moved to speed up the count to permit setting the hours and minutes. The seconds are set by inhibiting the carry FF and letting the time catch up with the clock.

The 9601 one shot, in conjunction with the octal counter, and the diode AND gate forms a phase-lock loop that locks the 9601 to 480 Hz, to provide correct time. For a 50 Hz line frequency, the timing components of the 9601 must be changed to give 400 Hz, and another gate must be added to the G_1, G_2 pair to force the second digit to count by 5.

This basic scheme makes it easy to count by a different base. By using a decade counter instead of the octal counter, and adding two 9300 4-bit registers, and a few gates, the system will count down from a 1 MHz crystal.

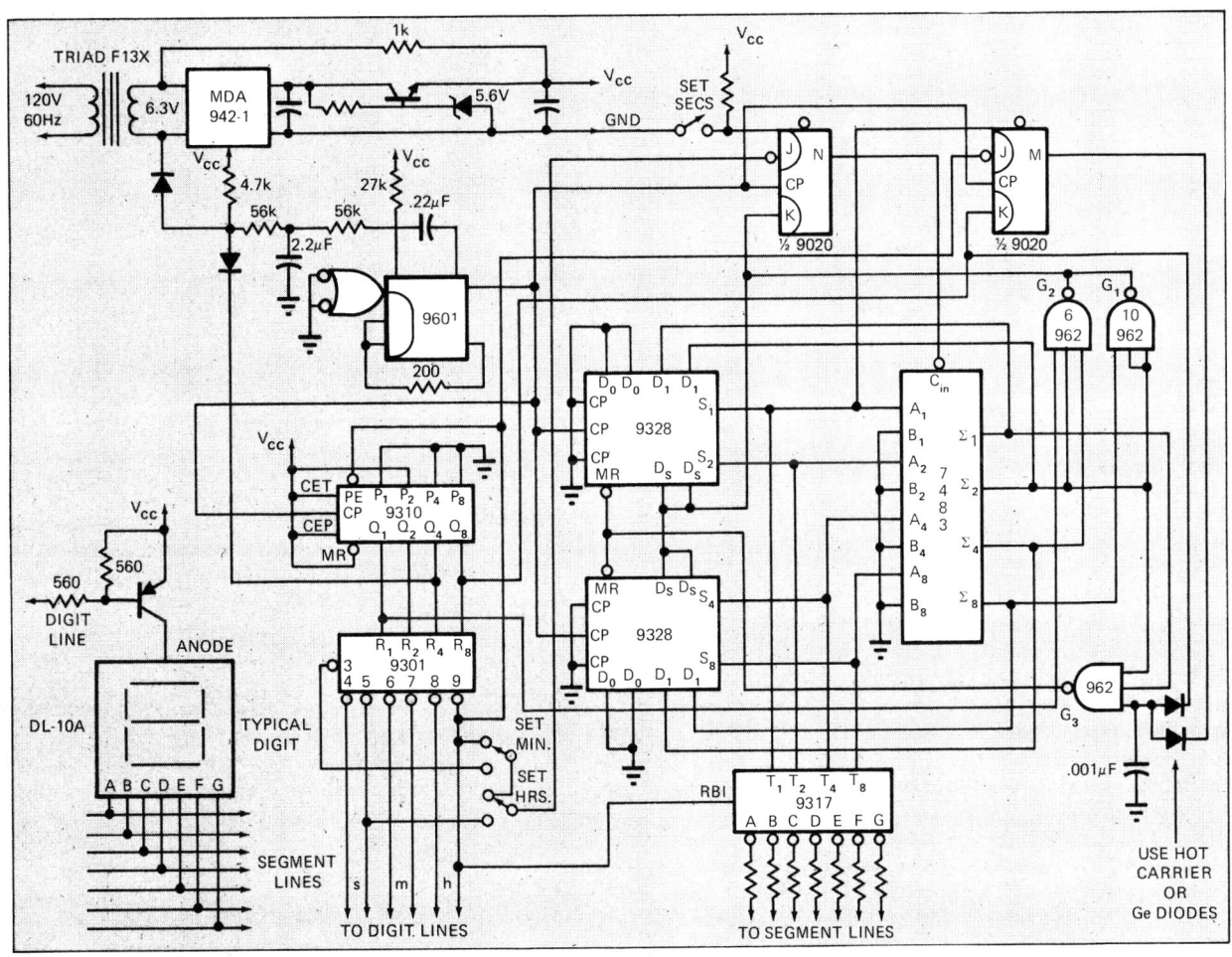

Digital clock scheme uses TTL MSI circuits to minimize package-count. A further savings in package count is achieved by multiplexing the LED readout.

60-Hz frequency discriminator

by Richard Burwen
Analog Devices
Cambridge, Mass.

THIS CIRCUIT delivers a dc output voltage proportional to the frequency deviation of the input voltage from a nominal center frequency. With the values shown, the center frequency is 60 Hz. This circuit is useful for frequency measurement or motor speed control. In addition, the phase shifter and multiplier portion of the circuit can be used for second harmonic generation without producing a dc output.

An input signal is fed directly to the Y input of the multiplier and also through a 90°-phase shifter A_1 to the X input of the multiplier. The output of the multiplier is the second harmonic of the input and a dc component. This signal is filtered and amplified by buffer A_2, A_2's dc gain is 10. Output voltage is proportional to frequency deviation from the frequency at which A_1 has 90°-phase shift. At 7 Vrms input, the dc output is 500 mV per percent of center frequency change. R_1 is adjusted for 0 Vdc at the center frequency.

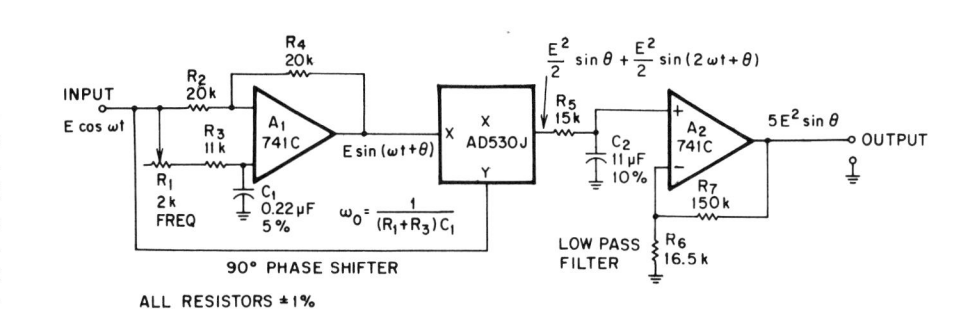

Two IC op-amps and a multiplier form a frequency discriminator circuit.

Convert wrap-around code to sign-plus-magnitude

Jerry N. Phillips
Litton Systems, College Park, MD

Usually angular information is available as a 0° to 360° "wrap-around" code (such as S/D converters, shaft encoders, etc.). However, when interfacing with a display, it is often desirable to use a 0° to ±180° sign-plus-magnitude code. The circuit shown provides a means for making this conversion.

From the truth table it is easy to see that for $\theta_{in} < 180°$ ($\omega_{180} = 0$), the output equals input. For $\theta_{in} \geq 180°$ ($\omega_{180} = 1$), the output equals the complement of the input plus one. Note that ω_{180} can serve as the output sign bit. The exclusive-OR gates are used to complement the input and the 5483 4-bit adder adds one to the result only when ω_{180} is true.

This same approach is applicable to any word length.

Truth table

θ_{IN}	Wrap-around code				θ_{OUT}	Sign-plus-magnitude code				
	W_{180}	W_{90}	W_{45}	$W_{22.5}$		Sign	M_{180}	M_{90}	M_{45}	$M_{22.5}$
+337.5°	1	1	1	1	−22.5°	1	0	0	0	1
+315.0°	1	1	1	0	−45.0°	1	0	0	1	0
+292.5°	1	1	0	1	−67.5°	1	0	0	1	1
+270.0°	1	1	0	0	−90.0°	1	0	1	0	0
+247.5°	1	0	1	1	−112.5°	1	0	1	0	1
+225.0°	1	0	1	0	−135.0°	1	0	1	1	0
+202.5°	1	0	0	1	−157.5°	1	0	1	1	1
+180.0°	1	0	0	0	−180.0°	1	1	0	0	0
+157.5°	0	1	1	1	+157.5°	0	1	1	1	1
+135.0°	0	1	1	0	+135.0°	0	1	1	1	0
+112.5°	0	1	0	1	+112.5°	0	1	0	1	1
+90.0°	0	1	0	0	+90.0°	0	1	0	0	0
+67.5°	0	0	1	1	+67.5°	0	0	0	1	1
+45.0°	0	0	1	0	+45.0°	0	0	0	1	0
+22.5°	0	0	0	1	+22.5°	0	0	0	0	1
+0.0°	0	0	0	0	+0.0°	0	0	0	0	0

For $\theta_{in} < 180°$, the output of this converter equals the input. For $\theta_{in} > 180°$, the output is the complement of the input plus one.

Fail-safe temperature sensor

By George J. Granieri
American-Standard
New Brunswick, N.J.

MANY INDUSTRIAL control systems require a temperature-sensing circuit that must be fail safe. The circuit must not only give an output pulse when the temperature reaches a predetermined critical value, but must also give an output pulse if the sensor element opens or shorts. This output pulse then shuts down the system.

The circuit shown here, meets all the above specifications. It was designed for use with a PTC thermistor having a resistance of 30 Ω from 32°F to 160°F and 800 Ω at the critical temperature of 200°F. At 200°F the circuit gives a negative output pulse. The circuit distinguishes between a shorted thermistor and a normal one with 30-Ω resistance.

A wide range of different operating temperatures can be obtained by choosing a suitable temperature sensor and selecting the appropriate values for R_2, R_3 and R_4.

The μA710C IC operates as a differential comparator. When the thermistor resistance increases to a value such that the voltage across it is greater than the voltage across R_4, the IC produces a pulse at pin 7. Transistors Q_1 and Q_2 condition this pulse, to give an output with correct amplitude, impedance and phase. Components R_5 and C_2, filter out spurious pulses triggered by input noise.

Normally, transistor Q_3 is biased off. This stage initiates an output pulse if the thermistor is either open circuited or shorted.

If the thermistor opens, the emitter-base reverse-breakdown voltage of Q_3 is exceeded. The transistor then clamps pin 3 of the IC to around 7.5 V. This protects the IC from damage and also generates a pulse at the output.

If the thermistor resistance is less than 30 Ω, the base-emitter junction of Q_3 is forward biased and the transistor conducts. This turns off Q_1, causing a pulse at the output.

With the components specified, the circuit has an operating-point stability of ±0.5°F for ambient temperatures in the range 0-60°C. This is achieved by using 1% resistors for R_2, R_3 and R_4. These resistors should have a temperature coefficient better than ±50 ppm/°C.

Resistors R_{13} and R_{14} should also be low-tempco types to minimize temperature drift of the base voltage of Q_3. This ensures that the fail-safe circuit works reliably with the specified thermistor variation of 30 Ω. Component values are selected as follows:

Resistor R_1 is dictated by the resistance curve of the thermistor. Resistor R_3 should be as low as possible to minimize variations of IC leakage current with temperature. But the minimum value of R_3 is determined by the maximum-allowable input voltage for the IC. Of course, R_2 must be equal to R_3. The values for the voltage divider R_{13}, R_{14}, are selected to ensure that Q_3 turns on when the thermistor resistance is less than 30 Ω.

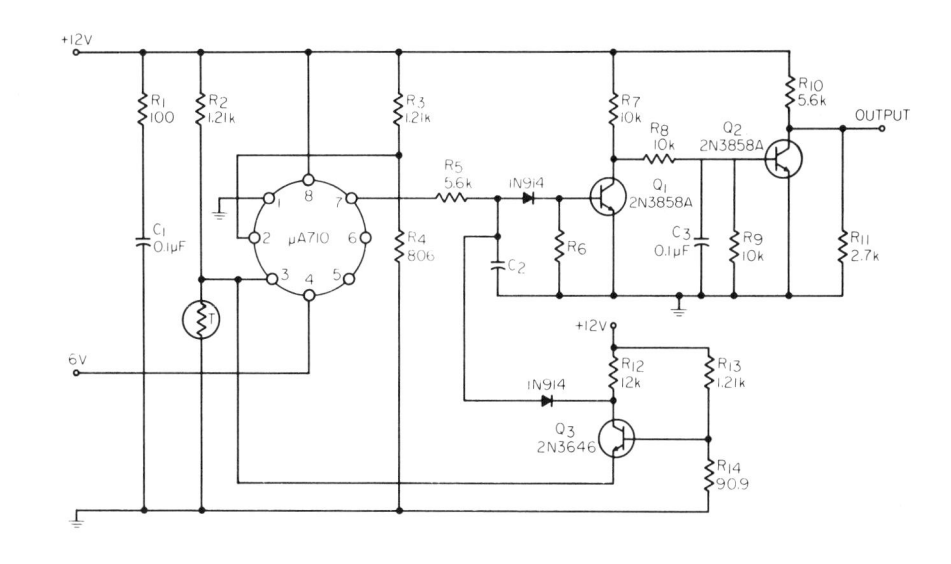

Temperature-sensing circuit gives an output pulse if the thermistor is shorted or open circuited.

Phase-Interlocked Gyro Motor Driver

By Eiji Koda
Systron-Donner Corp.
Concord, Calif.

PHASE SPLITTING to drive a hysteresis synchronous gyro ordinarily is obtained with a capacitor in series with one of two motor windings. But since the winding impedance is variable due to environments and manufacturing tolerance, it is not always possible to have the best suitable capacitor value. Further, the required capacitor usually is a tantalum type and its reliability can be doubtful at extremes of environments.

The logic diagram shown thus is a substantial improvement over the capacitor-type circuit.

The 90-deg phase relation can be established with logic circuitry to be entirely independent of load impedance. The flip-flops and logic can be made extremely reliable in adverse environments. Furthermore, since all transistors are used in a switching manner, the power consumption is considerably lower compared with a class AB or class B amplifier that normally is used to drive a gyro motor.

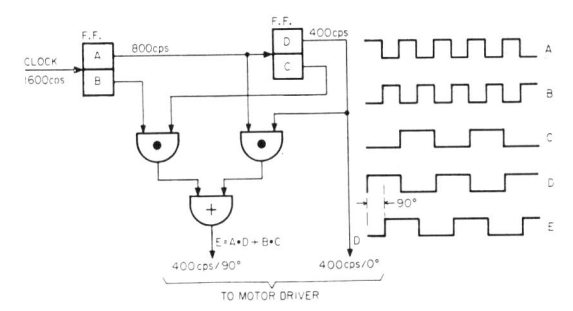

Phase-interlocked gyro motor driver.

Miscellaneous 187

Thirty-second timer uses IC one-shot

Entry by Frederick R. Shirley
Sanders Associates, Nashua, N.H.

Commercially available one-shot ICs are usually limited to applications requiring relatively short delay periods of a few microseconds or less. However, by adding just four external components, the 9602 improved dual one-shot IC can be used to generate timing periods of up to one minute. The major advantages of the circuit shown (**Fig. 1**) are as follows:
  Long time-constant
  Only five components
  Small size
  Low cost
  Standard TTL levels
  Full military temperature range
  Optional non-retriggerable operation
  High reliability.

With the component values shown in the schematic, the circuit has a delay period of 30 sec. The period can be extended to 1 minute by increasing the value of R from 270 kΩ to 560 kΩ. Or the period can be reduced to as little as 10 msec by decreasing R to 30 kΩ and decreasing C to 1 μF.

The circuit shown is non-retriggerable (i.e. it will ignore input pulses during the timing period). It can be made retriggerable by deleting the feedback connection between pins 7 and 5 of the IC. In this mode, the one-shot will be reinitiated by input triggers during the timing period.

The timing cycle is initiated by applying a positive-going trigger pulse at the input. Current from the 5V supply then charges capacitor C via resistor R. This yields a timing period proportional to the RC product.

When the voltage across C reaches the threshold voltage (approximately 1V) the capacitor discharges through Q1. The transistor has two functions. It serves as a current amplifier to permit values of greater than 50 kΩ for resistor R, and it provides level shifting to protect the electrolytic capacitor from the negative standby voltage (-0.7V) between pins 1 and 2 of the IC.

Resistor R1 does not affect timing, but must be included to provide the correct standby current for the IC. During standby, the IC saturates Q1 and holds the capacitor voltage at the discharge level until the arrival of another trigger input.

Maximum length of the timing period is limited by three main factors: the product of capacitance and leakage resistance for the capacitor, the β gain and the I_{CBO} leakage of the transistor.

The 330 μF tantalum capacitor used in this circuit has a maximum leakage of 5 μA at the maximum circuit voltage (1V) and with the maximum ambient temperature. This means that the value of R can be no greater than 600 kΩ (i.e. 3V/5 μA).

The 2N2484 transistor has a minimum β of 20 (at Ic = 10 μA) for the worst-case temperature of −55°C. So, at the steady current of 200 μA, the worst-case β will be about 37. To meet the requirements of the IC, this

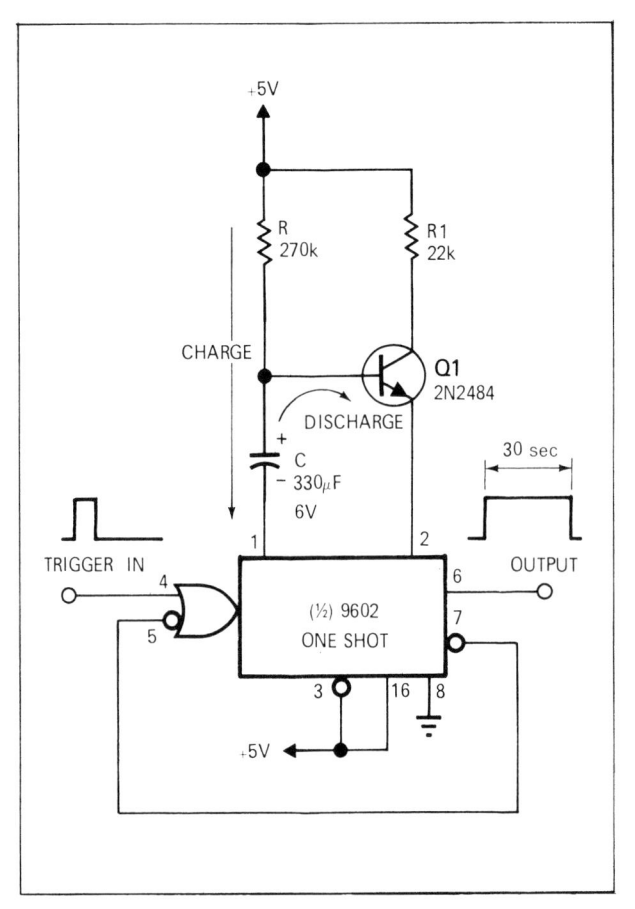

Fig. 1—**Long-delay timer** uses an IC one-shot plus four external components to generate delays of up to one minute.

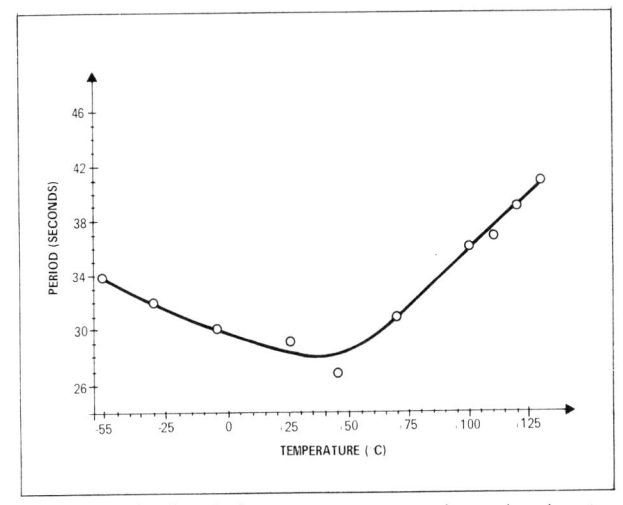

Fig. 2—**Graph of period versus temperature** shows that the circuit of **Fig. 1** is useable over the full military temperature range. At low temperatures, performance is limited by the reduction in transistor gain. At high temperatures, leakages in the capacitor and transistor are the limiting factors.

means the maximum value of R is 560 kΩ (i.e. 0.7 x 37 x 22 kΩ). The maximum transistor leakage is approximately 2 μA at 125°C. This allows a maximum value for R of 1.5 MΩ (i.e. 3V/2 μA). Thus, taking the three constraints into account, the maximum permissible value for the charging resistor is 560 kΩ, when using a CSR13 capacitor with a value of 330 μF.

Delay period for the timer can be calculated from the following equation:

$$T = RC/3$$

where T is the period in seconds, R is in kilohms and C is in farads.

The graph (**Fig. 2**) shows period plotted against ambient temperature for the 30-sec version of the circuit. This curve shows that the circuit operates satisfactorily from −55°C (where β is the limiting performance factor) to +125°C (where I_{CBO} and capacitor leakage are limiting factors). As can be seen, period varies from 28 to 41 sec over the temperature range.

Super-stable reference-voltage source

by Leonard Accardi
Kollsman Instrument Corp.
Elmhurst, N.Y.

With conventional reference-voltage circuits, the problem is not so much the stability of the temperature-compensated zener used, but the bothersome trimming or trial-and-error selection of the components that supply the zener current, and the often nebulous stability of the current-determining circuitry.

With the circuit described here, however, the current through the zener diode is truly independent of the power-supply voltage — which may be as low as 10V. The current is determined by the zener itself, thus one avoids the trimming and component selection needed for other circuits.

To understand how the circuit works, temporarily ignore components CR_3, CR_2, Q_1, R_4, R_5 and R_6. Let's assume also that zener CR_1 is in the breakdown region. Assume a voltage, $-V_1$, at the junction of R_1 and R_2. Then the output of A_1 is $-(V_1 + V_z)$. But resistors R_1 and R_2 form a voltage divider, therefore,

$$-(V_1 + V_z)\frac{R_2}{R_1 + R_2} = -V_1$$

If we select R_1 equal to $10R_2$, then,

$$V_1 = \frac{V_z}{10}$$

Therefore the zener current is given by the following equation:

$$\frac{V_1}{R_3} = \frac{V_z}{10R_3}$$

We can then calculate the value of R_3 needed to supply the specified zener current of 7.5 mA.

Now let's examine the functions of the remaining components. Resistor R_6 is included to sink a current of around 7.5 mA because most low-cost op amps can provide only 5 mA without exceeding their output rating. The remainder of the auxiliary components insure that the circuit assumes the current stable state when power is turned on.

Temperature-compensated zeners have internal forward-biased diodes, so CR_3 is included in this circuit to clamp the output voltage of the undesired positive-output stable state to about 1V. In this state, Q_1 is ON, and a negative potential, determined by R_1, R_2, R_4, R_5, $-V$ and the ON resistance of Q_1 (100Ω), appears at the noninverting input of A_1. This causes the circuit to revert to the desired negative-output stable state, turning off Q_1 and effectively removing the auxiliary components from the circuit in normal operation.

With the specified components, output stability approaches that of the zener itself. A μA741 amplifier typically introduces less than 1 mV of output-voltage variation over a temperature range of 100°C. Using premium versions of the 741, such as the Sprague 2151D or Precision Monolithics SSS741, output variations due to the op amp can be reduced to 150 μV over the same 100°C temperature range. The op amp used in this circuit should be frequency compensated for unity-gain operation since the impedance of the zener diode is low and thus provides almost 100% ac feedback.

Output-current capability can be increased, with no sacrifice in stability, by inserting a booster transistor inside the feedback loop of A_1. Also, the circuit can be built as a positive reference supply by using a p-channel FET, reversing the diodes, and changing the current-sinking voltage to $+V$.

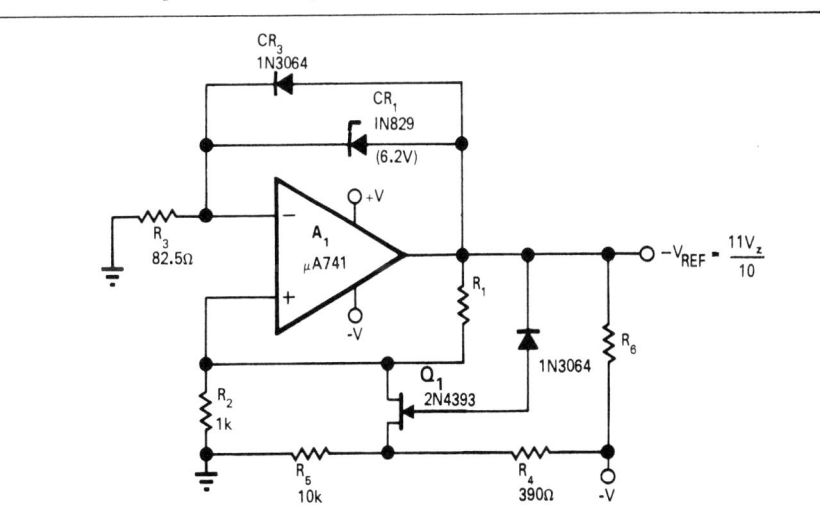

In this reference-voltage circuit, the current through the zener diode is stabilized by the diode itself.

Adjustable, low-impedance zener

by Karl Karash
Raytheon Co.
Waltham, Mass.

THE CIRCUIT in Fig. 1 performs the functions of a zener diode and is adjustable so that the voltage across its terminals is defined by

$$E_z = \frac{0.5\, R_1 + 0.5\, R_2}{R_1}$$

The circuit operates at any voltage above approximately 0.8 V and it has a zener impedance of 3 Ω max below E_z = 5 V. It's particularly useful in the 1- to 3-V region where normal zeners are not generally available and have considerably higher impedances.

The operating point is established by the voltage divider R_2/R_1, such that when the base-to-emitter voltage of Q_1 is greater than approximately 0.5 V, it causes Q_1 to pour current into R_3. When the voltage across R_3 becomes greater than approximately 0.5 V, Q_2 turns on quickly.

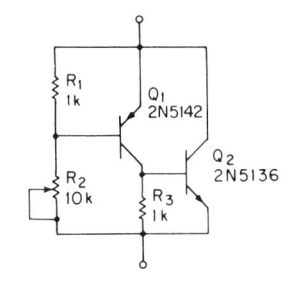

This circuit is particularly useful in the 1- to 3-V range, where conventional zeners have much higher impedance.

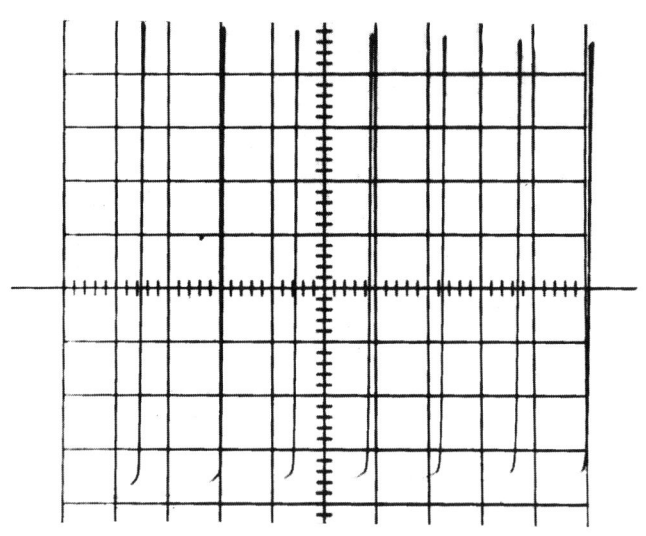

Fig. 2. Zener curves for various settings of R_2 from 0 to 10 kΩ. Vertical scale is 2 mA/div, horizontal is 0.5 V/div.

The high voltage gain of Q_1 and the low drive impedance provided by R_3 make the zener knee sharp, as in Fig. 2.

A digital deadband comparator

Richard W. Wilkens,
Electronic Comm. Inc., St. Petersburg, FL

The circuit shown here expands the use of the 54/7485 to a magnitude comparator with a deadband of ±2 bits. It illustrates a comparison of two 8-bit words which can be expanded to any length word where the compared accuracy must be ±2 bits.

The table illustrates several specific examples of the comparison of two 8-bit words. The inputs are listed, and the alterations to each word are shown as they input to the comparators. The two least-significant bits of each word are dropped at the input to the comparators and are indicated by (x). A new most-significant bit is shown in brackets which has a higher order of value than the most significant bits of either input word. This is a provision for the overflow generated when a ONE is added to an all-ONEs word. The outputs of the circuit are listed as A=B, A>B and A<B. These outputs actually mean that A is equal to B within ±2 bits or A is greater than B by more than 2 bits, etc.

In the logic, the three least-significant bits are examined for a numerical difference of two or three and a comparison of the third LSB. Where the difference between the numbers is three and the third LSBs agree, a ONE is added to the third LSB of the larger of the two inputs. Where the difference between the inputs is two and the third LSBs disagree, a ONE is added to the smaller of the two inputs to make the third LSBs agree going

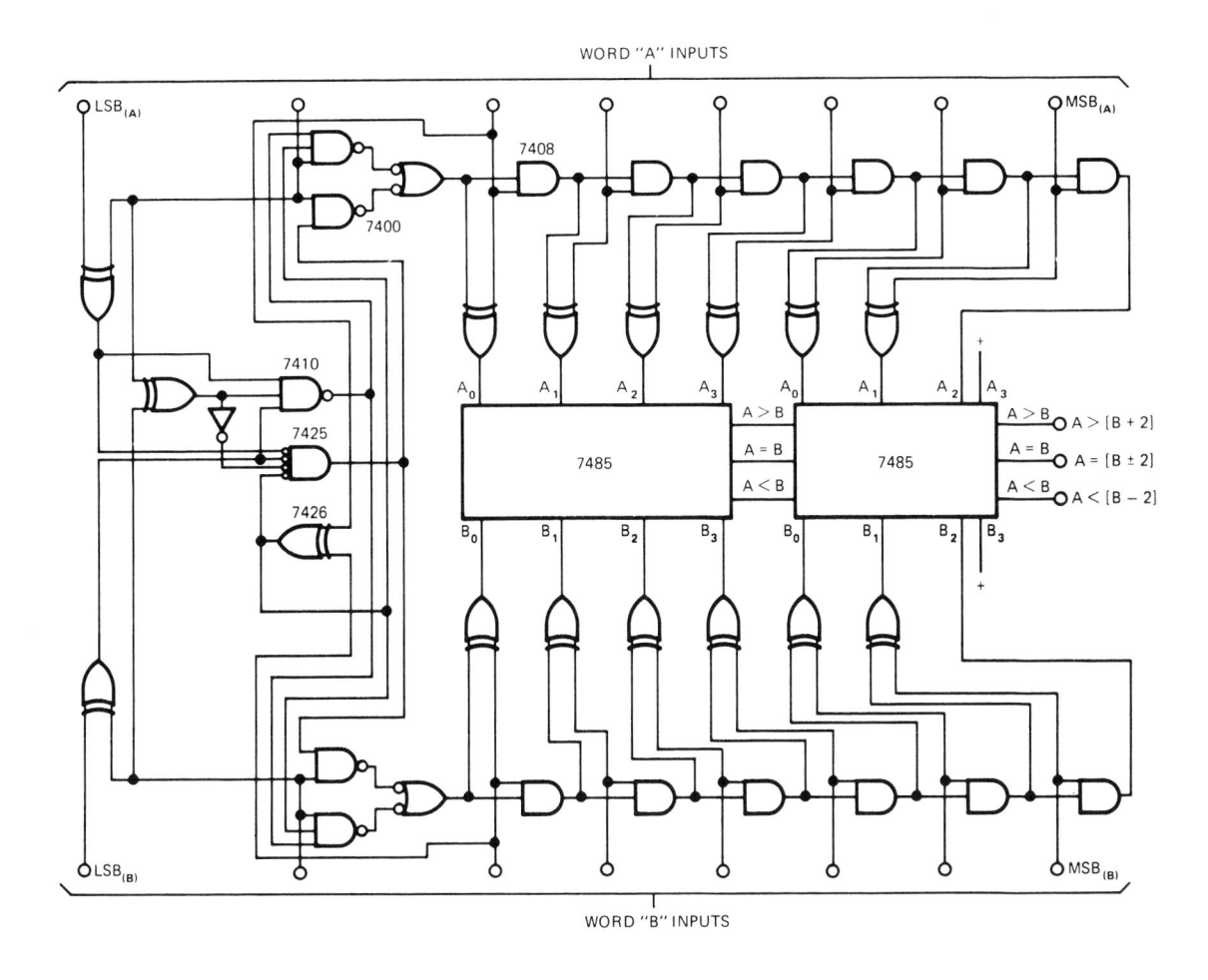

Basic 7485 magnitude comparators can be used to provide deadband comparisons for 2 words of any desired length.

TABLE 1
EXAMPLES OF TWO AND THREE BITS OF ERROR

INPUTS	INPUTS TO COMPARATOR	OUTPUTS		
LSB MSB	LSB MSB	A = B	A > B	A < B
A 00000000 B 01000000	x x 000000 (0) x x 000000 (0)	1	0	0
A 00000000 B 11000000	x x 000000 (0) x x 100000 (0)	0	0	1
A 10000000 B 11000000	x x 000000 (0) x x 000000 (0)	1	0	0
A 10000000 B 00100000	x x 000000 (0) x x 100000 (0)	0	0	1
A 01000000 B 00100000	x x 100000 (0) x x 100000 (0)	1	0	0
A 01000000 B 10100000	x x 000000 (0) x x 100000 (0)	0	0	1
A 11000000 B 10100000	x x 100000 (0) x x 100000 (0)	1	0	0
A 11000000 B 01100000	x x 100000 (0) x x 010000 (0)	0	0	1
A 00100000 B 01100000	x x 100000 (0) x x 100000 (0)	1	0	0
A 00100000 B 11100000	x x 100000 (0) x x 010000 (0)	0	0	1
A 10100000 B 11100000	x x 100000 (0) x x 100000 (0)	1	0	0
A 10100000 B 00010000	x x 100000 (0) x x 010000 (0)	0	0	1
A 01100000 B 00010000	x x 010000 (0) x x 010000 (0)	1	0	0
A 11111111 B 00111111	x x 000000 (1) x x 111111 (0)	0	1	0

into the comparator.

When the difference between the two LSB pairs is one and the third LSBs disagree while the second LSBs agree and are both ZERO, no alterations are made at the input to the comparator. However, when the second LSBs are both ONE, in this case, a ONE is added to both of the third LSBs. The resultant answer in both of these situations is identical since both number sets differed by three.

The advantage of this circuit is that it is static and requires no clocked operations for add and shift. It serves as an excellent monitor in a digital-to-analog scheme with analog-to-digital feedback. Here it may serve as a monitor to indicate performance or it may be used to add incremental changes to the D/A for quantized corrections.

Miscellaneous 191

Automatic scaling circuit for optical measurements

by Robert E. Keil
Honeywell Inc.
Hopkins, Minn.

One problem that occurs with precision optical measurements is a possible drift in light intensity due to thermal cycling of the lamp filament, dirty optics and other causes. Another problem that occurs, even with differential measurements, is the gain variation between photodetectors and between amplifiers. Both of these problems can be minimized by using an automatic scaling circuit to allow frequent and swift recalibration.

This circuit automatically produces a scale factor (K) and then computes the desired product (Z = KX) for a particular input reference level (e. g., when a known portion of the photodiode's active area is illuminated). Thus the circuit is essentially a very linear (1%) agc circuit that automatically corrects its scale factor on receiving a reset pulse at point B.

Comparator A_2 compares the output of the multiplier to a preset reference voltage on R_1. The output of A_2 switches from 15V (positive saturation voltage of A_2) to zero or vice versa, depending on the value of A_2's input voltage at pin 2.

If this input voltage is greater than the reference on pin 3, then the output will switch to zero and remain there until C_1 has discharged enough to lower the output of A_1, and, hence, the output of the multiplier, to a voltage less than the reference on pin 3 of A_2.

If on the other hand the input at pin 2 of A_2 is less than the reference on pin 3, then the output of A_2 will switch to 15V. The output of the multiplier will then be adjusted upwards until the voltage on pin 2 of A_2 again exceeds the reference on pin 3.

From the preceding explanation we can see that the output of A_2 is continually switching from 15V to zero during the entire reset or scaling period. This is the only time that Q_1 is turned on.

Capacitor C_1 and FET Q_1 together form a track-and-hold circuit. The voltage across C_1 is buffered and presented to the multiplier as the scale factor K. The scale factor is effectively the average value of the output from A_2.

After the reset pulse at point B has disappeared, the scale factor K remains constant during the measuring period.

Timing of the reset pulse depends on the actual system application. Normally one would initiate a reset pulse between each measurement. The circuit operates quite rapidly and a reset-pulse duration of around 20 msec should be adequate in most applications.

For a differential measuring system, one would, of course, include a scaling circuit in each leg to ensure a linear output over the entire measurement range. This technique allows accurate measurement of a part (i.e., photo-optical diameter or length measurement) even when it is not centered exactly on the two photodetectors.

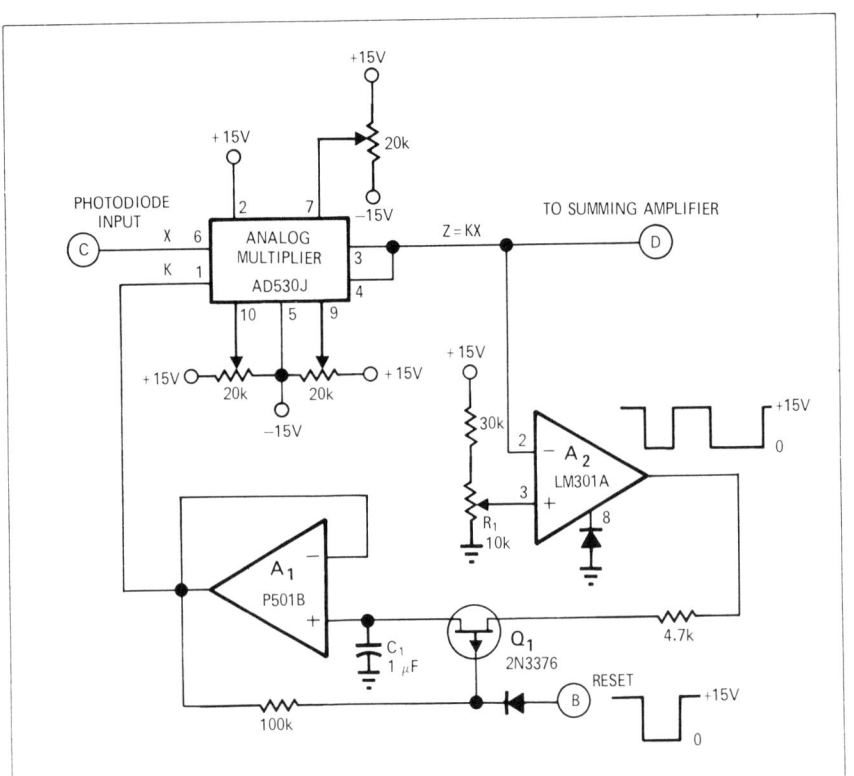

A reset pulse at point B commands this scaling circuit to automatically recalibrate itself. The circuit has important applications in optical measuring systems.

SCR's form electronic combination lock

Edwin R. DeLoach, Astro-Dynamics Electronics,
New Orleans, Louisiana.

This circuit may be employed the next time your design calls for a lock switch. Its operation is analogous to the mechanical combination lock. The prototype was constructed using nine push-button switches, grouped together like an adding machine keyboard. And the circuit provides a four-digit combination controlled by four SCR's-arranged in whatever order you choose.

The combination required to unlock the circuit shown in **Fig.** 1 is A-D-C-H. It is first necessary to press the 'A' push-button. This applies B+ through R_1 to the gate of SCR_1. This positive gate signal switches the SCR ON allowing current to flow through Q_2, SCR_1 and R_3 and applies B+ potential to the anode of SCR_2. The next step similarly applies an ON signal to SCR_2 through push button switch 'D'. This SCR switches ON providing current to R_4 which holds the SCR ON, applying B+ at the anode of the next SCR, SCR_3. If the correct combination is followed through, each SCR in succession with switch on until power is applied to the load relay or solenoid.

An error in dialing the correct combination will turn off all the SCRs and reset the entire circuit. To illustrate, assume SCR_1 and SCR_2 are ON and the next correct digit is 'C' but switch 'H' is pressed instead. No power exists at the anode of this SCR (SCR_4), so it cannot turn on. The low impedance of the gate, which appears as a forward biased diode, turns transistor Q_1 OFF. In turn Q_2 is switched OFF, which results in loss of power to the SCR's. All SCR's switch to the OFF state, which resets the lock. Thus the entire combination must be repeated to unlock the circuit.

The unused push-button switches are tied to ground, so their use also switches off the transistors, resetting the circuit. Transistors Q_1 and Q_2 do not turn off when the SCR's are fired in correct sequence because the SCR cathodes quickly rise to B+ potential to reverse bias the gate and provide a high "ON" impedance to transistor Q_1.

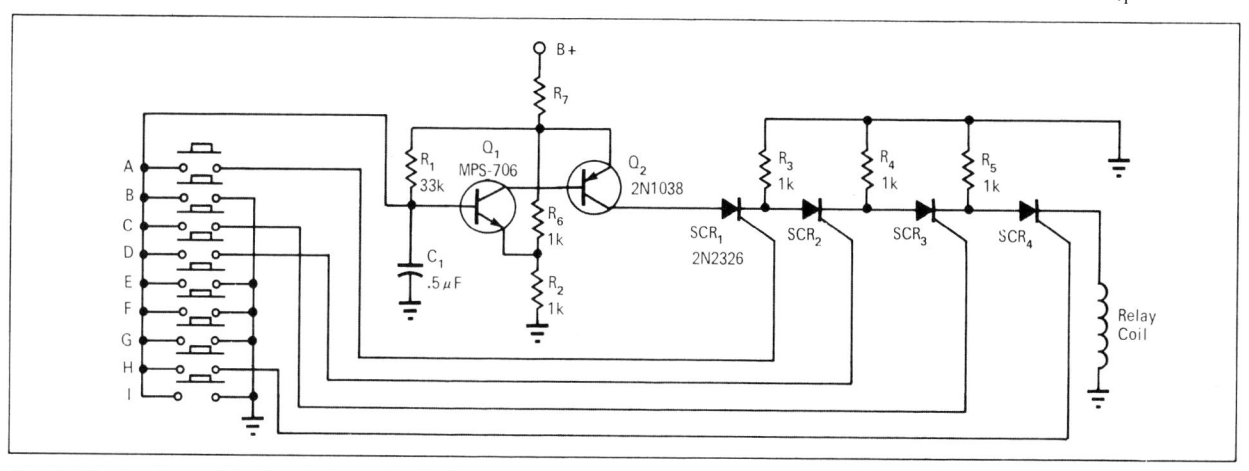

Fig. 1—Electronic combination lock can be built into electronic equipment or used separately with a solonoid. Changing of codes is simple. An incorrect or out of sequence combination will reset the lock, requiring the entire sequence of codes to be re-dialed.

Pulse catcher uses two ICs

S. J. Cormack
General Electric Co., Eau Gallie, FL

Ever need to select just one pulse of a wavetrain? The circuit shown will output one complete input pulse over a wide range of frequencies and pulse widths.

In operation, an enable input allows FF_1 to clock on the leading edge of an input pulse and FF_2 to clock on the trailing edge. This time period represents the input pulse width, which is delivered as an output by gate B. The edge-triggering characteristics of the type D flip-flops prevent operation if they are enabled during an input. In this case the following input pulse would then be delivered as an output by gate B, preserving pulse-width fidelity.

It should be noted that the enabling pulse width must exceed the input pulse width.

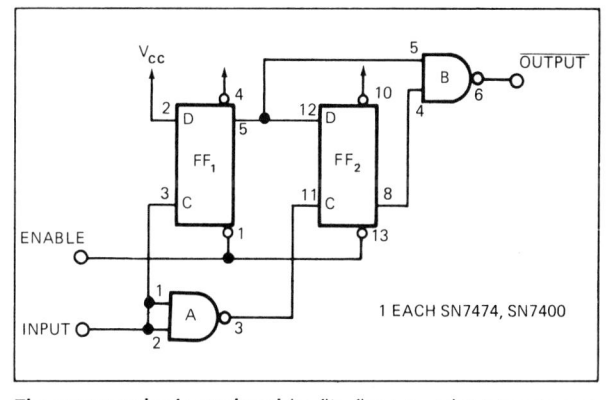

1 EACH SN7474, SN7400

The output pulse is produced by flip-flops 1 and 2 triggering on the leading and trailing edges, respectively, of the input pulse.

Pulse-Level Discriminator and Fault Indicator

By Richard L. Sazpansky
Honeywell, Inc.
St. Petersburg, Fla.

THIS CIRCUIT was devised to monitor and indicate gyro-wheel output faults. It sounds an alarm if the gyro wheel is locked up, as indicated by an input signal that remains at a high current (or voltage) level for a period of time longer than a preset interval. The meter-relay upper "set point" is adjusted to 120 ma (the meter relay is a non-locking type). As the input signal reaches 120 ma, (t_1), the meter relay latches and K_1 is pulled in. Plus 28 volts is applied to point A. SCR_1 is held off because its minimum breakover voltage (V_{BO}) is 50 v. Capacitor C_1 charges toward +28 v. If the input drops to 95 ma (good wheel curve), after 3 sec (t_2), the meter relay will move off the set point and K_1 will be de-energized, removing +28 v from point A. The unijunction transistor (Q_1) has not been gated on because the RC time constant, determined by R_1, R_2 and C_1, has been set to 3 sec.

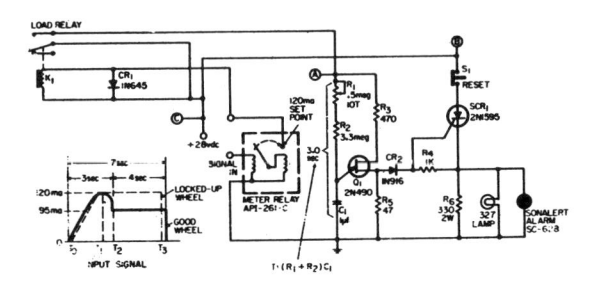

Pulse-height discriminator.

If, however, the wheel locks up, the input signal remains at 120 ma and now is present at point A for a a period long enough (3 sec) to permit capacitor C_1 to charge enough to fire the unijunction Q_1. The SCR will now be gated on by the output from Q_1 and the alarm circuit actuated. At t_3, where the input signal goes to zero, the dc voltage is again removed from point A. The alarm current can now be turned off by pushing reset switch S_1.

Automatic indication of a fault and resetting of the alarm circuit for continuous signal pulses can be achieved by removing the wire from B to C and connecting points A and B together.

Low-pass digital filter

Thomas H. Haydon
B.F. Goodrich, Troy, OH

An SN7400 quad 2-input NAND gate, an SN7431 dual 4-input Schmitt trigger, two 1N457 diodes and two capacitors are connected as shown to form a low-pass digital filter. This arrangement is useful in retrieving a pulse train from "outside world" noise generated by contact bounce, arcing, SCR spikes and inductive surges.

A Schmitt trigger, diode and capacitor comprise a pulse-delay circuit with delay on the positive-going transition. One gate of the SN7400 is used as an inverter to drive a second pulse delay operating on the negative-going transition of the input signal. The delayed pulse trains toggle a flip-flop to generate an output frequency equal to the input, but phase delayed.

Any additional pulses occurring during the delay circuit time-out will reset the delay time with no toggle action at the output. This limits throughput frequency to 1/ (2 × delay time). A higher frequency will latch the flip-flop, resulting in a zero output frequency. Rolloff is extremely sharp with cutoff frequency variation depending mostly on the diode thermal coefficient.

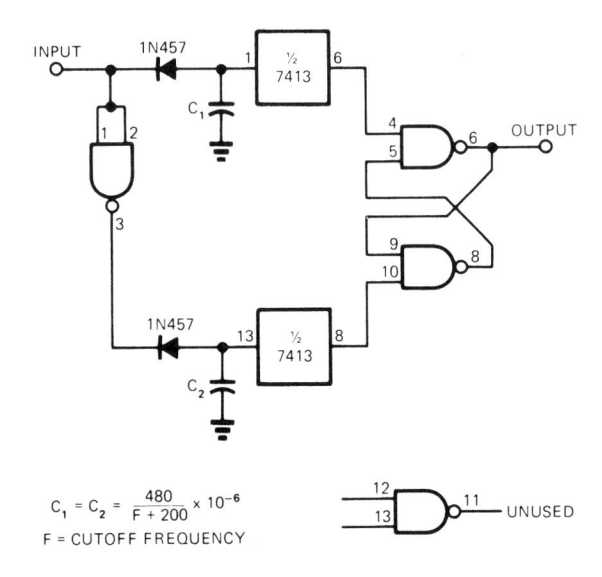

$$C_1 = C_2 = \frac{480}{F + 200} \times 10^{-6}$$

F = CUTOFF FREQUENCY

This low-pass digital filter, using only two TTL ICs, can retrieve pulse train data from very noisy signal lines.

Thermistor circuit senses air temperature and velocity

By Harold Greene
Electronic Associates
West Long Branch, N. J.

A THERMISTOR CIRCUIT can be used, not only to sense temperature changes, but also to sense changes in air velocity. Thermistor temperature depends on ambient temperature and also on heating caused by electrical power dissipated within the thermistor. Temperature rise from the latter source is governed by air flow past the thermistor.

The type of circuit described is useful for power-supply protection. Normally, forced-air-cooled supplies are protected by a pressure-actuated switch in the air duct and by temperature sensing devices at critical points in the circuit. But satisfactory protection can be achieved by a single solid-state circuit — thus eliminating electro-mechanical components which may be unreliable.

With careful design of the circuit, and of the thermistor enclosure, this approach can be used to sense any desired linear range of air velocity and temperature. A typical performance curve is shown in Fig. 1, where two set points of air velocity and temperature (v_1, T_{AF1}) and (v_0, T_{AF0}) are chosen as firing points for a control circuit. For best linearity, the thermal resistance of the thermistor should vary linearly with air velocity in the given enclosure.

Figure 2 shows the basic sensing circuit. In general, a resistance element R_2 is placed in series with a thermistor and a constant-voltage source V_T. We sense the voltage V_2 across resistor R_2. The thermistor produces the firing voltage V_{2F} when its resistance reaches some value R_F corresponding to a thermistor temperature of T_F.

In a practical circuit, element R_2 may be a relay which fires at a constant current I_F, or it may be a resistor across a semiconductor device whose input current is much less than I_F. Of course, the thermistor's voltage, current or resistance can be considered constant depending upon whether $R_2 \gg R_F$, $R_2 \approx R_F$, or $R_2 \ll R_F$, respectively.

Referring again to Fig. 1, the difference in temperature between the two set points depends on the difference in thermal resistance $(\theta_0 - \theta_1)$ of the thermistor at the two different air velocities. The change of temperature also depends on the constant electrical dissipation W_F in the thermistor. Then,

$$T_F - T_{AF1} = \theta_1 W_F \quad (1)$$
$$T_F - T_{AF0} = \theta_0 W_F \quad (2)$$

Therefore,

$$T_{AF1} - T_{AF0} = (\theta_0 - \theta_1) W_F \quad (3)$$

From the equations, we can see that to achieve maximum sensitivity to change in air velocity, with any given enclosure, we should make W_F as high as possible. We can decrease the lower operating temperature T_{AF0} by insulating the enclosure on the outside (while allowing forced air inside to cool the thermistor). To decrease the higher temperature T_{AF1}, we can impede the flow of forced air (while keeping the still-air insulation unchanged).

Figure 3 shows how this type of temperature-velocity sensor is used to protect an EAI power supply. The circuit interrupts the ac line if the fan fails, if the air flow is obstructed, or if the incoming temperature is higher than 50 degrees C at full load.

Because, in this supply, the transformer connections are adjustable for 115 or 230 volts, it was found convenient to operate the sensor circuit on an unregulated output V_{LN1} whose voltage is always somewhere in the range 38 to 50 volts dc. When the sensor fires, the breakdown diode conducts and triggers the SCR. The resulting overload then blows the fuse in the ac line circuit.

The sensor monitors exhaust air, and it triggers the SCR when the thermistor temperature at rated velocity v_1 is above $T_{AF1} = 65°C$, or when its temperature at zero velocity v is around $T_{AF0} = 30°C$.

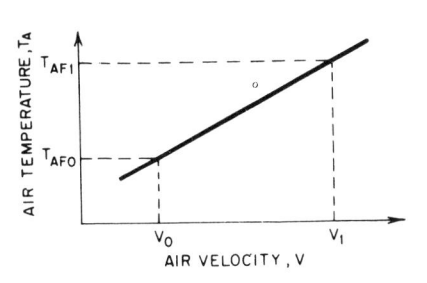

Fig. 1. Thermistor circuit triggers at two different preset points. With zero air velocity it fires at T_{AF0} and with normal air velocity it fires at a higher temperature T_{AF1}.

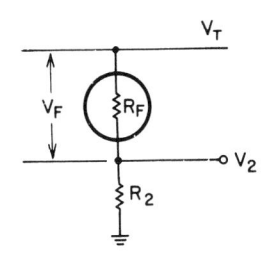

Fig. 2. Basic thermistor sensing circuit. Output V_2 can be used to trigger a breakdown diode, or R_2 can be replaced by a relay.

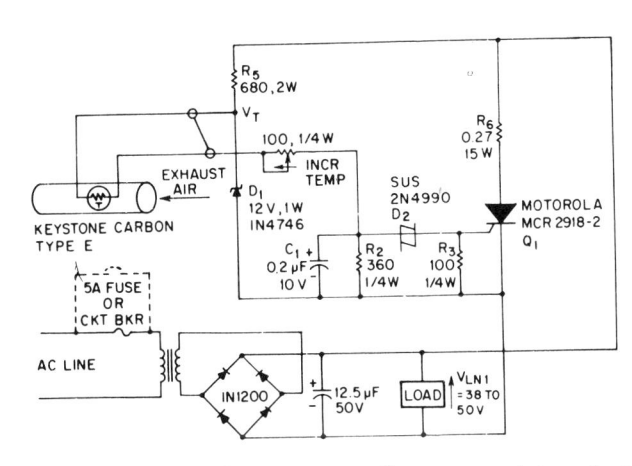

Fig. 3. One of EAI's power supplies uses a temperature-velocity circuit for protection. The SCR fires, blowing the fuse, if ambient temperature becomes excessive or if the forced-air supply fails.

DPM makes self-contained digital thermometer

Maxwell Strange, NASA,
Greenbelt, MD

Either of these circuits can be built into, and powered by, an inexpensive digital panel meter to produce a compact, accurate electronic thermometer.

The circuit in **Fig. 1** uses a Yellow Springs Instrument Co. linear thermistor network whose resistance is a linear function of temperature. The op amp biases this sensor at a constant current, so that the output voltage is proportional to temperature. Any 3-1/2-digit DPM having a 100 mV range will indicate temperature directly in degrees F with 0.1-degree resolution.

The Yellow springs 44203 sensor network is precalibrated to be interchangeable and linear to within 0.27°F over the entire range from −22°F to +122°F. The network actually uses 2 thermistors to achieve this linearity and a simple equation supplied on the spec sheet sent with each sensor shows the networks resistance value for any temperature within its range. This equation is based on temperature in degrees Centigrade, and that information has been converted to degrees Fahrenheit here. Thus, the circuit can be set up simply by replacing the sensor with a resistance decade box set for two known temperatures. There will be no interaction between these adjustments if the offset is set first with the "0° set" potentiometer R_4 and the decade box set for 0°F ($R_t = 14435\Omega$). Next the gain should be set for 100°F with the 100° set potentiometer R_6 and the decade box set for 100°F ($R_t = 7374\Omega$). Typical meter overrange capability provides readings well beyond the 0 to 100°F range.

This circuit was built with a Weston 1290DPM that had ±6.2V references available internally which could easily support the extra load of about 2 mA. Other voltages can be used (with appropriate resistance changes) as long as V_{ref} is stable. Sensor current should be less than 0.5 mA to minimize the self-heating error.

An "economy" circuit which uses the linear temperature coefficient of a forward-biased silicon diode for sensing is shown in **Fig. 2**. The diode is biased at about 0.3 mA. The current through R_2, R_3 and R_4 produces a voltage drop across R_3 which cancels the diode drop at 0°F to zero the DPM. The Weston 1290's internal adjustment is used to set the gain; a voltage divider may be required for other meters. This circuit is best calibrated by alternately dipping the sensor in crushed ice (32°F) and warm water monitored by a mercury thermometer.

In use, the two circuits behave identically, but the thermistor has somewhat better long-term stability and is easier to calibrate. These thermometers can replace thermocouples in most test setups, and provide better accuracy and readability. The diode circuit used with an inexpensive 2-digit DPM makes an ideal desk-top indoor/outdoor thermometer.

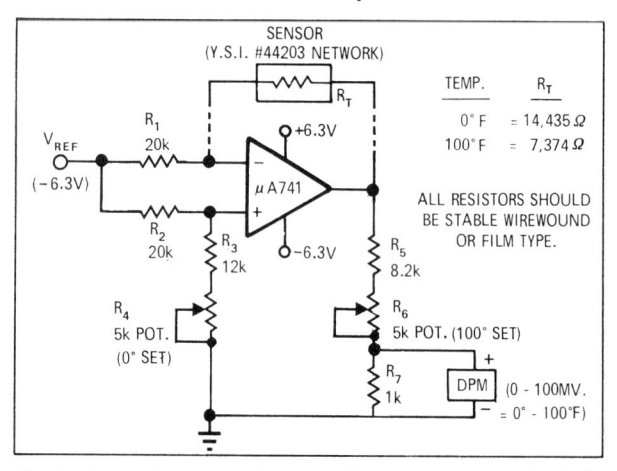

Fig. 1 — **Accurate temperature-to-voltage converter** can be built into a panel DPM. The sensor is a linear thermistor connected in the resistance mode for easy calibration.

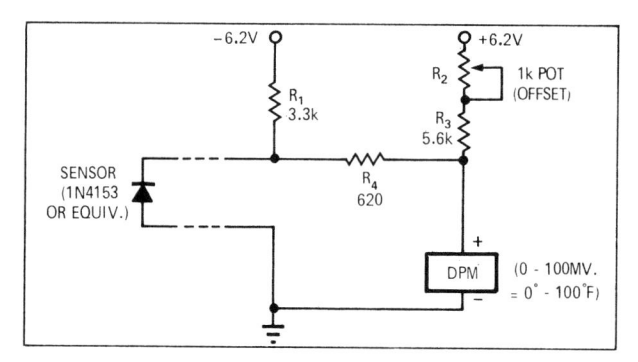

Fig. 2 — **Economy version** of the thermometer circuit uses silicon diode as a temperature sensor. Internal DPM gain adjustment is used for overall gain calibration

E-cell controls battery charging rate

by J. P. Yang
Digitronics
Albertson, N. Y.

THIS CIRCUIT uses an E-cell or electrochemical integrator to monitor the state-of-charge of a battery. An E-cell actually integrates the amount of charge (coulombs/sec) flowing in and out of a storage cell. This method is more accurate than sensing the terminal voltage's value as is done in conventional battery chargers.

Charge and discharge current is sensed by a small resistor (0.0015Ω ammeter) in series with the battery. When the battery discharges, current flows from B to A, making the voltage at B higher than A. This voltage difference is amplified 150 times by a CA 3029 op amp. Since V_B is higher than V_A, the output voltage of the op amp is negative. So current flows through the E-cell from cathode to anode. This action causes silver to be plated on the anode making the voltage across the E-cell nearly zero. The amount of silver plated on the anode is proportional to the charge that the battery has lost.

When the battery is in its charging mode, normally-closed relay K_1 shorts the 0.3 Ω resistor. This charges the battery at a high rate. Charging current now flows from A to B reversing the polarity of differential voltage V_{AB}. This is again amplified by the op amp whose output polarity goes positive. This in turn causes a current flow from anode to cathode in the E-cell which deplates silver away from the anode. As long as deplating continues, voltage across the E-cell is nearly zero, which prevents the SCR from firing.

When all of the silver has been plated off the anode (the battery has had its charge replaced), the voltage across the E-cell will rise to about 0.8 V firing the SCR. The SCR turns on the 2N2374 which drives K_1's relay coil. The relay contact switches a 0.3 Ω resistor into the charging path, making the system trickle charge.

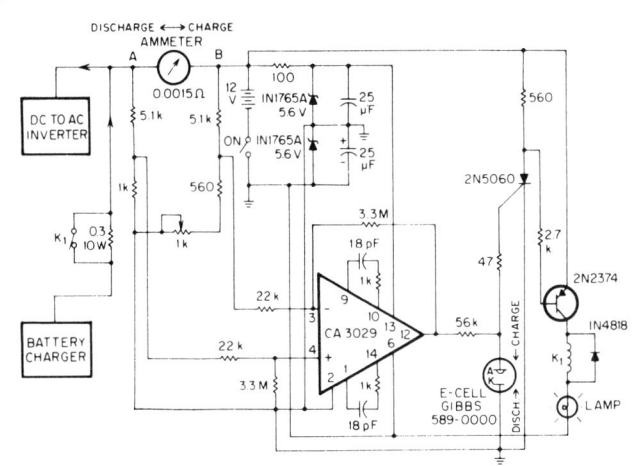

A differential amplifier and an E-cell are used to construct a battery-charger rate controller.

Proportional oven-temperature controller

by Robert L. Wilbur
Southwest Research Institute
San Antonio, Texas

This circuit provides proportional temperature control of an oven to within 1°C for temperatures from 75 to 250°C. The circuit uses an IC voltage regulator, Type 823D, which has a quiescent current of 1.5 mA. This particular IC operates well on the same 28V power source as the oven.

Bridge resistors R_1 and R_2 should be 1% carbon-film versions. To achieve adequate resolution of the temperature setting, the potentiometer should be a ten-turn wirewound type. Since the power transistor Q_1 operates either saturated or almost cut off, no heat sink is required and operational power derating is unnecessary.

The same basic circuit can be used to control temperatures over other operating ranges, and can be easily modified by selecting a suitable thermistor and bridge resistors. Ovens of different power ratings than indicated can be controlled by suitable selection of components R_3 (to provide the correct current drive), CR_1 and Q_1.

Low-power heater elements (where heater current does not exceed 15 mA) can be driven directly from the output (pin 6) of the regulator IC.

This circuit was developed in conjunction with the NASA-sponsored Technology Utilization Program, Contract No. NASW-1867.

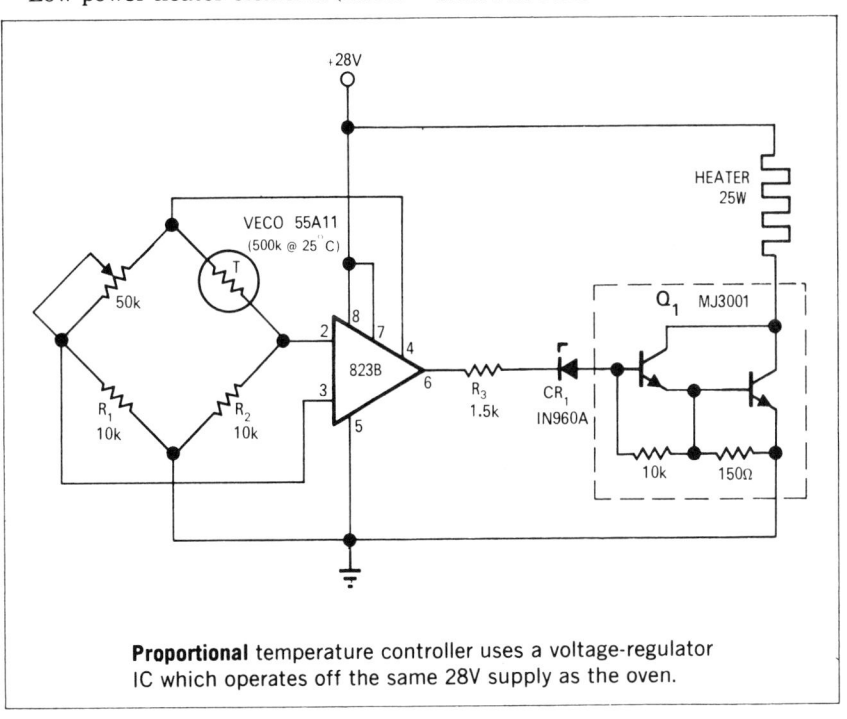

Proportional temperature controller uses a voltage-regulator IC which operates off the same 28V supply as the oven.

TTL/DTL interface to FET analog switch

by Walter G. Jung
MTI Div. of KMS Industries
Cockeysville, Md.

IT'S USUALLY difficult to control different types of analog-switch FETs, with their differing cutoff voltages and input levels (up to ±10V), from the 0 and +5-V logic levels of DTL and TTL circuitry. The circuits shown here make it easy and economical to switch different MOS and junction FETs from DTL, TTL or RTL logic.

We can see the problems if we look at either circuit shown here and consider an n-channel JFET like the 2N5459/MPF-105, whose maximum gate-source cutoff is −8 Vdc. This dictates a −V supply of $-V_{in} -8 -|V_{satQ2}| -|V_{D3}|$. The positive supply must be high enough (about $+V_{in}$) to back bias D_3 ($V_{gs}=0$) at the maximum positive peak of the input waveform.

We see that the required magnitudes of ±V can vary widely. So we need a constant-current base drive to insure full saturation or cutoff of Q_2, regardless of the value of −V, since Q_1's V_{cb} variations must not alter the value of I_c that drives Q_2 via the voltage developed across R_2.

In the circuits, Q_1 is a grounded-base level shifter that converts the "1" level emitter drive to a constant-current drive for the base of Q_2, independent of −V, which must equal or be less than V_{off} of the FET (assuming an enhancement-mode device). Q_2 is a simple inverter with positive and negative supplies of sufficient amplitude to control the gating FET.

The drive to Q_1 can take either of two forms. For current-sinking logic (Fig. 1), a dual-diode gate and a resistor to +5 V are used. A +5-V "1" level back biasis D_1 which turns on Q_1 through $R_1 - D_2$. A "0" level at D_1 deprives Q_1 of emitter current which, in turn, gates the FET on.

For current-sourcing logic (Fig. 2), a 2.4-kΩ resistor from a 3.6-V "1" level turns Q_2 on, gating the FET off. This circuit can be built for about 60 cents at single-piece pricing.

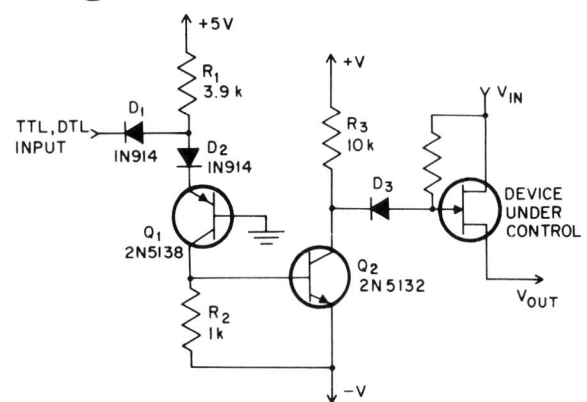

Fig. 1. TTL, DTL current-sinking control. A dual-diode gate is used at the input.

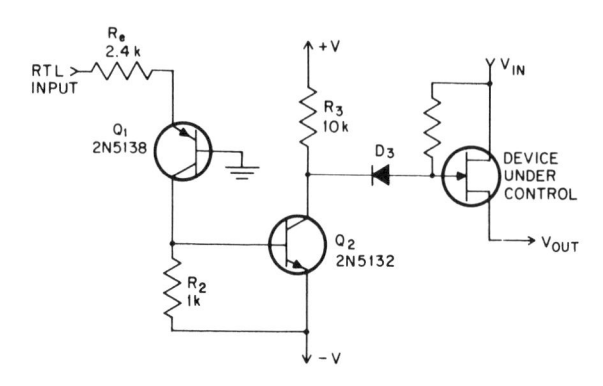

Fig. 2. RTL current-sourcing control uses a simple 2.4-kΩ resistor at the input.

FET provides automatic meter protection

By Ralph L. Charnley
Pako Corp.
Minneapolis, Minn.

MOST METER-PROTECTION circuits use some sort of non-linear device to shunt the meter and thus divert high currents. Though these circuits do protect the meter, many of them have the disadvantage that they also load down the external circuitry, causing errors. Errors can be minimized by using a series resistor and by recalibrating the meter together with its protection circuit. But, of course, this increases the cost.

Other protection circuits employ fuses or circuit breakers, but these are often slow and must be manually replaced or reset after each overload.

The circuit shown here has none of the drawbacks mentioned, yet it requires only two inexpensive components. It was developed for an application where the usual voltages of interest were in the range 2 to 3 volts, but where accidental overload voltages up to 14 volts could occasionally occur. With the components shown, there is no noticeable effect on meter accuracy with voltages to full scale. Yet with 16 volts applied, only 6.7 volts appear across the meter.

Maximum allowable voltage depends on the breakdown voltage of the FET. Circuit action is such that, below the onset of conduction in the zener diode, only the FET "on" resistance is in series with the meter. This resistance is negligible compared to the meter resistance. When the zener diode conducts, it applies gate voltage to the FET, cutting it off and preventing excessive meter current.

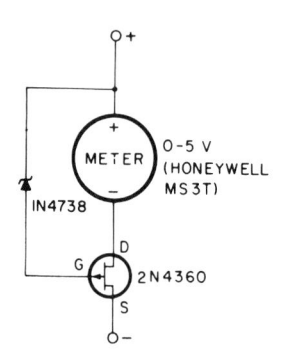

This simple and effective meter-protection circuit uses only two inexpensive components.

Two TTL gates drive very long coax lines

Robert W. Stewart
Control Data Corp., Santa Ana, Calif.

Need a great TTL coax driver and receiver? This circuit has all the good features you want and haven't been able to find. It will transmit information via coax or twisted pair over long lines at bit rates exceeding 10 Mb per sec. And the best part is its cost, that of standard logic gates, plus a few resistors.

The distance over which this driver/receiver will work is a function of the channel used (coax or twisted pair) and the data bit rate, due to both rolloff and phase distortion in the channel.

Safe operating maximums for two types of coax can be derived from **Fig. 1**. Twisted pair typically has a characteristic impedance in the range of 50 to 120Ω and has approximately the same loss characteristics as RG 174 below 10 MHz. For a convenient rule of thumb, the channel attenuation should not exceed 10 db for the data clock frequency. That is, if the clock frequency is 5 MHz, the overall channel attenuation should not be greater than 10 db at 5 MHz to support a data rate of 5 Mb per sec.

In **Fig. 2**, the device types have been chosen to provide two drivers or two receivers per IC package. The unused input in the receiver could also be connected as a strobe.

Success of the circuits depends upon operating about the threshold point of the receiving gate. This threshold is very accurately determined by connecting output to input of an adjoining gate on the same IC chip. The diffusion process by which the ICs are fabricated insures that the bias gate and the receiving gate have identically the same threshold and will track each other with temperature and voltage. Being a linear feedback amplifier, the bias gate, G_3, exhibits a low output impedance and can easily terminate the channel load resistor. Operation of the receiver is enhanced by forcing the driver to operate symmetrically about the receive threshold voltage. When the input to the driver is high, the driver output becomes saturated at near ground potential and when the input is low, feedback around the driver limits its output to twice the driver threshold voltage. Although different from the receiver threshold, close matching is not necessary and only becomes important for channel losses greater than 20 db. The logic input to the driver circuit should be a standard TTL gate with a positive swing of 3.0V or better and a saturated zero level of about 0.25V.

If you're at all concerned about operating these gates in their linear mode, a quick check of the open collector gate circuits will show that power dissipation is being kept within reasonable bounds and that nothing catastrophic can happen.

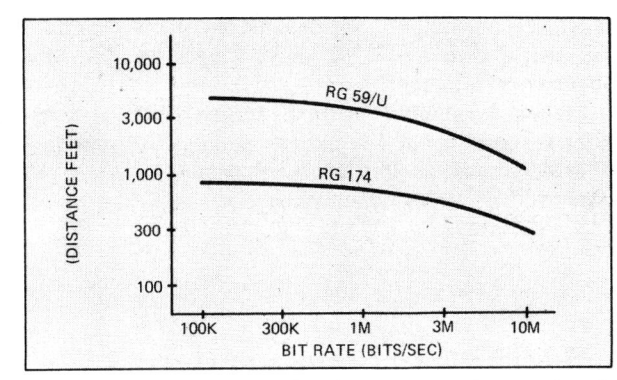

Fig. 1 — Performance curves for driver circuit when used with RG59 or RG174 coax. Twisted pair lines approximate the performance of RG174 at data rates below 10 MHz.

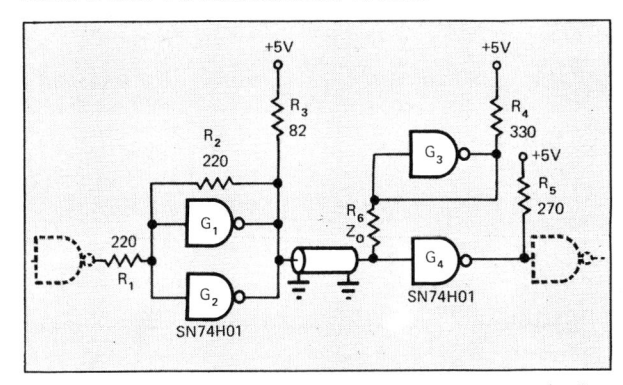

Fig. 2 — Driver/receiver pairs require only 4 TTL gates, and a few resistors. R_6 is the coax termination resistor.

Voltage- or pot-variable 400-Hz delay

by Louis L. Pechi
Master Specialties
Costa Mesa, Calif.

THE CIRCUIT in Fig. 1 provides delays variable from 0 to 2.5 ms as a function of control voltage or potentiometer adjustment. The delay is controlled by the R_1-R_7 divider or by a dc control voltage in its place.

The circuit senses and delays the positive excursion of an input 400-Hz square wave. The delayed signal triggers a 1.25-ms one-shot so the output looks like the original square wave, delayed.

When the input goes positive, Q_2 turns on and discharges C_2. Negative excursions are bypassed by CR_1. When the input goes low at the end of the positive excursion, voltage comparator A_1 senses the low voltage and its output goes low. C_2 begins to charge through the variable current source ($R_1 R_7 R_2 Q_1$), which determines the slope of the charging curve.

When the voltage across C_2 reaches the trigger level of A_1, its output goes high. The high-going signal turns on Q_3, which triggers the 1.25-ms one-shot.

The same cycle is repeated, producing an apparent delay of the input signal. The transistors used are fast switches. The µA710 is a differential comparator with fast response and high accuracy. The one-shot uses standard SUHL gates.

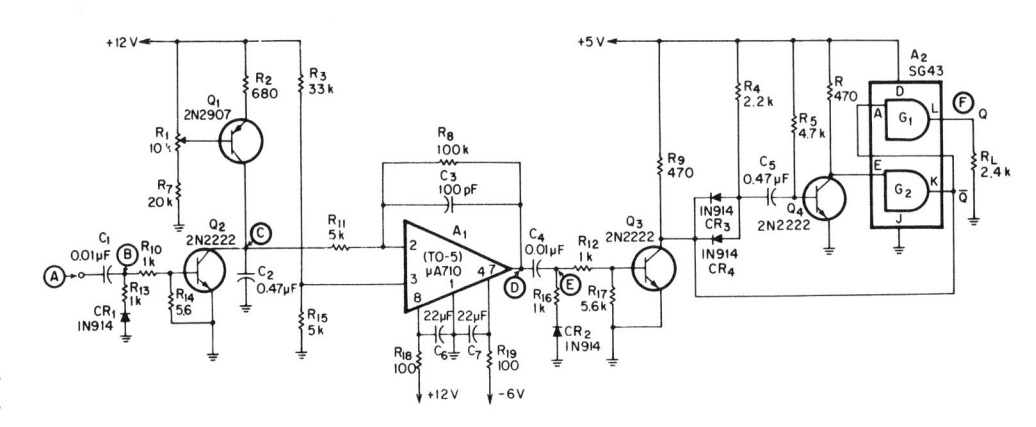

Fig. 1. This circuit provides variable apparent delay for a 400-Hz square wave, but it actually "reconstructs" the square wave.

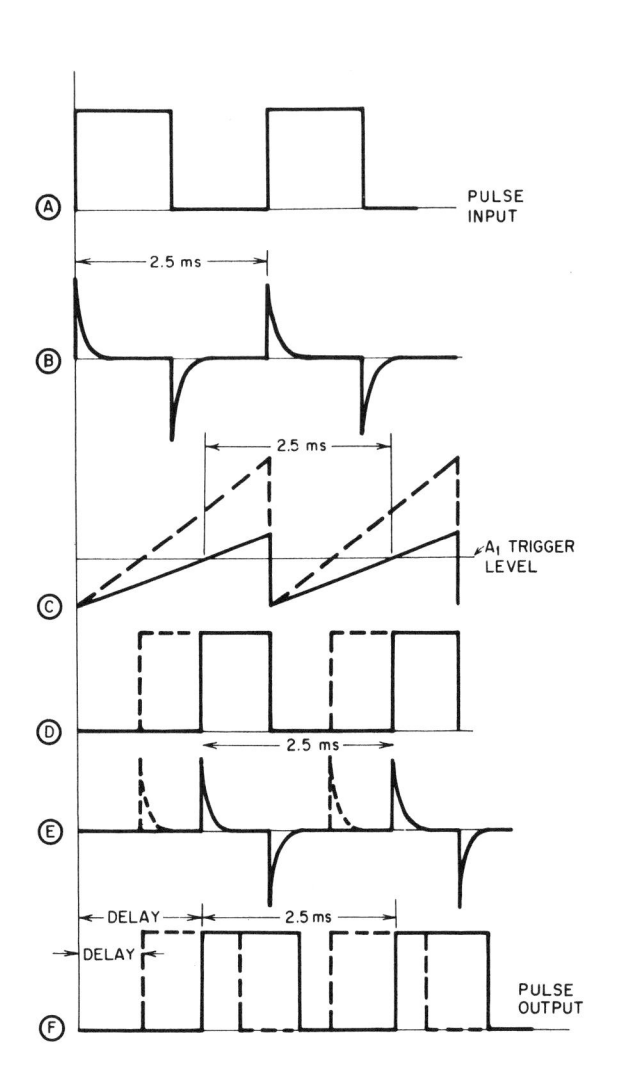

Fig. 2. Waveforms at key points in the delay circuit

Double duty photo alarm

by John F. Kingsbury
RCA Graphic Systems
Dayton, N. J.

THIS CIRCUIT is useful as an alarm for sensing either missing light pulses or for checking for the absence of an object on a moving conveyor belt. A CA-3062, light sensor and amplifier, detects light pulses synchronized to the 60-Hz line. With SW in position A, each pulse momentarily grounds diode CR_1. This continuously resets the 20-ms timing network of the UJT (2N2646) every 16.7 ms, thus preventing the UJT from firing.

If an object interrupts the light beam, the UJT is allowed to fire. Its output pulse triggers an SCR (2N3529) which in turn actuates a warning light or an audible alarm.

When SW_1 is in position B, the circuit detects interruptions in a steady light beam. Under this condition, an alarm is sounded only when an interruption does *not* occur.

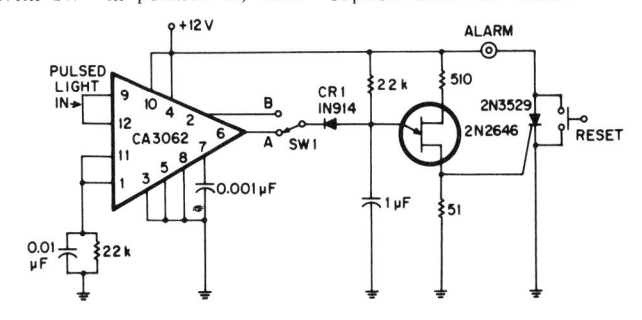

A combined light sensor and amplifier (CA30662) senses the interruption of a pulsed light source or the lack of interruptions of a steady beam of light.

200 Circuit Design Idea Handbook

Current-Pulse Generator for LED's

by James Dimitrios
Zenith Radio
Chicago, Ill.

It is often necessary to pulse a light-emitting diode (LED) with high peak currents of substantial pulse width with low time-averaged current drain. For example, the resolving power of some low-light-level (LLL) systems improves when the scene is illuminated with high peak radiation (as contrasted with constant illumination having the same time-averaged radiated power). Also, for portability, many LLL systems require battery operation of LED circuitry. In addition, pulsed operation of LEDs is useful in applications where timing of the radiated light pulse in relationship to a cyclic mechanical motion is under study.

The circuit shown in **Fig. 1** can generate peak currents in excess of 1A with pulse widths greater than 10 msec at repetition rates of 12 kHz with greater then 90% efficiency (100-mA dc current drain from 2.5V dc). Rise time of the pulses is 0.2 msec.

Circuit action can be described as follows: Assume Q_1 is saturated by a current through R_B. The battery voltage is then impressed across the primary inductance, L, with a consequent linear increase in Q_1's collector current, I_c. The Q_1 base current, I_{b1}, rides on a dc level, V/R_B, and is a rectangular pulse of current. Collector current I_c increases until it reaches $\beta_1(2V/R_B)$, where β_1 is the beta of Q_1. At this time, Q_1's collector voltage starts to increase from its saturated value. The base-emitter junction of Q_1 becomes back-biased due to the phase reversal of the transformer, and the transistor goes into the cut-off mode. Pulse repetition period is determined by L, R_B and β_1.

Since the transformer has stored energy, $1/2 LI_c$, and Q_1 is switched open at time T, a large voltage spike develops across the transformer windings. The current through R_b turns on the transistor, initiating the next cycle. The large voltage pulse is stepped down to drive the base-emitter junction of Q_1. A large current is needed to saturate Q_2 which has the LED in its emitter circuit. (The LED has a dynamic impedance of roughly an ohm.)

Pulse width and rise time are determined by base current drive and the saturation and switching characteristics of the transistor selected for Q_2. The peak current is of course determined primarily by the battery voltage and dynamic impedance of the LED. Larger battery voltages will require greater base drive to Q_2 to obtain saturation. This, in turn, will necessitate greater stored energy in the transformer, with larger Q_1 base drive and, therefore, a larger I_c, before Q_1 switches to the cut-off state. Fortunately, all this is provided automatically by the increase in battery voltage.

As an alternative, the transformer feedback winding can be deleted, and the base of Q_1 returned to a trigger source which provides the saturation of Q_1 needed to initiate the cycle. Then the pulse rate of the current pulse becomes variable – a desirable feature for testing LLL systems.

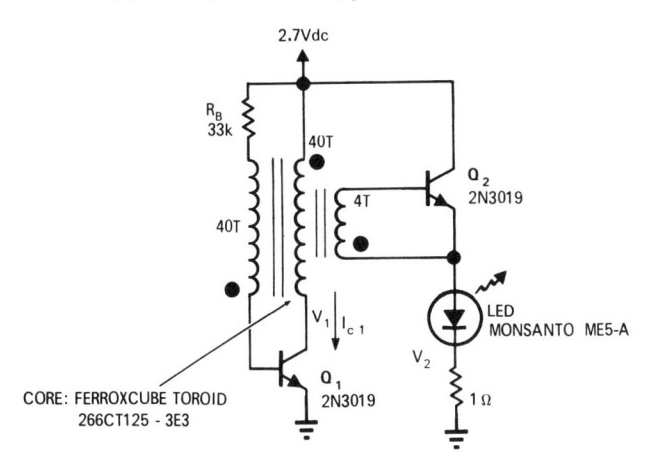

Fig. 1 – The LED in the emitter circuit of Q_2 is subjected to a train of rectangular current pulses instead of a continuous dc voltage. Pulsed light is useful in low-light-level TV systems or for strobing cyclical mechanical equipment.

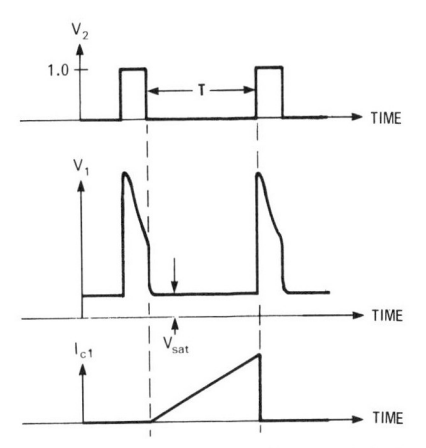

Fig. 2 – Typical waveforms at various points in the pulse-generator circuit. The top waveform shows the rectangular current pulses through the LED, while the bottom waveform shows the ramp of current through the transformer primary which determines pulse-repetition frequency.

Photocells allow a free pendulum to operate silently

Robert B. Bohlken
Honeywell, Hopkins, MN

Using optoelectronic techniques, an electronic pendulum can be built that delivers a 1-pulse/sec output suitable for driving a digital display.

Mechanical details of the pendulum are shown in (**A**). At center position, both photocells PC_1 and PC_2 are dark, as the flag attached to the pendulum, P, covers both cells. As P swings to the right, PC_1 is illuminated and turns Q_1 OFF. This turns OFF driver D_2 through its gate circuit (**B**) and enables driver D_1. As P swings to the left, both cells are dark; and as P passes the center position, both inputs to driver D_1 are enabled allowing coil D_1 to attract the iron slug on P. As the swing continues, photocell PC_2 is illuminated, turning D_1 OFF.

The excursion of the pendulum is self-regulating, since a longer excursion causes the time that both PC_1 and PC_2 are dark to be shorter, thus reducing the drive pulse. DS_1 and DS_2 indicate drive-coil operation.

The audio circuit allows a choice of clock-like tick-tock or silent operation.

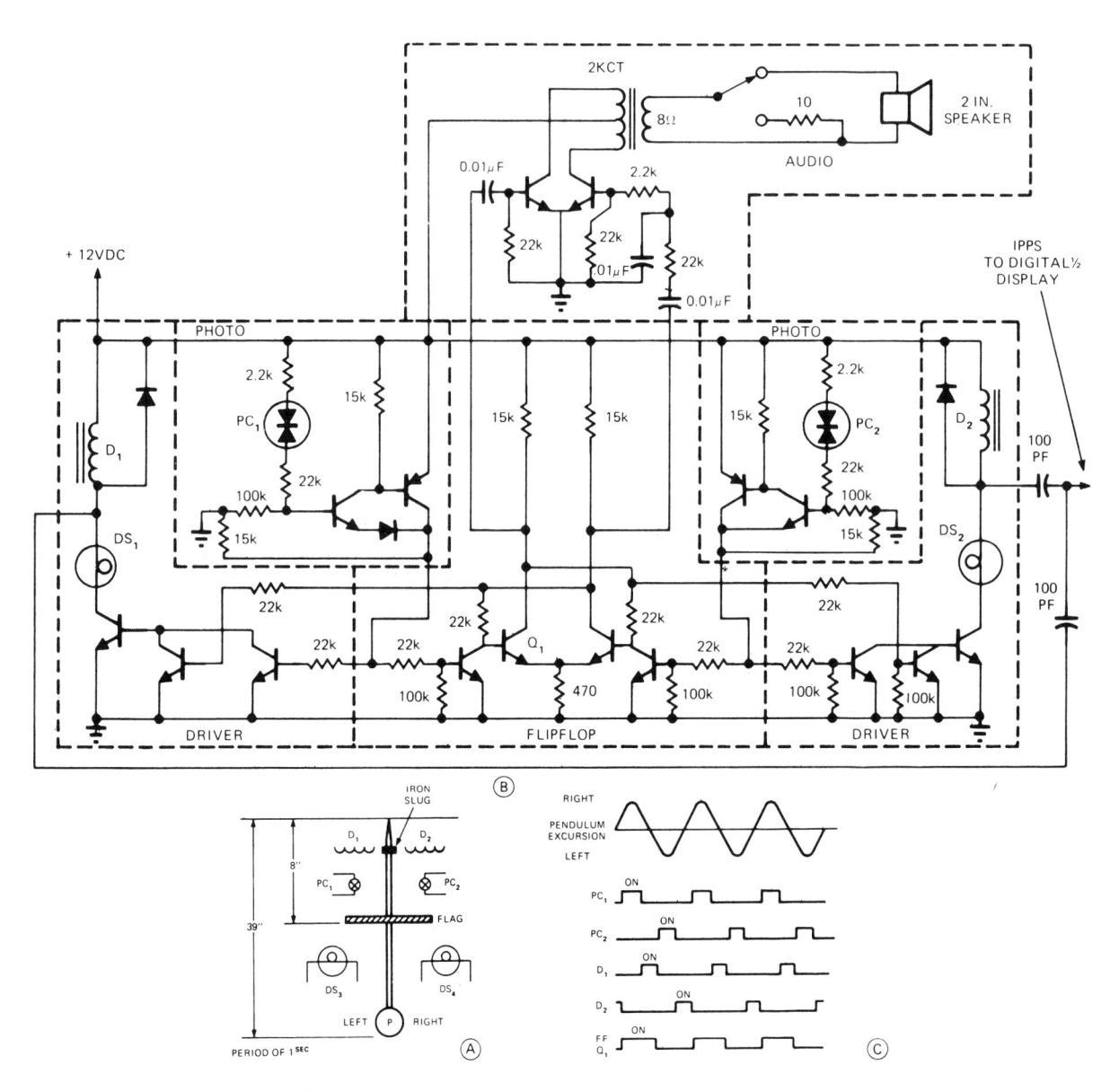

Alternate illumination of photocells by swinging pendulum (A) is converted to a 1-pulse/sec signal for driving a digital display (B).

202 Circuit Design Idea Handbook

Low-cost, long-delay timer

Fractional-hour time delays are provided at low cost by bipolar transistors in this solid-state timing circuit. This design avoids the use of more expensive FETs and UJTs. The resulting circuit is comparable in cost to mechanical timers but with greater reliability. Timing period can be varied over a 1- to 15-minute range.

The timer consists of a ramp circuit and a comparator circuit. When input A_1 is up, capacitor C_1 charges through R_1 and D_1 to about 11.4 volts. When the input goes down, D_1 becomes reverse biased and capacitor C_1 charges through the discharge resistor network and the base of Q_1. The discharge resistor network, composed of R_{13} through R_{19}, allows adjustment of the timing period. The discharge resistance is varied by shorting out pin positions 1 through 10. Resistors R_2 and R_3 provide a voltage division to bias the negative side of C_1 to 6 volts to prevent C_1's 6-V breakdown voltage from being exceeded.

When C_1 has discharged to the reference level maintained by R_8, D_3 and R_9, comparator transistors Q_1 and Q_2 change state. During the timing period Q_1 is on. Emitter current of Q_1 is held constant at 300 μA by current source Q_3. When the capacitor discharges to the reference voltage, Q_1 turns off and Q_2 turns on.

When Q_2 turns on, its collector voltage turns on Q_4. The current flow from Q_4 turns on Q_5 and the output level drops. Diode D_4, in the emitter of Q_4, ensures that Q_2 turns on ahead of Q_4.

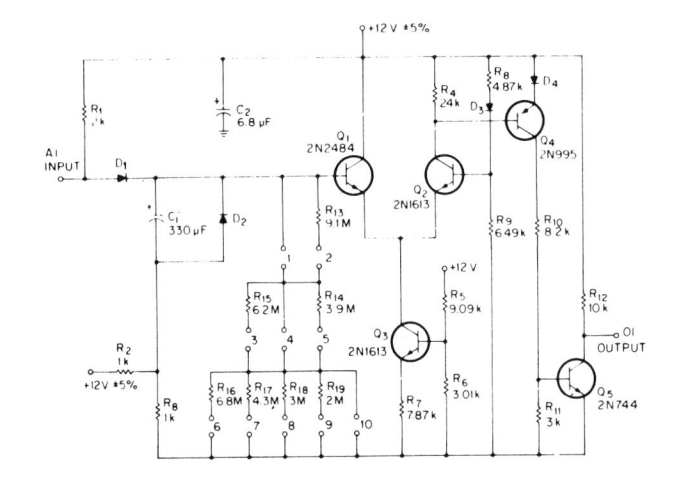

Once the timer has completed its period and the output has dropped, the output will remain down as long as the input is down. If the input level goes up, output will go up resetting the ramp. Reset time is 10 seconds.

Critical components are C_1 and Q_1. Capacitor C_1 is 330-μF with a leakage current of 1000 nA at 60°C and +5 Vdc. Transistor Q_1 has a minimum h_{FE} of 400 and an I_{CBO} of 5nA at V_c = 10Vdc and T = 25°C.

Leakage testing of diode and JFETs

David Dilatush
National Camera, Englewood, CO

A very sensitive circuit for testing diodes and FETs for leakage can be made using an FET, a capacitor and a resistor.

The FET to be tested for leakage and a 22 kΩ resistor are connected as a source-follower, with a capacitor across the input from the gate to ground. The leakage of the FET charges the capacitor at a rate which is directly proportional to leakage and inversely proportional to the capacitance. The quantities are related mathematically as follows:

$$I \text{ (in amperes)} = \frac{Q \text{ (in coulombs)}}{T \text{ (in seconds)}} \text{ and}$$

$$V \text{ (in volts)} = \frac{Q \text{ (in coulombs)}}{C \text{ (in farads)}}$$

A suitable value of C is 0.01 μF. With this capacitance, each volt of potential change across C indicates a stored charge of 10^{-8} coulomb. This can be interpreted as current if the time for the voltage on the capacitor to rise 1V is known.

Example: $C = 10^{-8}$ F, $\frac{dv}{dt} = \frac{1V}{38.7 \text{ sec}}$

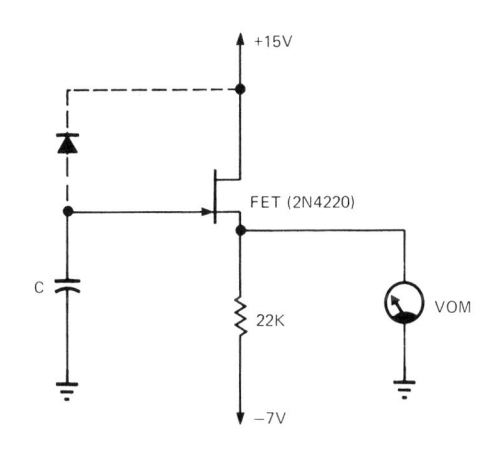

JFET leakage can be checked with a voltmeter and a stopwatch or timer in this simple test setup.

$$I = \frac{dQ}{dt} = \frac{10^{-8} \text{ coul.}}{38.7 \text{ sec.}} = 2.58 \times 10^{-10} \text{ A} = 0.258 \text{ nA}.$$

Diodes are tested in the same way, except that a "good" FET is used as a follower and the diode provides almost all of the charging current.

Universal Transformers

By Richard Hofheimer
Non-linear Systems Inc.
Del Mar, Calif.

HERE IS A NOVEL way to obtain a wide variety of output voltages from a single transformer, using a minimum number of windings. The trick is to use a number of secondary windings giving output voltages in ratios of powers of three. Every voltage step from zero to the sum of the secondary voltages can then be obtained by switching the winding connections. The voltage increments will be determined by the smallest winding voltage.

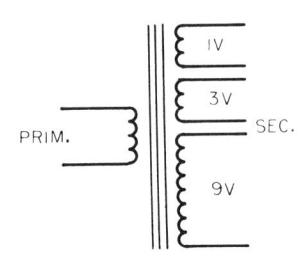

Windings required for transformer giving 1 to 13 V in 1 V steps.

The example shown in the figure has three secondaries of 1 V, 3 V, and 9 V. In this case, the smallest voltage obtainable is 1 V and the largest is 13 V. But it is also possible to obtain all the intermediate voltages in steps of 1 V. For example: 2 V is obtained by connecting the 3-V winding and the 1-V winding in series opposition; 3 V is obtained by using the 3-V winding alone; 4 V is obtained by connecting the 3-V winding and the 1-V winding in series aiding; etc.

The table shows the possible connections for the transformer described in the example. "+" denotes a winding in phase with the required output, "−" denotes a winding in antiphase with the output, and "0" denotes a winding not used. Note that redundant windings, denoted by "0" in the table, are simultaneously available for driving a separate load if required.

There is a considerable difference in regulation for different output voltages. Winding impedances are always additive, regardless of phase.

OUTPUT VOLTS	1–VOLT WINDING	3–VOLT WINDING	9–VOLT WINDING
1	+	0	0
2	−	+	0
3	0	+	0
4	+	+	0
5	−	−	+
6	0	−	+
7	+	−	+
8	−	0	+
9	0	0	+
10	+	0	+
11	−	+	+
12	0	+	+
13	+	+	+

Connections for Transformer with Three Secondaries.

Thus a 13-V and a 5-V connection will both give the same secondary impedance, as both connections use all three windings. Note however that the percentage regulation will be different due to the different total voltages.

The technique described can be extended to any number of windings. Thus adding a fourth winding with a voltage of 3^3 or 27 V will give all voltages up to 40 V in steps of 1 V. The relationship is given by the equation

$$S = \sum_{x=0}^{(n-1)} 3^x$$

where S—number of steps, and n—number of secondary windings.

Low cost IR system detects intruders

Alexander Liu, Fairchild Microwave
& Optoelectronics Division, Palo Alto, CA.

This IR intruder detection system offers the following advantages: both transmitter and receiver can be packed into boxes; the IR beam path is invisible; and cost of the system is low.

The transmitter is comprised of three sections: a 500-Hz clock generator, a clock shaper, and a driver. The clock generator is made from three hex-inverter gates (μA 9016) and provides a square wave of approximately 500 Hz.

The clock shaper, consisting of two dual-input gates and one gate from the hex-inverter, is a one-shot which generates approximately 20-μs pulses triggered by the positive-going edge of the clock. This is a very simple way to generate a 500-Hz, 1% duty-cycle clock.

The driver is a Fairchild FLD 100 (a gallium arsenide infrared emitting diode), which emits an intense narrow band of radiation peaking at approximately 900 nm (non-visible). The most efficient way to operate the IR diode is with high peak current and low duty cycle pulsing. For this application, 2A peak current operation is suitable for the FLD 100. The FLD 100 is physically placed at the focal point of the transmitter lens.

The receiver consists, basically, of a Fairchild FPT 100 (a phototransistor whose spectral response peaks at 900 nm) and its associated circuitry. This phototransistor is mounted at the focus point of the receiving lens and receives a continuous stream of IR pulses. The delay time of the 9601 is 4.5 msec (0.36 R4C3) and was selected to be a little bit over the period T (2 msec) of the clock frequency. The 9601 is retriggered continuously by the incoming pulses with its output remaining at a "HIGH" level. Once the IR beam path is interrupted by an intruder for a minimum of one clock period (2msec), the 9601 times out. The result is Q;"LOW", which sets the latch (μA9936). This, in turn, triggers the SCR to activate the alarm circuit. To clear the alarm the RESET button is pressed; this clears the latch, and recycles the SCR.

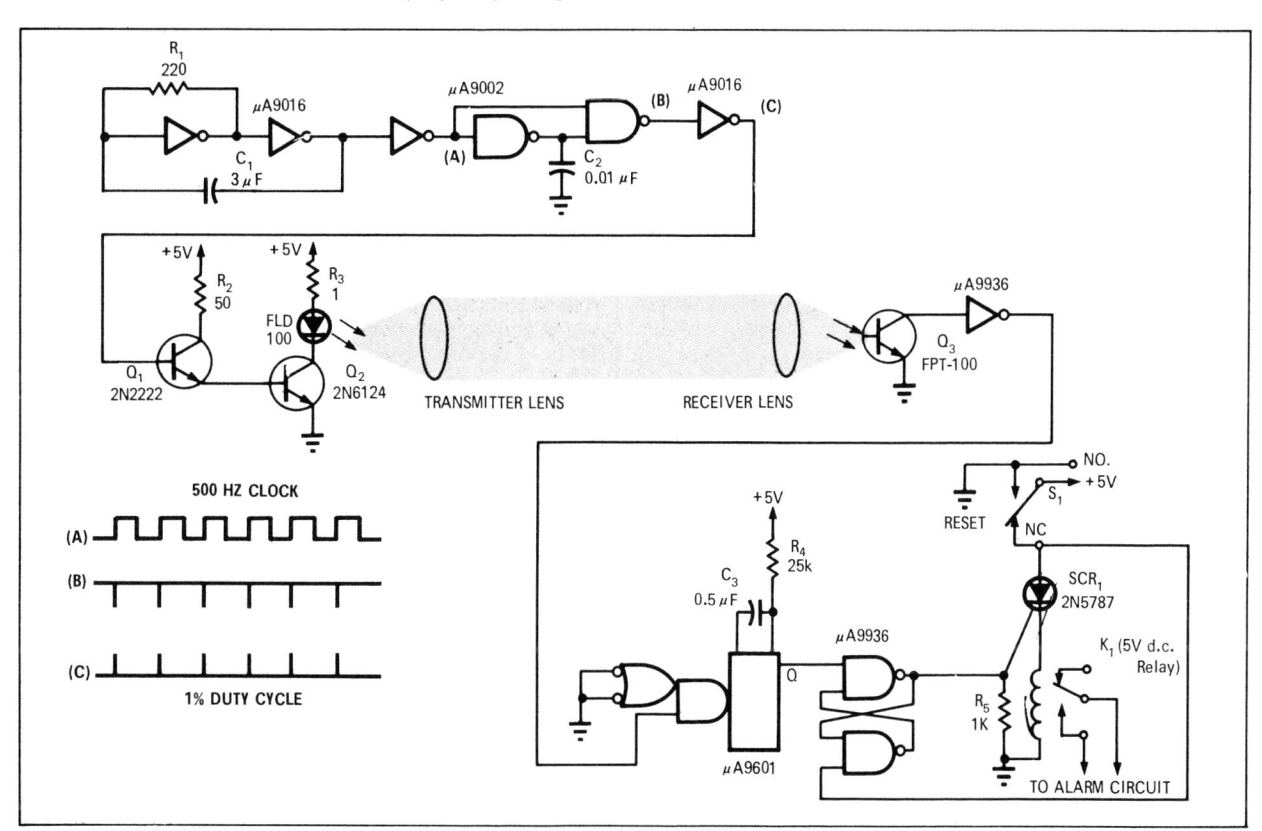

Fig. 1.—The invisible infrared radiation from the LED is formed into a beam across the protected area. Any interruption of this beam for longer than 4-1/2 msec will activate the alarm circuit which remains on until switch S_1 is reset.

Optimum zener decoupling

by Raymond J. D'Auteuil
Avco Missile Systems
Wilmington, Mass.

Decoupling capacitors are often used across zeners, as in Fig. 1a, to filter out conducted interference. Because the zener's internal resistance $R_\approx$ is small compared to the bias resistor R_s, the value of $R_\approx$ determines the filtering, as shown in the equivalent circuit of Fig. 1b. The effective filter cutoff frequency f_c equals $1/(2\pi R_\approx C_p)$.

The small change in Fig. 2a can improve performance significantly. The dc regulation remains the same but, for decoupling, the value of $R_\approx$ is negligible and the network, whose equivalent circuit appears in Fig. 2b, has the maximum possible value of series resistance, $R_s/4$. The cutoff frequency, now equal to $4/(2\pi R_s C_p')$ is reduced by a factor of $R_s/4R$ for circuits with $C_p = C_p'$.

As an example, let's assume a 6.2-V zener regulator operates from a 29-V supply with conductive interference up to 1000 Hz. A 3-kΩ resistor R_s provides 7.5-mA bias current to the

$$C_p' = \frac{1}{2\pi \times 1000 \times 5}$$
$$= 31.8\ \mu F$$

For the revised circuit, however, we can use

$$C_p = \frac{1}{2\pi \times 1000 \times 750}$$
$$= 0.21\ \mu F$$

5-Ω zener. For the shunt capacitor of Fig. 1a we need for a reduction in capacitor size by more than two orders of magnitude.

The proposed circuit would use two 1/8-watt resistors and a 0.2-μF capacitor operating at about 18 V. The less effective network has a 1/4-watt resistor and a 30-μF capacitor operating at 6.2 V. The component trade-off favors the newer circuit.

Load circuits with internally generated noise may still require the conventional decoupling. In that case, the proposed circuit, if used with a capacitor across the zener, effectively decouples residual noise from the power line.

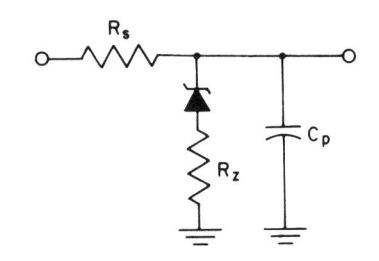

Fig. 1a. Basic circuit of zener shunted by decoupling capacitor used to filter out conducted interference.

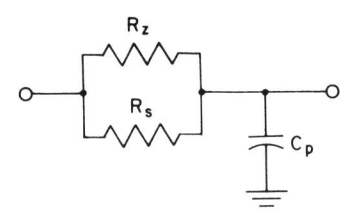

Fig. 1b. Equivalent circuit of Fig. 1a.

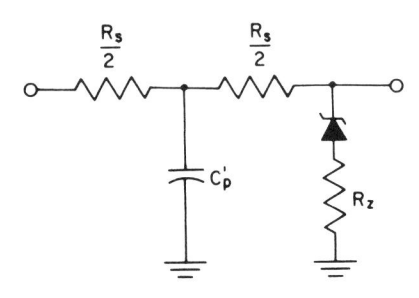

Fig. 2a. Slight modification of earlier circuit gives significant improvement.

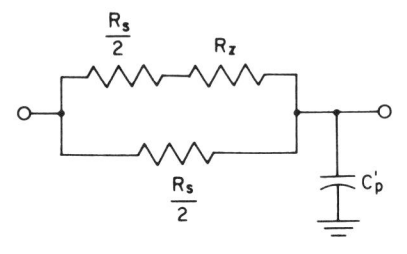

Fig. 2b. Equivalent circuit of Fig. 2a.

Index

Amplifiers
   audio, 20
   gain control, 32, 34
   high voltage, 11
   instrumentation, 18
   non-linear, 20, 27, 28, 33
   operational, 3, 5, 8, 11, 13, 17, 18, 19, 20, 21,
      22, 23, 27, 28, 31, 32, 37, 38, 42, 44, 65,
      102, 105, 107, 108, 110, 114-115, 120,
      122, 124, 125, 126, 128, 130, 133, 134,
      138, 172, 176, 189
   oscilloscope, 14
   preamps, 9
   pulse, 87
   sampling, 21
   solar cells, 29
   transistor, 8, 9, 25, 29
   video, 42
Analog
   arithmetic unit, 5, 17
   counters, 45
   divider, 5
   monitor, 18
   multiplier, 203
   signal selection, 7
   switches, 16, 17, 40
   squaring, 12

Battery chargers, 197
Binary
   coded decimal, 51, 56, 57, 61, 64, 69
   counters, 93

Capacitance testing, 176, 177
Clamping circuits, 38, 75

Clippers, 3
Clock, digital, 185
Code conversion circuits, 61, 72, 186
Combinational logic, 69
Comparators
   analog, 10, 23, 46, 67
   digital, 91, 94, 97
Converters
   current-to-voltage, 8
   dc to dc, 62
   D/A, 32
   period-to-voltage, 42
Counters
   linear, 45
   pulse, 84
   reversible, 45
Current
   boosters, 27
   constant, 13, 19, 39, 43, 87, 177
   sinks, 25
   switches, 16

Data
   acquisition, 60, 81
   processing, 71, 86, 89
Delay
   analog, 135, 200
   digital, 49, 70, 92, 96, 136, 138, 203
Detector
   level, 38, 54-55, 59, 71, 74, 78-79, 83, 88,
      96, 194
   limit, 85
   optical, 65
   peak, 14, 73
   phase difference, 22, 43

product, 9
pulse, 70, 81, 84, 90, 179, 194
synchronous, 39
WWV, 35
zero-crossing, 22, 76, 156, 164
Discriminator
frequency, 6, 91, 186
pulse-height, 73
pulse-width, 22, 95, 97
Dividers digital, 50, 62, 94
Drivers
coax, 29, 59, 199
clock, 79, 85
LED, 88, 201
lamp, 190, 204
motors, 165, 187

E cells, 197
Electronic lock, 193
Error stabilization, 5

FETs, 9, 17, 38, 39, 41, 56, 63, 102, 104, 116, 124, 161
Filters
active, 22, 37, 38
audio, 4
bandpass, 37
digital, 88
tone, 26
Flip-flop circuits, 50, 53, 62, 74, 77, 89, 194
Frequency
detectors, 94
doublers, 34

Generators
clock-pulse, 110, 117, 121, 130, 132, 135
constant current, 19, 21, 34, 39
dc offset, 43
pulse, 60, 63, 70, 92, 135
ramp, 42, 102, 110, 111, 116, 120, 128, 133, 135
raster, 33
staircase, 106, 111
sweep, 112, 133
triangle, 101, 107, 108, 130
vertical sweep, 36

High voltage
converters, 146
source follower, 8, 11
triacs, 155

Indicators, 85, 171, 174, 179, 180, 181, 186, 190, 201, 104

Integrators, 5, 24
Interface circuits, 191, 198

Latch, set-reset, 8
LED
driver, 96, 201
modulator, 31
Lock-electronic, 193
Logic gates, 7, 60, 76, 87, 146, 161
Logic probes, 171, 174, 179, 180, 181

Metering technique, 172, 177, 193
Microwave, 40
Mixers
diode, 41
double-balanced, 41
Modems, 7
Modulators
LED, 31
linear, 31
phase, 37
pulse, 26, 28
MOS circuit, 56, 85
Multiplexing, 41, 159
Multivibrators
astable, 58-59, 101, 114-115, 117, 118, 119, 122, 123, 128, 129, 130, 133, 136, 137, 138, 139
bistable, 50, 53, 62, 74, 77, 94, 98
monostable, 42, 51, 52, 55, 56, 63, 65, 66, 68, 73, 75, 77, 82, 83, 84, 92, 93, 96, 136, 138, 165, 188-189

One-shot multivibrators, 42, 51, 52, 55, 56, 63, 65, 66, 67, 68, 73, 75, 77, 82, 83, 84, 90, 92, 93, 96, 165, 188-189
Opto-electronic devices, 31, 65, 160, 200, 201, 205
Oscillators
gated, 108, 118, 130, 133
microwave, 40
phase locked, 35, 58-59, 64
relaxation, 103, 109, 113, 119, 125, 126
sine wave, 103, 104, 105, 108, 115, 116, 122, 124, 125, 126, 127, 132, 134
sweep, 110, 112
voltage controlled, 40, 91, 101, 125, 132
Oscilloscope circuits, 174

Peak-holding, 73
Phase
detectors, 22, 30, 94
locking, 6, 23, 35, 53, 58-59, 64, 187
modulators, 37

shifters, 82
Photo transistors, 65, 92, 160, 200, 205
PIN diodes, 40
Power switching circuits, 143, 144, 146, 151, 155, 157, 158, 160, 161, 162, 163, 165, 166, 167, 176
Protective circuits, 38, 143, 147, 150, 156, 161, 165, 167, 173, 181, 190, 194, 198, 213
PUTs, 103, 113, 125, 132, 135, 160

RMS Measurements, 15

Sample and hold circuits, 11, 17, 42
Sampling amplifiers, 21
Scaling circuits, 192
Schmitt triggers, 66, 67, 68, 71, 78-79, 80, 83, 96
SCR and triacs, 101, 103, 143, 150, 155, 157, 158, 165, 193, 194
Shift register, 79
Signal switching, 157, 159, 161, 163
Squaring circuits, 34

Tachometer circuits, 191
Temperature
   coefficient, 178
   control, 197
   sensing, 187, 195, 196, 197
Timers, 51, 55, 56, 75, 101, 113, 151, 165, 175, 186, 188-189, 203
Tone burst gates, 21
Transformers, 204

UJTs, 33, 101, 112, 126

VCOs, 127
Video
   signal processing, 3, 8, 27, 33, 70, 89
   sweep generators, 36, 133, 135
Voltage
   converters, 144, 146, 149
   references, 13, 25, 43, 148, 189, 190
   regulators, 143, 144, 146, 147, 149
   sensing, 143, 145, 147, 149, 150
Voltage controlled resistors, 32
Voltmeter, 172

Waveshaping circuits, 107, 122, 175

Zener diodes, 13, 108, 112, 130, 165, 190, 206
Zero-crossing switches, 156, 164